THE MARINE ELECTRICAL AND ELECTRONICS BIBLE

Other Books by John C. Payne

The Fisherman's Electrical Manual

The Great Cruising Cookbook: an International Galley Guide

Motorboat Electrical and Electronics Manual

Understanding Boat Batteries and Battery Charging

Understanding Boat Communications

Understanding Boat Corrosion, Lightning Protection and Interference

Understanding Boat Diesel Engines

Understanding Boat Electronics

Understanding Boat Wiring

The purchase of this book
was made possible by a
generous donation from
The Friends
Of
The Blake Library,
in Stuart, Inc.

THE MARINE ELECTRICAL AND ELECTRONICS BIBLE

Third Edition

JOHN C. PAYNE

SHERIDAN HOUSE

This edition first published 2007 by
Sheridan House Inc.
145 Palisade Street,
Dobbs Ferry, NY 10522
www.sheridanhouse.com

Library of Congress Cataloging-in-Publication Data
Payne, John C.
The marine electrical and electronic bible /
John C. Payne. — 3rd ed.
p. cm.
Includes bibliographical references and index.
ISBN-13: 978-1-57409-242-4 (alk. paper)
ISBN-10: 1-57409-242-1
1. Boats and boating—Electronic equipment. I. Title.

VM325.P39 2007
623.8'503—dc22 2006029733

Illustrations by Paul Checkley and Grahame MacCleod

Printed in China

ISBN 10: 1-57409-242-1
ISBN 13: 978-1-57409-242-4

CONTENTS

SECTION ONE ELECTRICAL SYSTEMS

SECTION TWO ELECTRONICS SYSTEMS

INTRODUCTION

Both cruising and racing yachts continue to evolve in many aspects, from hull design and exotic materials to the sailing rig and an increasing array of sophisticated electronic equipment. Quite often this equipement equals and is even superior to that found on many modern merchant vessels. One of the shortcomings is the frequent failure of the basic electrical systems that provide the power. Equipment and systems still often remain poorly planned, installed, and maintained. While the ABYC in the United States and the Recreational Craft Directive (RCD) in the UK/Europe set out standards that most manufacturers comply with, there are still many boat owners who fail to keep them up to these standards.

The Marine Electrical and Electronics Bible is specifically written to meet the real and practical requirements of cruising and racing yacht electrical and electronics systems. Electrical theory is explained only to allow the proper consideration, selection, installation, operation, maintenance, and troubleshooting of systems. I have deliberately set out to correct the dangerous illusion that vessel and automotive systems are alike except for the voltage levels. This third edition has a revised section on boat wiring, engine starting and battery charging and takes a "big ship" reliability based approach. Electrical problems are not an inevitable part of cruising and racing, and an acceptable level of reliability is possible, and in fact is necessary.

Marine electronics and systems technology still continues to advance at a fast pace. The greatest impact in the last few years has been the introduction of the Global Maritime Distress and Safety System (GMDSS). In this new edition, the marine communications chapters have been revised to reflect this important change and the ramifications are profound for all boaters. It will result in a much safer and more efficient distress and safety system afloat. Satellite telephones as well as e-mail and Internet are now an affordable reality along with electronic charting systems and network based instrument and data systems. One significant feature of the new equipment is the ability for boaters to take care of the installation themselves as most systems are plug-and-play.

An inherent danger with electronic navigation is the surrendering of seamanship skills. Far too many people sail offshore without the basic knowledge or ability to survive a loss of electronic navigation aids or the ability to maintain the equipment to keep them running. These high technology systems are only aids to navigation, and there is no substitute for seamanship, a well-found vessel, up to date charts, and the ability to find out where you are, using traditional methods.

This book captures and distills my 30 years of professional marine electrical experience on merchant vessels and in the offshore oil industry as a marine electrical engineer and surveyor, as well as my ownership of both cruising yachts and power vessels. In this book, I hope to answer many of the frequently asked questions and information requests that I get. I cannot overstress the importance of adopting a keep-it-simple approach to electrical systems and to marine electronics. Successful and trouble-free cruising depends on simplicity.

John Payne
www.fishingandboats.com

SECTION ONE

ELECTRICAL SYSTEMS

Batteries

1.0 Battery Selection. In any sailing yacht the battery has a primary role as a power storage device, and a secondary one as a buffer, absorbing power surges and disturbances that arise during charging and discharging. The foundation of a reliable and efficient power system is a correctly specified and rated battery. The following chapters explain all the factors essential to the installation of a reliable power system. After recent incidents it is recommended that all yachts consider sealed batteries. This chapter covers some basic factors that should be considered, and the following battery types are examined:

- **a. Lead Acid Batteries.** The lead acid battery is used in the majority of marine installations and the principles are explained.
- **b. Low Maintenance Batteries.** The viability of these batteries for cruising is considered.
- **c. Gel Cell Batteries.** Gel cell batteries are a relatively new battery type and their suitability for cruising applications will be analyzed.
- **d. AGM Batteries.** Absorbed Glass Mat (AGM) batteries are now a viable and efficient battery technology and the advantages will be discussed.
- **e. Nickel Cadmium Batteries.** These batteries are usually found on larger cruising vessels and are an alternative to lead acid batteries.

The house or service battery capacity should be based on calculation of the vessel's power consumption for a 24-hour period. The power calculations should include all the equipment and systems that will run continuously or intermittently during the calculated period.

House or service loads draw current over long periods. Equipment in this category includes lights, navigation instruments, communication radios, radar, autopilots, inverters and entertainment systems. The deep cycle battery is normally used for these applications. Calculations are based on the maximum power consumption over the longest probable period between battery recharging. Typically this is 24 hours.

A separate battery should be provided for the propulsion engine. The engine start battery capacity should be based on the provision of 10 consecutive start attempts of 5 seconds with a 30 second period between each attempt, at an ambient temperature of 5°C. The engine manufacturer's recommendations should be the minimum battery specification for a starting battery.

Starting loads require large current levels for relatively short time periods. Loads in this category include the engine starter motor, the diesel engine pre-heating, the anchor windlass, electric sail handling deck winches, bow thrusters and also electric toilets. The starting battery is normally used for all these applications. The battery rating should always allow for worst case starting scenarios, which are in cold temperatures where battery efficiency is lowered and engine starting requires greater power due to increased oil viscosity etc. It should also allow for problems where multiple start attempts are required or cranking for diesel system bleeding.

1.1 Battery Ratings. Manufacturers use a range of ratings figures to indicate battery performance levels. When selecting a battery it is essential to understand the ratings and how they apply to your own requirements. The various ratings are defined as follows.

a. Amp-hour Rating. Amp-hour rating (Ah) refers to the available current over a nominal period until a specified final voltage is reached. Rates are normally specified at the 10- or 20-hour rate. This rating is normally only applicable to deep cycle batteries. For example, a battery is rated at 84 Ah at 10 hour rate, with a final voltage of 1.7 volts per cell. This means that the battery is capable of delivering 8.4 amps for 10 hours, when a cell voltage of 1.7 volts will be attained. (Battery Volts = 10.2 VDC). Where a battery is discharged faster than the nominal rating the available capacity also decreases. This is called the Peukert Effect and the decline follows a logarithmic curve.

b. Reserve Capacity Rating. This rating specifies the number of minutes a battery can supply a nominal current at a nominal temperature without the voltage dropping below a certain level. This rating, normally applied to automotive applications, indicates the power available when an alternator fails and the power available to operate ignition and auxiliaries. Typically, the rating is specified for a 30-minute period at 25°C with a final voltage of 10.2 volts.

c. Cold Cranking Amps (CCA). This rating defines the current available at -18°C for a period of 30 seconds, while being able to maintain a cell voltage exceeding 1.2 volts per cell. This rating is only applicable for engine starting purposes. The higher the rating, the more power available, especially in cold weather conditions.

d. Marine Cranking Amps (MCA). This rating defines the current available at 0°C for a period of 30 seconds, while being able to maintain a cell voltage exceeding 1.2 volts per cell. Again, this rating is only applicable for engine starting purposes. If you are in cold climate area (UK/Europe and USA) then CCA is more relevant.

e. **Plate Numbers.** Data sheets state the number of positive and negative plates within a cell. The more plates installed, the greater the plate material surface area. This increases the current during high current rate discharges and subsequently improves cranking capacity and cold weather performance.

f. **Casing Type.** Battery casings are either a rubber compound or plastic. Where possible always select the rubber types, as they are more resilient to knocks and vibration.

g. **Marine Battery.** This often misused sales term applies to certain constructional features. Plates may be thicker than normal or there may be more of them. Internal plate supports are also used for vibration absorption. Cases may be manufactured with a resilient rubber compound and have carry handles fitted. Filling caps may be of an anti-spill design. These days, batteries are of a similar design with very little to distinguish between the automotive types except the label. In many cases you are paying a premium for a label.

1.2 Battery Rating Selection. It is important to select suitable batteries for use in service or house power roles. The majority of power problems arise from improper battery selection. Battery bank capacities are either seriously underrated with resultant power shortages, or overrated so that the charging system cannot properly recharge them. This results in premature failure of the batteries due to sulfation in lead acid batteries. Initially, it is essential that all the equipment on board be listed along with power consumption ratings. Ratings can usually be found on equipment nameplates or in equipment manuals. It is recommended that the ratings, usually expressed in watts, be converted to current in amps. To do this, divide the power by your system voltage. Calculate the current consumption for a 12-hour period while in port or anchored. The calculation assumes that the engine will not be operated, and no generator with battery charging will be operational. While motoring all power is being supplied from engine alternators, and when batteries are charged the alternator effectively supplies all power. Tables 1-1 and 1-2 illustrate the typical power consumptions, with space for you to insert and calculate the data for your own vessel.

a. **Load Calculation Table.** To calculate the total system loading, multiply the total current values by the number of hours to get the amp-hour rating. If equipment uses 1 amp over 24 hours, then it consumes 24 amp-hours.

b. **Capacity Calculation.** Select the column that matches the frequency of your charging periods. The most typical scenario is one with the boat at anchor or on a mooring and operating the engine every 12 hours to pull down refrigerator temperatures and also charge the battery.

e.g. Total consumption is 120 Ah over 12 hours = 10 amps/hour

c. **Capacity De-rating.** As we wish to keep our discharge capacity to 50% of nominal battery capacity, we can assume that a battery capacity of 240 amp-hours is the basic minimum level. In an ideal world, this would be a minimum requirement, but certain frightening realities must now be introduced into the equation. The figures below typify a common system, with alternator charging and standard regulator. Maximum charge deficiency is based on the premise that boat batteries are rarely above 70% charge and cannot be fully recharged with normal regulators, and there is reduced capacity due to sulfation, which is typically a minimum of 10% of capacity. The key to maintaining optimum power levels and avoiding this common and frightening set of numbers is the charging system.

Nominal Capacity		240 Ah
Maximum Cycling Level (50%)	Deduct	120 Ah
Maximum Charge Deficiency (30%)	Deduct	72 Ah
Lost Capacity (10%)	Deduct	24 Ah
Available Battery Capacity		**24 Ah**

d. **Amp-hour Capacity.** It is important to discuss a few more relevant points regarding amp-hour capacity as it has significant ramifications on the selection of capacity and discharge characteristics:

(1) **Fast Discharge (Peukerts Equation).** The faster a battery is discharged over the nominal rating (either 10- or 20-hour rate) the less real amp-hour capacity there is. This effect is defined by Peukerts Equation, which has a logarithmic characteristic. This equation is based on the high and low discharge rates and discharge times for each to derive the Peukert coefficient *'n'*. Average values are around 1.10 to 1.20. If we discharge a 250 amp-hour battery bank, which has nominal battery discharge rates for each identical battery of 12 amps per hour at a rate of 16 amps, we will actually have approximately 10-15% less capacity. Battery discharge meters such as the E-Meter incorporate this coefficient into the monitoring and calculation process.

(2) **Slow Discharge.** The slower the discharge over the nominal rate the greater the real capacity. If we discharge our 240 amp-hour battery bank at 6 amps per hour we will actually have approximately 10-15% more capacity. The disadvantage here is that slowly discharged batteries are harder to charge if deep cycled below 50%.

e. **Battery Load Matching.** The principal aim is to match the discharge characteristics of the battery bank to that of our calculated load of 10 amps per hour over 12 hours. Assume that we have a modified charging system so that we can recharge batteries to virtually 100% of nominal capacity. The factors affecting matching are as follows:

(1) **Discharge Requirement.** The nominal required battery capacity of 240 Ah has been calculated as that required to supply 10 amps per hour over 12 hours to 50% of battery capacity. In most cases, the discharge requirements are worst for the night period, and this is the 12-hour period that should be used in calculations. What is required is a battery bank with similar discharge rates as the electrical current consumption rate. This will maximize the capacity of the battery bank with respect to the effect defined in Peukerts Coefficient.

(2) **Battery Requirements.** As the consumption rate is based on a 12-hour period, a battery bank that is similarly rated at the 10-hour rate is required. In practice you will not match the precise required capacity, therefore you should go to the next battery size up. This is important also as the battery will be discharged longer and faster over 12 hours, so a safety margin is required. If you choose a battery that has 240 amp-hours at the 20-hour rate in effect you will actually be installing a battery that in the calculated service has 10-15% less capacity than that stated on the label, which will then be approximately 215 Ah, so you are below capacity. This is not the fault of the supplier, but simply a failure to correctly calculate and specify the right battery to meet system requirements.

f. **Battery Capacity Formulas.** A range of formulas are frequently used to determine battery capacity. These are as follows:

(1) **Four-Day Consumption Formula.** One of the more unrealistic formulas states that you should be able to supply all electrical needs over four, complete, 24-hour periods without recharging batteries. Given that an average 10 amps per hour is a typical consumption rate, the four-day formula tells us we'll use 960 amp-hours. If we only discharge to 50%, that translates into an incredible 2000 amp-hour battery capacity. In addition, the recharging period, which requires an additional 20%, must replace about 1200 amp-hours. Even with a fast-charge device, a 100-amp alternator, the finite charge acceptance rate of the battery will force you to spend at least 12 hours charging.

(2) **75/400 System.** This was included in a magazine article as one of three formulas for various sized vessels, and was for a 40 to 45-foot yacht. This was the nearest I have seen to a rational set of numbers, based on a 75-amp consumption over 24 hours. Though perhaps too conservative, the formula is based on a 130- or 150-amp alternator with fast-charge device to recharge half of a 400 amp-hour battery bank.

(3) **My Personal Formula (240/460 System).** My own personal formula is based on the worst-case consumption of 240 amp-hours over 24 hours. This entails the installation of two banks of 230-hour batteries, each bank made up of two 6-volt batteries. The batteries each supply a split switchboard, with electronics off one bank and with pumps and other circuits off the other, limiting any interference. Charging is from an 80-amp alternator with a fast charge regulator (Ample Power, Adverc etc) through a diode isolator. It's simple and able to cope with all load conditions. Charging is relatively fast, and at a rate similar to the battery's ability to accept it. In reality, there are no easy formulas. Each vessel has different requirements, and systems must be tailored to suit.

1.3 Sailing Load Calculations. It is essential that all equipment on board be listed along with power consumption ratings. Ratings can usually be found on equipment nameplates or in equipment manuals. Insert your own values into the Actual column. Calculate power used for 12 hours. To convert power (in watts) into current (in amps) simply divide the power value by your system voltage. Add up all the current figures relevant to your vessel and multiply by hours to get an average amp-hour consumption rate. Space is reserved to add in specific values. Most of these items will be on when anchored or moored, but many will not be relevant if the vessel is at a marina connected to a battery charger.

Table 1-1 DC Load Calculation Table

Equipment	Typical	Actual	12 Hours	Other
Radar- Transmit	4.5 A	______	______	______
Radar – Standby	0.5 A	______	______	______
SSB – Receive	0.5 A	______	______	______
VHF – Receive	0.5 A	______	______	______
Satcom	1.0 A	______	______	______
Weatherfax	0.5 A	______	______	______
GPS/LORAN	0.5 A	______	______	______
Navtex	0.5 A	______	______	______
Fishfinder	1.0 A	______	______	______
Instruments	0.3 A	______	______	______
Stereo	0.5 A	______	______	______
Gas Detector	0.5 A	______	______	______
Inverter – Standby	1.0 A	______	______	______
Anchor Light	1.0 A	______	______	______
Refrigeration	4.0 A	______	______	______
Interior Lights	5.0 A	______	______	______
Computer	1.0 A	______	______	______
Television	2.0 A	______	______	______
Video	1.0 A	______	______	______
TOTAL				

1.4 Additional Load Calculations. Other basic load characteristics have to be factored in to load calculations. Add up all the current figures relevant to your vessel and multiply by expected run times to get an average amp-hour consumption rate.

a. **Intermittent Loads.** It is often hard to quantify actual real current demands with intermittent loads. My suggestion is simply to use a baseline of 6 minutes per hour, which is 0.1 of an hour.

b. **Motoring Loads.** Certain loads are only applicable when motoring. Loads must be subtracted from charge current values, and actually may impact on charging system efficiency at low speeds. Loads include navigation lights, refrigeration electric clutch, water-maker clutch and ventilation fans.

Table 1-2 DC Load Calculation Table

Equipment	Typical	Actual	12 Hours	Other
Other Loads				
Bilge Pump	3.5 A	________	________	________
Shower Pump	3.5 A	________	________	________
Water Pump	3.5 A	________	________	________
Wash Down Pump	3.5 A	________	________	________
Toilet	3.5 A	________	________	________
Macerator	3.5 A	________	________	________
SSB - Transmit	3.5 A	________	________	________
VHF -Transmit	3.5 A	________	________	________
Spot Light	3.5 A	________	________	________
Extraction Fan	3.5 A	________	________	________
Inverter	3.5 A	________	________	________
Cabin Lights	3.5 A	________	________	________
		________	________	________
		________	________	________
		________	________	________
Sub Total Table 2				
Sub Total Table 1				
LOAD TOTAL				

1.5 Lead Acid Batteries. The fundamental theory of the battery is that a voltage is developed between two electrodes of dissimilar metal when they are immersed in an electrolyte. In the typical lead-acid cell the nominal generated voltage is 2.1 volts. The typical 12-volt battery consists of 6 cells, which are internally connected in series to make up the battery.

a. Cell Components. The principal cell components are:

(1) Lead Dioxide (PbO2) — the positive plate active material.

(2) Sponge Lead (Pb) — the negative plate material.

(3) Sulfuric Acid (H2SO4) — the electrolyte.

b. Discharge Cycle. Discharging of the battery occurs when an external load is connected across the positive and negative terminals. A chemical reaction takes place between the two plate materials and the electrolyte. During the discharge reaction, the plates interact with the electrolyte to form lead sulfate and water. This reaction dilutes the electrolyte, reducing the density. As both plates become similar in composition, the cell loses the ability to generate a voltage.

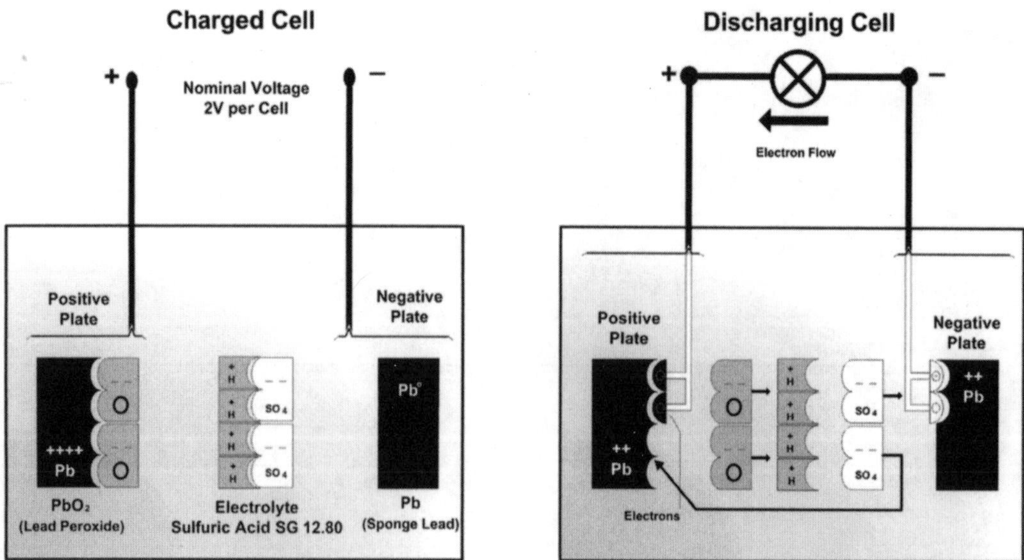

Figure 1-1 Lead Acid Charge and Discharge Reaction

c. **Charge Cycle.** Charging simply reverses this reaction. The water decomposes to release hydrogen and oxygen. The two plate materials are reconstituted to the original material. When the plates are fully restored and the electrolyte is returned to the nominal density, the battery is completely recharged.

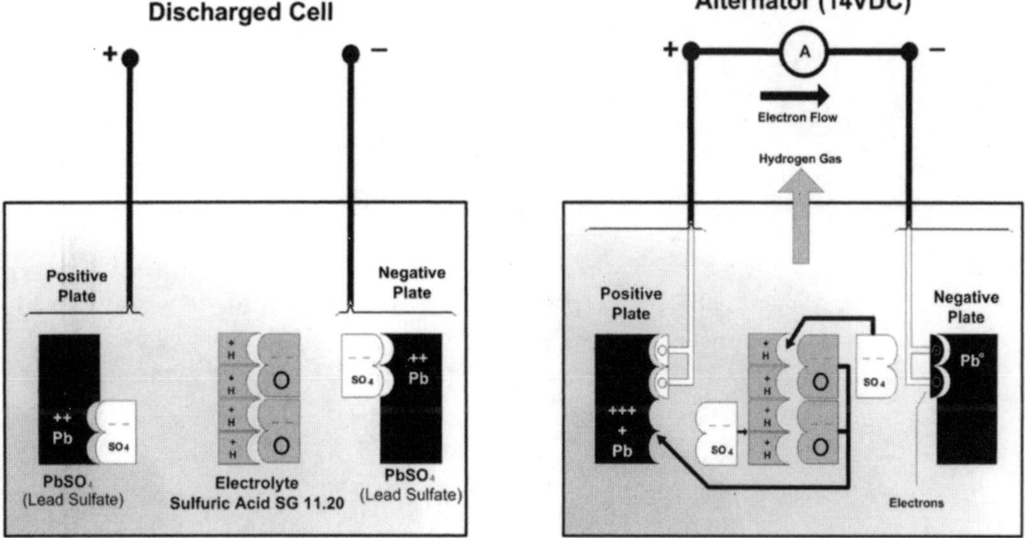

Figure 1-2 Lead Acid Discharge and Charge Reaction

1.6 Battery Electrolyte. The cell electrolyte is a dilute solution of sulfuric acid and pure water. Specific Gravity (SG) is a measurement defining electrolyte acid concentration. A fully charged cell has an SG typically in the range 1.240 to 1.280, corrected for temperature. This is an approximate volume ratio of acid to water of 1:3. Pure sulfuric acid has an SG of 1.835, and water a nominal 1.0. The following factors apply to battery electrolytes.

a. **Temperature Effects.** For accuracy, all hydrometer readings should be corrected for temperature. Ideally, actual cell temperatures should be used, but in practice ambient battery temperatures are sufficient. Hydrometer floats have the reference temperature printed on them and this should be used for calculations. As a guide, the following should be used for calculation purposes.

   (1) For every 1.5°C the cell temperature is *above* the reference value *add* 1 point (0.001) to the hydrometer reading.

   (2) For every 1.5°C the cell temperature is *below* the reference value *subtract* 1 point (0.001) from the hydrometer reading.

b. **Nominal Electrolyte Densities.** Recommended densities are normally obtainable from battery manufacturers. In tropical areas it is common to have battery suppliers put in a milder electrolyte density, which does not deteriorate the separators and grids as quickly as temperate climate density electrolytes.

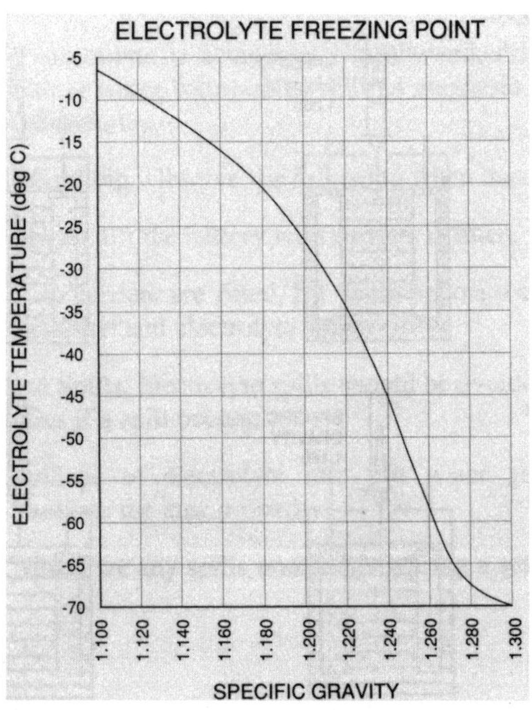

Figure 1-3 Electrolyte Temperature Effects

1.7 Battery Water. When topping up the cell electrolyte, always use distilled or de-ionized water. Rainwater is acceptable, but under no circumstances use tap water. Tap water generally has an excessive mineral content or other impurities that may pollute and damage the cells. Impurities introduced into the cell will remain, and concentrations will accumulate at each top up, reducing service life. Long and reliable service life is essential so the correct water must always be used. Water purity levels are defined in various national standards.

Table 1-3 Electrolyte Correction at 20°

Temperature	Correction Value
-5°C	deduct 0.020
0°C	deduct 0.016
+5°C	deduct 0.012
+10°C	deduct 0.008
+15°C	deduct 0.004
+25°C	add 0.004
+30°C	add 0.008
+35°C	add 0.012
+40°C	add 0.012

1.8 Plate Sulfation. Sulfation is the single greatest cause of battery failure, and occurs as follows:

a. During discharge, the chemical reaction causes both plates to convert to lead sulfate. If recharging is not carried out within a couple of hours, the lead sulfate starts to harden and crystallize. This is characterized by white crystals on the typically brown plates and is almost non-reversible. If a battery is only 80% charged, this does not mean that only 20% is sulfating, the entire plate material has not fully converted and subsequently sulfates.

b. The immediate effect of sulfation is partial and permanent loss of capacity as the active materials are reduced. Electrolyte density also partially decreases, as the chemical reaction during charging cannot be fully reversed. This sulfated material introduces higher resistances within the cell and inhibits charging. As the level of sulfated material increases, the cell's ability to retain a charge is reduced and the battery fails. The deep cycle battery has unfairly gained a bad reputation, but the battery is not the cause, improper and inadequate charging is. As long as some charging is taking place, even from a small solar panel, a chemical reaction is taking place and sulfation will not occur.

c. **Efficiency.** Battery efficiency is affected by temperature. At 0°C, efficiency falls by 60%. Batteries in warm tropical climates are more efficient, but may have reduced life spans, and batteries commissioned in tropical areas often have lower acid densities. Batteries in cold climates have increased operating lives, but are less efficient.

d. **Self Discharge.** During charging, a small quantity of antimony or other impurities dissolve out of the positive plates and deposit on the negative ones. Other impurities are introduced with impure topping up water and deposit on the plates. A localized chemical reaction then takes place, slowly discharging the cell. Self-discharge rates are affected by temperature, with the following results:

(1) At 0°C, discharge rates are minimal.

(2) At 30°C, self-discharge rates are high and the specific gravity can decrease by as much as 0.002 per day, typically up to 4% per month.

(3) The use of a small solar panel, or regular and *complete* recharging will prevent permanent damage as it can equal or exceed the self-discharge rate.

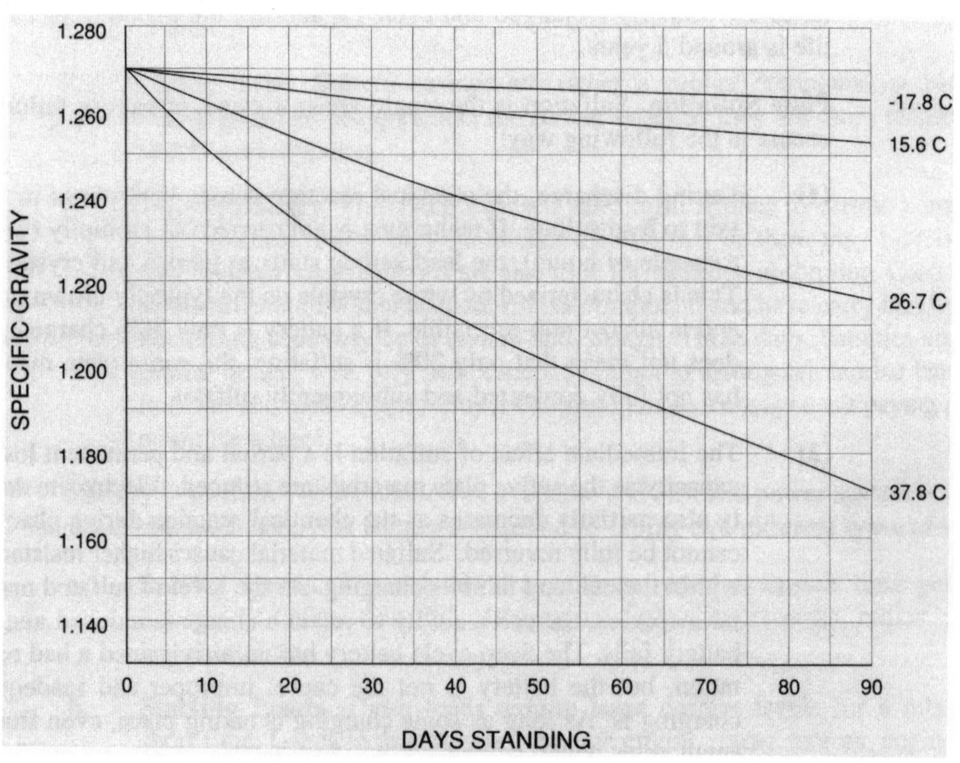

Figure 1-4 Self-Discharge Rates

1.9 Deep Cycle Batteries. Service loads require a battery that can withstand cycles of long continuous discharge, and repeated recharging. This deep cycling requires the use of the suitably named deep cycle battery. Top of the range Surrette (www.surrette.com), Trojan and Rolls (www.rolls.com) typify the quality deep cycle batteries. The deep cycle battery has the following characteristics.

a. **Construction.** The battery is typified by the use of thick, high-density flat-pasted plates, or a combination of flat and tubular. The plate materials may also contain small proportions of antimony to help stiffen them. Porous, insulating separators are used between the plates and glass matting is used to assist in retaining active material on the plates that may break away as plates expand and contract during charge and recharge cycles. If material accumulates at the cell base, a cell short circuit may occur, although this is less common in modern batteries. If material is lost the plates will have reduced capacity or insufficient active material to sustain the chemical reaction with resultant cell failure. Much has been done into developing stronger and more efficient plates. Rolls have their Rezistox positive plates. The grid design has fewer heavier sections to hold the high density active material. This is due to the dynamic forces that normally cause expansion and contraction with subsequent warping and cracking. Separator design has also evolved and Rolls use double insulated thick glass woven ones that totally encase the positive plate along with a microporous polyethylene envelope. This retains any material shed from the plates than cause cell short circuits.

b. **Cycling.** The number of available cycles varies between individual battery makes and models. Typically it is within the range of 800-1500 cycles of discharge to 50% of nominal capacity and complete recharging. Battery life is a function of the number of cycles and the depth of cycling. Batteries discharged to only 70% of capacity will last appreciably longer than those discharged to 40% of capacity. In practice you should plan your system so that discharge is limited to 50% of battery capacity. The typical life of batteries where batteries are properly recharged and cycle capabilities maximized can be up to 5–10 years.

c. **Charging.** The recommended charging rate for a deep cycle battery is often given as 15% of capacity. In vessel operations, it is not possible to apply these criteria accurately. Essentially, the correct charge voltage (corrected for temperature) should be used. Deep cycle battery charging characteristics are as follows:

(1) During charging, a phenomenon called "counter voltage" occurs. Primarily, this is caused by the inability of the electrolyte to percolate at a sufficiently high rate into the plate material pores and subsequently convert both plate material and electrolyte. As the battery resists charging, plate surface voltage rises artificially high and "fools" the regulator into prematurely reducing charging.

(2) To properly charge a deep cycle battery, a charge voltage of around 14.5 volts is required, corrected for temperature. If you do not fully recharge the battery, it will rapidly deteriorate and sustain permanent damage. A charge level of approximately 80% does not represent a fully-charged battery, and is not acceptable if you want a reliable electrical power system and reasonable battery life.

d. **Equalization Charge.** An equalization charge consists of applying a higher voltage level at a current rate of 5% of battery capacity. This is done to reactivate the plates. There is a mistaken belief that this will also completely reverse the effects of sulfation. There may be an improvement following the process, but it will not reverse long-term permanent damage. Equalization at regular intervals can increase battery longevity by ensuring complete chemical conversion of plates, but care must be taken.

1.10 Starting Batteries. The starting battery must be capable of delivering the engine starter motor with sufficient current to turn and start the engine. This starting load can be affected by engine compression, oil viscosity, and engine driven loads. Some loads such as an inverter, thruster or an anchor windlass under full load require similar large amounts of current. Starting batteries have the following characteristics.

a. **Construction.** The starting battery is characterized by thin, closely spaced porous plates, which give maximum exposure of active plate material to the electrolyte and offer minimal internal resistance. This enables maximum chemical reaction rates, and maximum current availability. Physical construction is similar to deep cycle batteries.

b. **Cycling.** Starting batteries cannot withstand cycling, and if deep cycled or flattened have an extremely short service life. Ideally they should be maintained within 95% of full charge.

c. **Sulfation.** In practice, sulfation is not normally a problem, as batteries are generally fully charged if used for starting applications only. If improperly used for deep cycle applications and undercharged they will sulfate.

d. **Self Discharge.** Starting batteries have low self-discharge rates and this is generally not a problem in normal engine installations.

e. **Efficiency.** Cold temperatures dramatically affect battery performance. Engine lubricating oil viscosities are also affected by low temperature and further increase the starting loads on the battery. The reduction in battery capacity in low temperatures, combined with the increased starting current requirements, amplifies the importance of having fully charged batteries. Table 1-4 illustrates the typical cranking power loss when temperature decreases from 27°C to 0°C using a 10W-30 multi-viscosity lubricating oil and the increased percentage of power required to turn over and start an engine.

Table 1-4 Battery Power Table

Temperature	Battery Level	Power Required
+ 27°C	100 %	100 %
0°C	65 %	155 %
- 18°C	40 %	210 %

f. **Charging.** Recharging of starting batteries is identical to deep cycle batteries. Additional factors to consider are:

(1) Discharged current must be restored quickly to avoid damage. Similarly temperature compensation must be made.

(2) Normally after a high current discharge of relatively short duration, there is no appreciable decrease in electrolyte density. The battery is quickly recharged, as the counter voltage phenomenon does not have time to build up and has a negligible affect on the charging.

g. **Battery Ratings.** Starting batteries are normally specified on the basis of engine manufacturers' recommendations, although I have found these to be imprecise. The following is given as a guide only. Table 1-5 shows recommended battery ratings and typical plate numbers for various diesel ratings as well typical starter motor currents:

(1) **Start Capability.** Calculate a good safety margin allowing for a multi-start capability. Some classification societies specify a minimum of six consecutive starts, and that should be the absolute minimum value.

(2) **Temperature Allowance.** Additional allowances should be made for the decreased efficiency in cold climates as a greater capacity and greater load current is required.

h. **Additional Starting Battery Loadings.** The starting battery should also be used to supply short duration, high current loads. Check with your engine supplier for the recommended battery rating, and then add a margin for safety. Also factor in the following:

(1) **Anchor Windlass.** The very heavy current loadings that electric windlasses demand require a much higher bank rating. The battery banks should be doubled up, so that two identical batteries are then parallel connected.

(2) **AC Generator.** In some cases the engine battery can be used for starting. Be careful if starting the engine while the generator is running as the small 10- to 15-amp alternators regularly suffer damage from the engine starting motor's high current load. It is better to have a separate battery.

(3) **Bow Thrusters.** DC powered thrusters are generally powered directly from the engine starting batteries. As the engine should be running much of the load is supplied directly from the alternator. It must be noted that as the engine is in slow or idle the full output is often not available, and considerable load will be taken from the battery if engine speed is too low.

(4) **Deck Winches.** The electric deck winch also has very heavy current loadings and will also require a much higher bank rating. The battery banks should be doubled up, so that two identical batteries are then parallel connected.

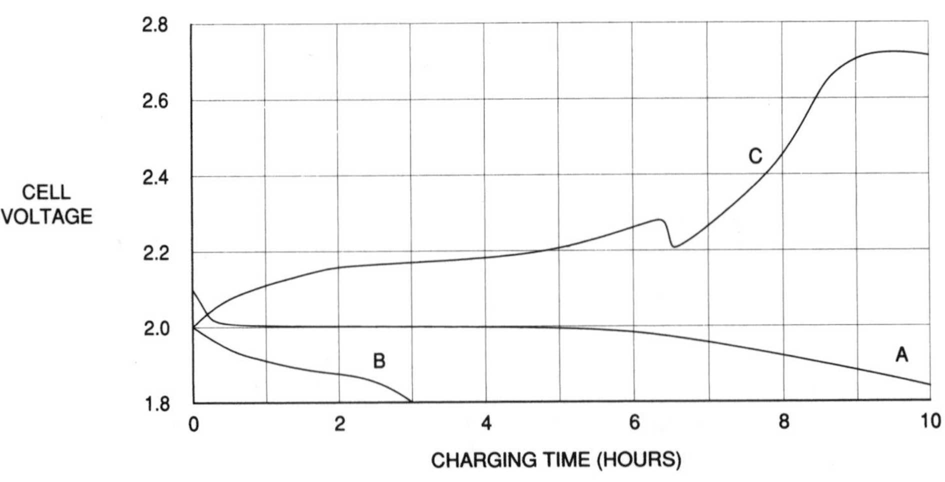

Figure 1-5 Lead Acid Battery Characteristics

Table 1-5 Battery Ratings Table

Engine Rating	Current Load	Battery CCA	Voltage
10 hp 7.5 kW	59 amps	375 CCA	12
15 hp 11 kW	67 amps	420 CCA	12
20 hp 15 kW	67 amps	420 CCA	12
30 hp 22 kW	75 amps	450 CCA	12
40 hp 30 kW	85 amps	500 CCA	12
50 hp 37 kW	115 amps	500 CCA	12
100 hp	115/60 amps	500 CCA	12/24
150 hp	150/75 amps	600 CCA	12/24
200 hp	120 amps	800 CCA	24

1.11 Battery Safety. The lead-acid battery is potentially hazardous and the following safe handling procedures should be used:

a. **Gas.** Battery cells contain an explosive mixture of hydrogen and oxygen gas at all times. An explosion risk is always possible if naked flames, sparks or cigarettes are introduced into the immediate vicinity. Always use insulated tools. Cover the terminals with an insulating material to prevent accidental short circuit. Watchbands, bracelets and neck chains can accidentally cause a short circuit.

b. **Acid.** Sulfuric acid is highly corrosive and must be handled with extreme caution. If there is ever a need to refill a battery with new acid on motorboats observe the following precautions:

- **(1)** Wear protective clothing and eye protection.
- **(2)** Avoid splashes or spills as acid can cause severe skin and clothing burns. If acid splashes into the eyes, irrigate with water for at least 5 minutes. Seek medical advice.
- **(3)** If electrolyte is accidentally swallowed, drink large quantities of milk or water, followed by milk of magnesia. Seek immediate medical attention.

c. **Manual Handling.** Always lift the battery with carriers if fitted. If no carriers are fitted, lift using opposite corners to prevent case distortion and electrolyte spills.

d. **Electrolyte Spills.** Electrolyte spills should be avoided, but observe the following measures:

- **(1)** Spilling electrolyte into salt water will generate chlorine gas, so ventilate the area properly.
- **(2)** Neutralize any spills immediately using a solution of baking soda.

1.12 Battery Voltage and Installation. Batteries must be installed correctly. There are a number of important criteria to consider when installing battery banks to make up the required voltage and capacity.

Cell Size. Battery banks may be installed either in cell multiples of 1.2V, 6V, 12V, 24V, 32V, 36V (42V) or 48V. Each configuration has advantages both physically and operationally:

(1) **1.2 Volts.** This is generally impractical because the batteries takeup so much space. The battery plates are generally more robust and thicker. This leads to increased service life, but it is an expensive option.

(2) **6 Volts.** This is the ideal arrangement. The cells are far more manageable to install and remove (Size GC2 and J250). Large capacity batteries are simply connected in series. Electrically they are better than 12-volt batteries, generally having thicker and more durable plates. Contrary to some opinions, a series arrangement does not necessarily reduce the available power range, nor does it require an equalization network, and these are rarely found. The one proviso is that batteries must be of the same make, model and age. If one battery requires replacement then the other should also be replaced simultaneously.

(3) **12 (14 Volts).** This is the most common arrangement. Physically batteries up to around 115Ah (Size 27 and 31) are easily managed. They are paralleled in banks of up to three and this is the most common arrangement. It is not rare to see traction or truck batteries of very large dimensions such as 8D sizes installed but this is very impractical from any service standpoint. If the battery space is constructed to take a 3-battery arrangement, it is relatively easy to replace one unit. Additionally if you have a multiple bank and lose one with cell failure, two will remain. Charging voltage is 14 volts.

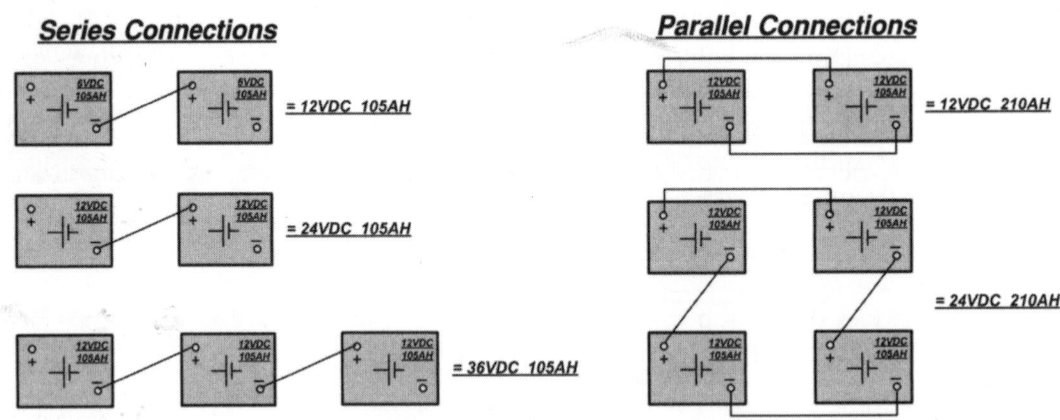

Figure 1-6 Cell and Battery Arrangements

(4) **24 (28 Volts).** Used in commercial vehicles it also is prevalent on larger vessels, and is a standard voltage in commercial shipping control and backup power supply systems. This is simply any of the above battery or cell sizes connected in series to get the 24 volts. Some boats have series-parallel systems. Charging voltage is 28 volts.

(5) **32 Volts.** This voltage level has been in service for many years and is comparatively uncommon. The voltage level is made up from 4 x 8 volt batteries, which some manufacturers still make available.

(6) **36 (42 Volts).** This relatively new automotive voltage requires the connection of 3 x 12 volt cells or 6 x 6 volt cells in series to make 36 volts. The charging voltage is 42 volts and probably will not appear on boats for quite a while.

(7) **48 Volts.** This voltage level is used for the powering of thrusters. This entails series battery connection of 4 x 12 volt units to get 48 volts. This gives rise to charging problems, and this is overcome by use of two 24-volt banks with a series switch when in use so that charging is still only at 28 volts.

1.13 Battery Installation Recommendations.

Batteries should be installed within a separate space or compartment that is located above the maximum bilge water level, and protected from mechanical damage.

The batteries should be installed in a lined box protected from temperature extremes. The preferred temperature range is 10°C–27°C. The box should be located as low down as possible in the vessel for weight reasons, but high enough to avoid bilge water or flooding. After a bad knockdown, and with water over the sole, many boats have compounded their problems by having the batteries contaminated with salt water. On multihulls the batteries should either be located centrally in the mast area to center the weight, or be divided into two banks with one bank in each hull. This effectively gives two separate house banks, plus two engine-start batteries. Sufficient natural light should be available for testing or servicing. If this is not possible, an ignition-protected light can be installed. Allow sufficient clearance to install and remove batteries. Ensure there is sufficient vertical clearance to allow hydrometer testing. Ideally, batteries should not be located within machinery spaces where they might be exposed to high ambient temperatures.

Batteries shall not be installed adjacent to any fuel tank, fuel pipe or parts of the fuel system.

The batteries should not be installed close to any source of ignition, such as fuel tanks and parts such as fuel filters, separators and valves. Any leak or accumulation of fuel represents a serious hazard and so any source of ignition should be removed.

Batteries should be installed within an enclosure, or have a tray that will contain any spills of electrolyte at all angles of heel or inversion.

The box should be made of plastic, fiberglass or lead lined to prevent any acid spills contacting with wood or water. Boxes should be at least the full height of the battery so that any spills will be contained at all times. PVC battery boxes are acceptable alternatives.

Batteries should be secured so that they do not move at any angle of heel or inversion.

Physically secure batteries with either straps or a removable restraining rod across the top. Batteries should be prevented from movement. Insert rubber spacers around the batteries to stop any minor movements and vibrations. The ideal configuration is to arrange the batteries athwartships. This offers marginally better protection against acid spilling under excess heel. Yet even in a fore-and-aft layout, I have not come across any adverse problems. On a friend's steel cruising yacht, I have seen a gimballed tray to prevent electrolyte spills when heeling, although this is probably excessive.

Battery terminals and connections shall be installed or protected against any accidental contact with metallic objects. Battery terminals should be coated with petroleum jelly or equivalent compound to prevent corrosion or interaction with electrolyte spray.

Battery box lids should be in place at all times and secured. PVC or other connection covers should be installed where accidental contact by tools or other items can cause a short circuit across the terminals. Terminals should be coated to limit the corrosive effects of acid.

Battery compartments shall not contain any electrical equipment liable to cause ignition of any generated gases or vapor.

Do not install any electrical equipment within the compartment or adjacent to batteries, as sparks may be generated and cause ignition of hydrogen gas generated after battery charging.

Where batteries are not of the sealed type, the battery compartment should have adequate ventilation to atmosphere of all generated gases.

The area should be well ventilated and vented to atmosphere. The use of an extraction fan is rarely required but should be considered if natural convection methods will be insufficient. If a fast charging device is installed, ensure that the ventilation remains sufficient to remove any generated gasses, and prevent them accumulating.

Where start and service batteries have an interconnecting switch for emergency power supply, the switch should be normally open.

In dual battery systems where an emergency bridging or paralleling switch is installed, it should be always be in the open position. Many people leave them permanently closed and discharge both batteries.

Start and service batteries shall be electrically separate and arranged so that service loads cannot discharge the engine start battery.

In most cases the battery negatives are bridged, and a separate negative for each should be installed. This requires a separate alternator charging negative and grounding negative to the same grounding point as the other battery. Any arrangement should ensure that the start battery could not be accidentally discharged. Where a solenoid system is used to parallel the batteries for charging it must always open when the charging ceases.

Battery interconnection cables should have the same rating as the main start circuit cables.

In dual battery system the cables connecting each battery negative or positive should be rated the same as main supply cables. Many are installed in smaller cable sizes.

Equipment having high current ratings such as thrusters and windlass systems should be installed to limit the disturbances or affects on the stability of the electrical system.

Where high current equipment can cause system disturbances such as large load surges and voltage droops, consideration should be given to installing a separate battery bank with the required characteristics to power the equipment. Alternatively ensure that if they are powered off the engine start battery that the battery is uprated to meet the additional loads.

1.14 Battery Commissioning. After installation, the following commissioning procedures should be carried out.

a. **Battery Electrolyte Level.** Check the electrolyte level in each cell:

(1) Cells with separator guard — fill to top of guard.

(2) Cells without guard — fill to 2mm above plates.

b. **Battery Electrolyte Filling.** If the level is low, and evidence suggests a loss of acid in transit, refill with an electrolyte of similar density. Specific gravity is normally in the range 1.240 to 1.280 at 15°C. If no evidence of spillage is apparent, top up electrolyte levels with de-ionized or distilled water to the correct levels.

c. **Battery Terminals.** Battery terminals are a simple piece of equipment, yet they cause an inordinate amount of problems:

(1) **Terminals.** Install heavy-duty marine grade brass terminals. Do not use the cheaper plated brass terminals, as they are not robust and fail quickly.

(2) **Clean Terminals.** Make sure that the terminal posts are clean, that they do not have any raised sections, and are not deformed, as a poor connection will result.

(3) **Replace Connections.** Replace the standard wing nuts on terminals with stainless steel nuts and washers. The wing nuts are very difficult to tighten properly without deformation and breakage. I have encountered many installations where the wing castings are broken.

(4) **Coat Terminals.** Coat the terminals with petroleum jelly.

d. **Battery Cleaning.** Cleaning involves the following tasks:

(1) **Clean Surfaces.** Clean the battery surfaces with a clean, damp cloth. Moisture and other surface contaminations can cause surface leakage between the positive and negative terminals.

(2) **Grease and Oil Removal.** Grease and oil can be removed with a mild detergent and cloth.

e. **Battery Charging.** After taking delivery of a new battery perform the following:

(1) **Initial Charge.** Give a freshening charge immediately.

(2) **Routine Charging.** Give a charge every week if the vessel is incomplete or not in service.

1.15 **Battery Routine Testing.** The following tests can be made on a weekly basis to monitor the condition of the battery. Battery status can be measured by checking the electrolyte density and the voltage as follows.

a. **Stabilized Voltage Test.** Voltage readings should be taken with an accurate voltmeter. Switchboards should incorporate a high quality meter, not a typical engine gauge charge indicator. The difference between fully charged and discharged is less than 1 volt, so accuracy is essential. A digital voltmeter is the ideal. Battery voltage readings should only be taken a minimum of 30 minutes after charging or discharging. Turn off all loads before measuring. Typical values at 15°C are shown in Table 1-6. Manufacturers have slightly varying densities so check with your supplier.

Table 1-6 Typical Open Circuit Voltages and Densities

Charge Level	SG Temperate	SG Tropical	Voltage
100%	1250	1240	12.75
90%	1235	1225	12.65
80%	1220	1210	12.55
70%	1205	1195	12.45
60%	1190	1180	12.35
50%	1175	1165	12.25
40%	1160	1150	12.10
30%	1145	1135	11.95
20%	1130	1120	11.85
10%	1115	1105	11.75
0	1100	1090	11.65

b. **Battery Electrolyte Specific Gravity.** A hydrometer should be used weekly to check acid density. The hydrometer is essentially a large syringe with a calibrated float. The calibration scale is corrected to a nominal temperature value, which is normally marked on the float. The following points should be observed during testing with a hydrometer:

(1) Never test immediately after charging or discharging. Wait at least half an hour until the cells stabilize. It takes some time for the pockets of varying electrolyte densities to equalize. Never test immediately after topping up the electrolyte. Wait until after a charging period, as it similarly takes times for the water to mix evenly.

(2) Ensure the float is clean and not cracked and the rubber has not perished. Keep the hydrometer vertical. Ensure that the float does not contact the side of the barrel, which may give a false reading. Draw sufficient electrolyte into the barrel to raise the float. Ensure that the top of the float does not touch the top. Observe the level on the scale. Disregard the liquid curvature caused by surface tension. Adjust your reading for temperature to obtain the actual value.

(3) Wash out the hydrometer with clean water when finished.

c. **Battery Load Test.** The load test is carried out only if the batteries are suspect. The load tester consists of two probes connected by a resistance and a meter. The tester is connected across the battery terminals effectively putting a heavy load across it. A load of approximately 275 amps at 8 volts is normal. Take your battery to your nearest automotive electrician or battery service center for a test.

1.16 Battery Maintenance. Battery maintenance is simple and is not the tedious chore that it is often made out to be. The following tasks should be carried out.

a. **Battery Terminal Cleaning.** (3-monthly). Remove battery terminals and ensure that terminal posts are clean and free of deposits. Refit and tighten terminals and coat with petroleum jelly, not grease.

b. **Battery Electrolyte Checks.** (monthly). Check levels along with density. Record each cell density so that a profile can be built up. Record the battery voltage as well. Top up cells as required with distilled or de-ionized water.

c. **Battery Cleaning.** (monthly). Wipe battery casing top clean with a damp rag. Moisture and salt can allow tracking across the top to ground or negative, slowly discharging the battery. This is a common cause of dead batteries, and mysterious but untraceable system "leaks."

1.17 Battery Additives. There are a number of additives on the market. The claims made by manufacturers appear to offer significant performance enhancement. The compounds are specifically designed to prevent sulfation or dissolve it off the plate surfaces. If you read the fine print on one brand, it is not recommended for anything other than new or near-new batteries. If the additive is to dissolve sulfates on battery plates, it will be only on the surface, as plate sulfation occurs through the entire plate, so only a partial improvement is achieved. Recently a friend of mine returned after an extended Pacific cruise and called over a charging problem. I had installed a smart regulator three years previously and he had managed the entire period as a liveaboard without a problem, until he put in an additive. My advice is to leave such additives alone, and your battery electrolyte should remain untouched. Just make sure the battery is properly charged and topped off and you won't need to resort to such desperate measures.

Table 1-7 Lead-Acid Battery Troubleshooting

Symptom	Probable Fault
Will not accept charge	Plates sulfated Maximum battery life reached
Low cell electrolyte SG	Cell sulfated
Low battery SG value	Low charge level (regulator failure) Plates sulfated (undercharging problem)
Will not support load	Low charge level (undercharging problem) Plates sulfated
Cell failure	Improperly commissioned Electrolyte contaminated (impure water) Overcharging problem (regulator failure) Undercharging problem (regulator failure) Excess vibration and plate damage Cell internal short circuit
Battery warm	Plates sulfated Excessive charge current (regulator failure) Cells damaged

1.18 Low Maintenance Batteries. Sealed low maintenance batteries are not generally suited to cruising vessel applications. Frequently, they are installed without considering their performance characteristics or their various advantages and disadvantages.

a. **Low Maintenance Principles.** Basic chemical reactions are similar to the conventional lead acid cell. The differences are as follows:

(1) **Lead Acid Batteries.** In a normal lead acid battery, water loss occurs when water is electrically broken down into oxygen and hydrogen close to the end of charging. In any battery during charging, oxygen will develop at the positive plate at approximately 75% of full charge level. Hydrogen is generated at the negative plate at approximately 90% of full charge. These are the bubbles seen in the cells during charging. In normal batteries, the gases disperse to atmosphere, resulting in electrolyte loss that requires periodic water replacement.

(2) **Low Maintenance Batteries.** The low maintenance recombinational battery has different characteristics. The plates and separators are held under pressure. During charging, oxygen is only able to move through the separator pores from positive to negative, reacting with the lead plate. The negative plate charge is then effectively maintained below 90%, inhibiting hydrogen generation.

b. **Low Maintenance Battery Safety.** Batteries are totally sealed, but incorporate a safety valve. Each cell is also sealed, with a one-way vent. When charging commences, oxygen generation exceeds the recombination rate and the vents release excess pressure within the battery. Excessive charge rates can create internal pressure build-up. If the pressure exceeds the safety vent's discharge rate, an explosion can occur.

c. **Charging.** Low maintenance batteries must only be charged at recommended charging rates and charge starting currents. The result of any overcharging may be explosion.

d. **Advantages.** The following are advantages of low maintenance batteries:

(1) **Low Water Loss.** Low water loss is the principal advantage; however performing a routine monthly inspection and occasional topping up of a lead acid battery is not labor intensive or inconvenient. I am amazed that this factor is the main one put forward as the criterion for these batteries. If you are continually topping up, then you have a charging problem or a high ambient temperature.

(2) **Inversion, Heel and Self Discharge.** The batteries are safe at inversion or excessive heel angles without acid spilling, and have a low self-discharge rate.

e. **Disadvantages.** There are two major disadvantages that make low maintenance batteries unsuitable for cruising applications:

(1) **Over-Voltage Charging.** Low maintenance batteries are incapable of withstanding any over-voltage during charging. If they are subjected to high charging voltages (above 13.8 V), water will vent out and they have been known to explode. This means no fast charging devices should be installed to charge them.

(2) **Cycle Availability.** Cycle availability is restricted, and an approximate lifespan of 500 cycles to 50% of nominal capacity is typical. Any discharge to 40% of capacity or less makes recharging extremely difficult if not impossible, and requires special charging techniques.

1.19 Gel Cell Batteries. These battery types are known as Dryfit from Sonnenschein (www.sonnenschein.org) or Prevailer batteries. A quality deep cycle lead acid battery can have a life exceeding 2500 cycles of charge and discharge to 50%. A gel cell has a life of approximately 800–1000 cycles. They do have a much greater cycling capability than normal starting batteries, but not of good deep cycle or AGM batteries.

a. **Electrolyte.** Unlike normal lead-acid cells the gel cell has a solidified thixotropic gel as an electrolyte, which is locked into each group of plates. The gel electrolyte has a high viscosity and during the charge and discharge process develops voids and cracks. These can impede the flow of acid and cause capacity loss. During charging the gel also liquefies due to its thixotropic properties, and solidification after charging can exceed an hour as thixotropic gels have a reduced viscosity under stress. The newer types use phosphoric acid in the gel to retard the sulfation hardening rates. Water loss can occur in VRLA batteries, and the oxygen recombination cycle is used to minimize electrolyte loss. Sources of water loss are small for each source, however cumulatively they can cause failure, and the process is called dry-out. Water loss causes include reduced recombination efficiency, high charge voltages, corrosion of the positive grid, transpiration through the cell casing, typically in temperatures exceeding 40°C, and self discharge, in temperatures greater than 20°C. Batteries should be installed in cool areas wherever possible.

b. **Construction.** The plates are reinforced with calcium, rather than antimony, which reduces self-discharge rates, and they are relatively thin. This facilitates gel diffusion and improves the charge acceptance rate as diffusion problems are reduced. The separator provides electrical and mechanical isolation of the plates, and must have a high porosity to facilitate ion migration and electrolyte acceptance. Each cell has a safety valve to relieve excess pressure if the set internal pressure is exceeded. Typical values are 114 psi (100mbar). The valve recloses tightly to prevent oxygen from entering the cell. Lead acid batteries suffer from corrosion of the current collector grid. Typical grids in VRLA batteries are manufactured from lead-calcium-tin, lead-tin-alloys or lead-antimony-cadmium. Positive plate grids corrode due to the conversion of lead-to-lead dioxide. This corrosion effect doubles for every 10°C, as well as charge voltages and electrolyte density. As the lead dioxide requires a substantially higher volume, this causes mechanical stressing and deformation of the grid. The effect is to active plate material contact with the grid causing capacity loss, internal short circuits and even cell case ruptures, which increases with service time. The process also consumes water reducing the electrolyte causing lowered performance. VLRA cells also are prone to chemical degradation of the negative plate lugs and strap surfaces called sulfate rot. This is caused by the oxygen recombination reaction and electrolyte inorganic sulfate salts.

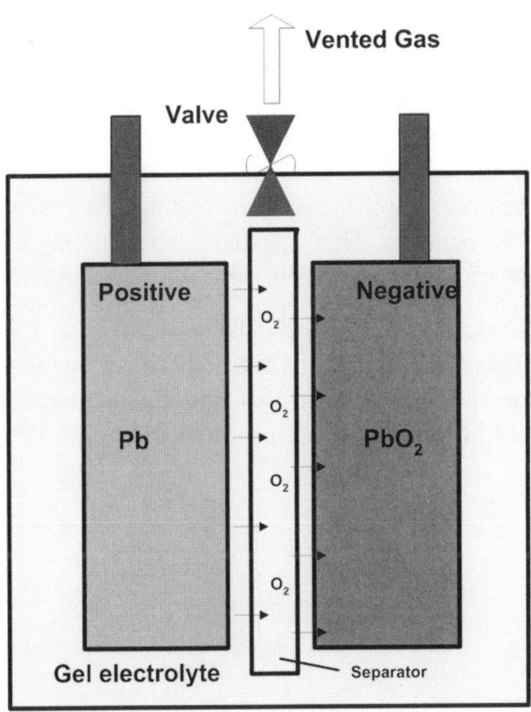

Figure 1-7 Gel Battery Characteristics

1.20 Gel Cell Charging. Batteries have a much higher charge acceptance rate, and therefore a more rapid charge rate is possible. A gel cell cannot tolerate having any equalizing charge applied and this over charge condition will seriously damage them. During charging the current causes decomposition of the water and the evolvement of oxygen at the positive plate. The oxygen diffuses through the unfilled glass mat separator pores to the negative plate, and chemically reacts to form lead oxide, lead sulfate and water. The charge current then reduces and does not evolve hydrogen gas. The end of charge voltage is typically 2.22 to 2.28 Vpc (Volts per cell). If recombination of hydrogen is incomplete during overcharge conditions, the gases may vent to the battery locker causing an explosion risk. Although accepting a higher charge rate than a lead-acid deep cycle battery, and consequentially charging to a higher value, there is at a certain point the problem of attaining full charge, and therefore capacity usage of the battery bank. As no fast charge devices can be safely used, a longer engine run time is required for complete recharging. While these batteries will accept some 30–40% greater current than an equivalent lead acid battery they are restricted in the voltage levels allowed. Typical open circuit voltage at 100% charge is 12.85–12.95 volts, 75% is 12.65 volts, 50% is 12.35 volts, 25% is 12 volts and 11.8 volts is flat. Gel cells are intolerant to over voltage charge conditions and can be damaged in over charge situations. The normal optimum voltage tolerance on Dryfit units is 14.4 volts. There are some minimal heating effects during charging, and this is caused by the recombination reaction. Where batteries are not kept within reasonable temperature ranges thermal runaway can occur. This normally occurs during charging when the temperature of the battery and charge current create a cumulative increase in temperatures that leads to battery destruction. Continuous over or undercharging of gel cells is the most common cause of premature failure. In many cases this is due to use of imprecise automotive type chargers.

1.21 Absorbed Glass Mat (AGM) or Valve Regulated Lead Acid (VRLA) Batteries. These batteries are installed on my boat and to date the performance has been excellent. There are variations to flat plate manufacturing techniques, and the Optima AGM batteries have a spiral cell, dual plate construction. Another important feature is a greater shock and vibration resistance than gel or flooded batteries. They also have extremely high CCA values of up to 800 amps at 0°F. Manufacturers include www.eastpenn-deka.com and www.concordebattery.com

a. **Electrolyte.** The electrolyte is held within a very fine microporous (boron-silicate) glass matting that is placed between the plates. This absorbs and immobilizes the acid while still allowing plate interaction. They are also called starved electrolyte batteries, as the mat is only 95% soaked in electrolyte.

b. **Recombinant Gas Absorption Principles.** In a normal lead-acid battery, water loss will occur when it is electrically broken down into oxygen and hydrogen near the end of charging. In a battery during charging, oxygen will evolve at the positive plate at approximately 75% of full charge level. Hydrogen evolves at the negative plate at approximately 90% of full charge. In normal batteries, the evolved gases disperse to atmosphere, resulting in electrolyte loss and periodic water replacement. These are the bubbles seen in the cells during charging. During charging the current causes decomposition of the water, and oxygen evolves on the positive plate. The oxygen then migrates through the unfilled pores of the separator matting to react with the negative plate and form lead oxide, lead sulfate and water. The charge current reduces and does not generate hydrogen. The low maintenance recombinational battery has different characteristics. The plates and separators are held under pressure. During charging, the evolved oxygen is only able to move through the separator pores from positive to negative, reacting with the lead plate. The negative plate charge is then effectively maintained below 90% so inhibiting hydrogen generation. They emit less than 2% hydrogen gas during severe overcharge (4.1% is flammable level).

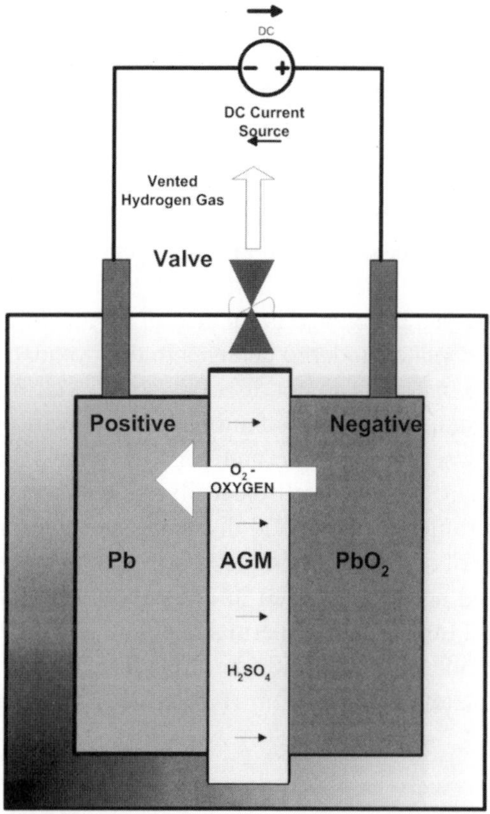

Figure 1-8 AGM Battery Characteristics

c. **Charging.** Charging of AGM cells has few limitations, and no special charge settings are required. Typical charge voltages are in the range 14.4 to 14.6 volts at 68°F (20°C). The batteries have a very low internal resistance, which results in minimal heating effects during heavy charge and discharge. They can be bulk charged at very high currents, typically by a factor of five over flooded cells, and a factor of 10 over gel batteries. They also allow 30% deeper discharges and recharge 20% faster than gel batteries, and have good recovery from full discharge conditions. Self-discharge rates are only 1%–3%. If you are a weekend, harbor or river cruiser, who does limited motoring periods, or leave the boat unattended for long periods, the AGM battery is a viable proposition, as it has very low self-discharge rates, and very high recovery rates from deep discharges. If a small solar panel is left on with a suitable regulator, they will recoup the annual costs of replacing deep cycle batteries by lasting several seasons, with the more important improvements in reliability. Typical charge voltage levels are: 100% is 12.8–12.9 volts, 75% is 12.6 volts, 50% is 12.3 volts, 25% is 12 volts and 11.8 volts is dead. At high temperatures AGMs (and gel cells) are unable to dissipate the heat generated by oxygen and hydrogen recombination and this can create thermal runaway. This will lead to gassing and the drying out of cells. A premature loss of capacity can occur when the positive plate and grids degrade due to higher operating temperatures. This is caused by the exothermic recombination process and higher charge currents. In addition negative plates also degrade due to inadequate plate conversion. The main failure modes are cell shorting and pressure vent malfunctions caused by manufacturing faults.

1.22 Alkaline Batteries. Alkaline cells are typified by the nickel cadmium (NiCad) and nickel iron (NiFe) batteries. The principal factors are cost, (typically 500% greater), greater weight and physically larger bank size. Normally these batteries will only be found in larger motor vessels for those reasons. They have completely different operating characteristics to the lead-acid cell. The obvious difference is the use of an alkaline electrolyte instead of an acid. Unlike lead-acid cells, plates undergo changes in their oxidation state, altering very little physically. As the active materials do not dissolve in the electrolyte, plate life is very long. The electrolyte is a potassium hydroxide solution with a specific gravity of 1.3. The electrolyte transports ions between the positive and negative plates and the alkaline solution is chemically more stable than lead-acid cell electrolytes. Unlike lead-acid cells, the density does not significantly alter during charge and discharge and hydrometer readings cannot be used to determine the state of charge. Electrolyte loss is relatively low in operation. Lead-acid and NiCad batteries should never be located in the same compartment as the cells will become contaminated by acid fumes causing permanent damage. The components of the NiCad cell are Nickel-Hydroxide (2Ni(OH)2)–the positive plate; Cadmium Hydroxide (Cd(OH)2)–the negative plate; and Potassium Hydroxide (KOH)–the electrolyte.

a. **Discharge Cycle.** Cells are usually characterized by their rate of discharge, such as low, medium, high or ultra high. Classification UHP is for starting applications and VP for general services. There is also a category for deep cycle applications. Discharge ratings are given at the five hour rate and typically they will deliver current some 30% longer than lead-acid equivalents. The amp-hour capacity rating remains stable over a range of discharge currents values. An over-discharge condition can occur when the cell has been driven into a region where voltage has become negative. A complete polarity reversal takes place. No long-term effects occur on occasional cell reversal at medium discharge rates. Discharge current reduces cell voltage from 1.3 volts to 1.0 volt over 10 hours.

b. **Charge Cycle.** During charging, the negative material loses oxygen and converts to metallic cadmium. The positive material gradually increases in the state of oxidation. While charging continues, the process will proceed until complete conversion occurs. Approaching full charge gas will evolve and this results from electrolysis of the electrolyte water component. NiCad cells can be charged rapidly with relatively low water consumption. The disadvantages are that cell imbalances may occur and this can cause thermal runaway. The NiCad cell will generally absorb maximum alternator current for about 85% of the cell charge period, so the alternator must be capable of withstanding this load and have adequate ventilation. Typical voltage regulator settings for a nominal 12-volt battery bank of 10 cells over a 2–4 hour period should be in the range of 15 to 15.5 volts. A NiCad battery accepts high charge currents and will not be damaged by them. At 1.6 volts per cell, a NiCad can absorb up to 400% of capacity from a charging source. In most cases it will accept whatever the alternator can supply. The problem with normal alternator regulators is that they fix the output at only 14 volts, which is far too low for proper charging. Absolute maximum charging rates require 1.6 to 1.8 volts per cell, which is 16–18 volts on a typical 10-cell battery bank. The typical 14-volt output of an alternator is only a float charge voltage level for a NiCad battery. Constant voltage charging is the only practical method of charging on vessels. The regulator setting should be around 15.5 volts for a 2–4 hour charge period. Higher voltages will increase current. The charging cell voltage is 1.5 times the 10-hour discharge current. Water additions should be made immediately after charging, and never after discharging.

1.23 NiCad Battery Characteristics. The open circuit voltage of a vented cell is around 1.28 volts. This depends on temperature and time interval from last charge period. Unlike a lead-acid cell, the voltage does not indicate the state of charge. The nominal voltage is 1.2 volts. This voltage is maintained during discharge until approximately 80% of the 2 hour rated capacity has been discharged. This is also affected by temperature and rate of discharge. The closed circuit voltage is measured immediately after load connection. Typically it is around 1.25–1.28 volts per cell. The working voltage is that observed on the level section of the discharge curve of a NiCad cell, voltage plotted against time. Typically the voltage averages 1.22 volts per cell. Capacity is specified in amp-hours. Normally it is quoted at the five-hour rate. The nominal rating is the amp-hour delivery rate over 5 hours to a nominal voltage of 1.0 volt per cell. Internal resistance values are typically very low. This is due to the large plate surface areas used and is why the cells can deliver and accept high current values.

1.24 Small Appliance Batteries. Standard disposable batteries are zinc carbon, long life alkaline, and super alkaline, lithium, silver-oxide and zinc-air batteries. Rechargeable batteries are far more economical. The small nickel cadmium (NiCad) can be recharged several hundred times. These should be completely discharged before recharging. More recent battery technology is the nickel metal hydride (NiMH) cell used in many cellular phones. These cells can withstand recharging up to 1000 cycles. They do not suffer with partial discharge and charge, although it is still good practice. Please do not dispose of batteries over the side into the sea.

Battery-Charging Systems

2.0 Battery-Charging Systems. An efficient battery-charging system is essential for optimum battery and electrical system performance. The principal charging systems on cruising vessels consist of the following:

a. **Alternators.** The alternator is the principal charging source on the majority of sailing yachts. In many cases, it is the only source utilized, even at the dock, due to the alternator's higher available charging currents.

b. **Alternative Energy Systems.** The following methods of alternative energy charging are available as options to supplement engine charging sources:

- Solar Panels
- Wind Generators
- Propeller Shaft Alternators
- Water-Powered Charging Systems
- Fuel Cell Charging Systems

c. **Battery Chargers.** When a vessel is in port, and particularly in liveaboard situations where the main power source is via a shore-powered charger, the battery charger has an important role in the power system.

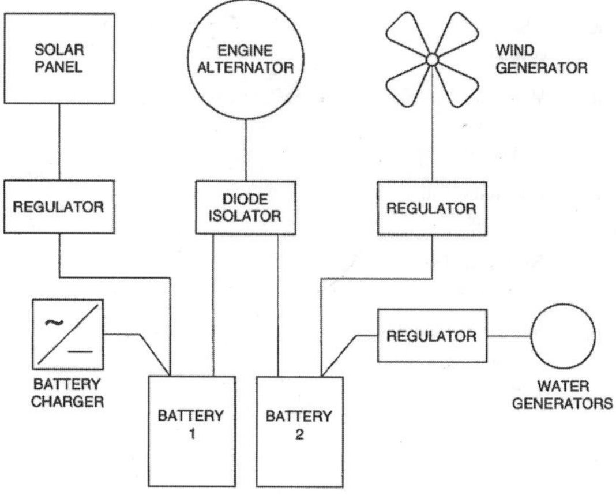

Figure 2-1 Charging Systems

2.1 Charging Cycles. There are four recognized parts of any charging cycle; understanding these parts is crucial to understanding charging systems and problems.

a. **Bulk Charge.** The bulk-charge phase is the initial charging period before the gassing point is reached. This is typically in the range 14.4 to 14.6 volts, corrected for temperature, though with a traditional alternator and regulator, output is fixed at 14 volts. The bulk charge rate can be anywhere between 25% and 40% of rated amp-hour capacity at the 20-hour rate as long as temperature rises are limited.

b. **Absorption Charge.** After attaining the gassing voltage, the charge level should be maintained at 14.4 volts until the charge current falls to 5% of battery capacity. This level normally should equate 85% of capacity. In a typical 300 amp-hour bank, this will be 15 amps.

c. **Float Charge.** The battery charge rate should be reduced to a float voltage of approximately 13.2 to 13.8 volts to maintain the battery at full charge.

d. **Equalization Charge.** A periodic charge rated at 5% of the installed battery capacity should be applied for a period of 3–4 hours until a voltage of 16 volts is reached. A suitable and safer way of equalizing is applying the unregulated output from the wind generator or solar panel once a month for a day.

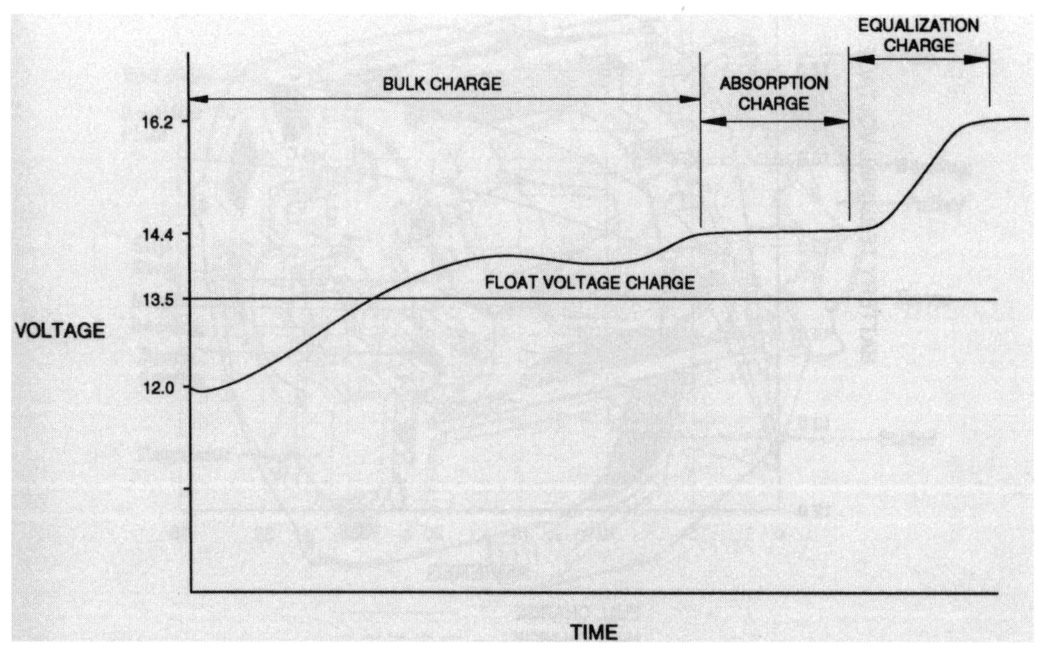

Figure 2-2 Charging Cycles

2.2 Charging Efficiency. Before any charging systems can be considered, a number of factors must be summarized and taken into account. Nominal capacities of batteries are specified by manufacturers, and the total capacity of the bank must be taken into consideration. Older batteries have reduced capacities due to normal in-service aging, and plate sulfation. Sulfation increases internal resistance and therefore inhibits the charging process. The electrolyte is temperature dependent, and the temperature is a factor in setting maximum charging voltages.

a. **State of Charge.** The state of charge when charging begins can be checked using the open circuit voltage test and electrolyte density. The level of charge will affect the charging rate. Also critical to the state of charge is the temperature. It has a dramatic effect on charge voltages as indicated in the curve below.

b. **Charging Voltage and Regulation.** Charging voltage is defined as the battery voltage plus the internal cell voltage drops. Cell volt drops are due to internal resistance, plate sulfation, electrolyte impurities, and gas bubble formation that occur on the plates during charging. These resistances oppose charging and must be exceeded to effectively recharge the battery. Resistance to charging increases as a battery reaches a fully-charged state and decreases with discharge. A battery is self regulating in terms of the current it can accept under charge. Over-current charging at excessive voltages, which many fast-charging devices do, simply generates heat and damages the plates, which is why some fast-charging devices are not recommended.

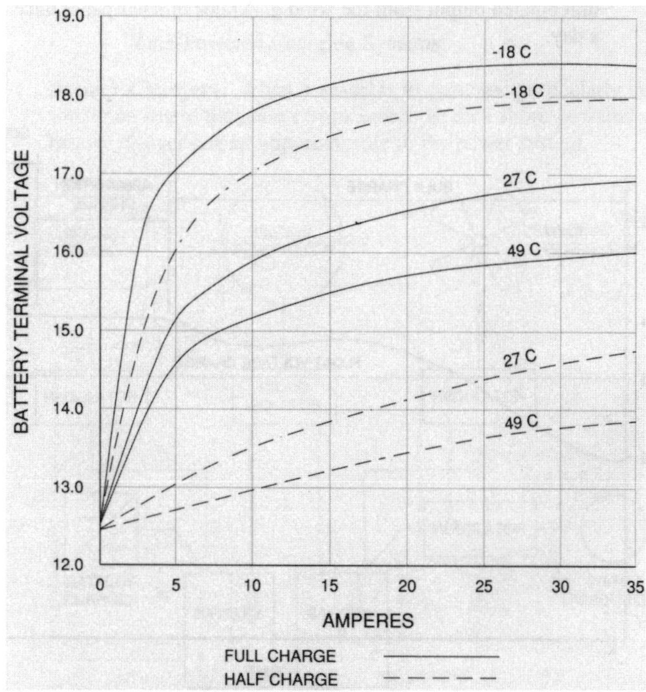

Figure 2-3 State of Charge/Temperature Characteristics

2.3 Charging System Configurations. There are three principal charging systems in use: the changeover switch, the relay and the diode isolator. The charging system on most engines uses the same cabling as the engine starter circuit. Basically, it consists of a switch with three positions and off. The center position parallels both battery banks. It is not uncommon to see both batteries left accidentally paralleled under load with flattening of both. Paralleling of a heavily discharged battery and a fully charged one during charging can also cause some instability in the charging as they both equalize.

a. **Switch Operation Under Load.** If a changeover switch is operated under load, the surge will probably destroy the alternator diodes. Most switches incorporate an auxiliary make before break contact for connection of field. This advanced field switching disconnects the field and therefore de-energizes the alternator fractionally before the opening of the main circuit. In reality this is rarely connected as most alternators have integral regulators and it is difficult to connect the switch into the field circuit.

b. **Surges.** If both batteries are paralleled during an engine start, sensitive electronics can be damaged by the surge.

c. **Circuit Resistance.** In most cases, the cables must run from the batteries to the switch location and back to the starter motor introducing voltage drops. Switches are notoriously unreliable and can introduce voltage drops into the circuit and total alternator or switch failure.

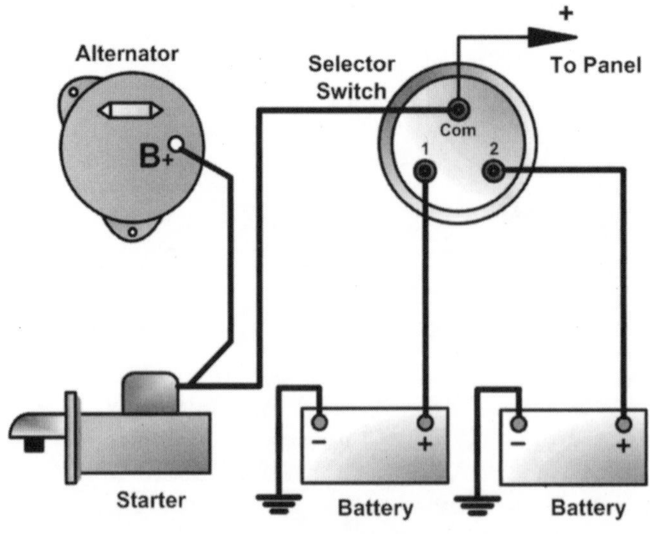

Figure 2-4 Single Engine Changeover Switch Charging System

2.4 Relay/Solenoid Configuration. This system improves on the switch system by separating the charging system from starting circuits. The relay or solenoid does offer a point of failure if incorrectly rated for the task. The relay interconnects both batteries during charging, and separates them when off, preventing discharge between the batteries. The relay-operating coil is interlocked with the ignition and energizes when the key is turned on. When modifying the system, it is necessary to separate the charging cable from the alternator to starter motor main terminal where it is usually connected. A cable is taken directly from the alternator output terminal to the relay as illustrated. Relay ratings should at least match the maximum rated output of the alternator. It is prudent to over-rate the relay. Relays are marketed in various forms, the most common being automotive solenoid types. Another system is the Voltage Sensitive Relay (VSR). The relay is open when the engine is started, and when the voltage rises to 13.7 volts it closes to parallel the two batteries, which then charge together. When the engine stops the relay opens to split the two batteries again. The AutoSwitch from PowerTap is also a similar device. The Isolator Eliminator from PowerTap (www.pwrtap.com and www.amplepower.com) has a similar function. It is a multi-step device; however it does not connect battery banks but charges from the higher housed bank under charge. It is temperature compensated like an alternator regulator.

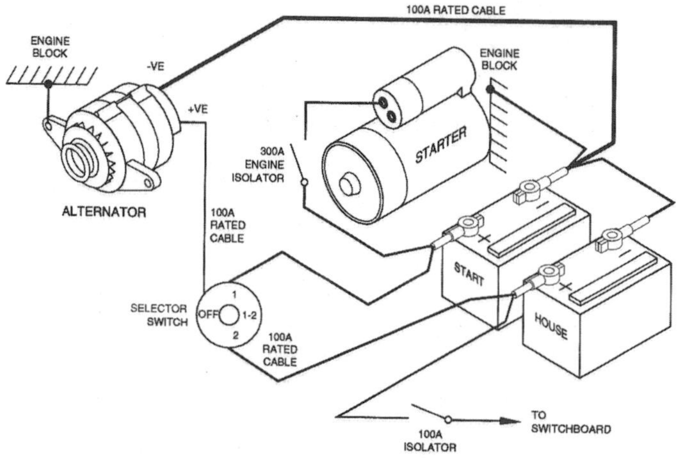

Figure 2-5 Single Engine Relay Charging System Configuration

2.5 Diode System. The diode system is the simplest configuration and the most reliable. A diode has an inherent voltage drop of typically 0.7–0.8 volt. This is unacceptable in a normal charging circuit. If the alternator is machine sensed and does not have any provision for increasing the output in compensation, the diode should not be used. Essentially a diode isolator consists of two diodes with their inputs connected. They allow voltage to pass one way only, so that each battery has an output. This prevents any back feeding between the batteries. They are mounted on heat sinks specifically designed for the maximum current carrying capacity and maximum heat dissipation. The diode isolators must be rated for at least the maximum rating of the alternator, and if mounted in the engine compartment must be overrated to compensate for the de-rating effect caused by engine heat. Heat sink units should have the cooling fins in the vertical position to ensure maximum convection and cooling. Do not install switches in the cables from each output of the diode to the batteries. A variation on this is the Cross Charge Diode from PowerTap, which uses a Schottky diode. These diodes have low voltage drops in comparison with normal diodes.

Diode Isolator Testing. With engine running the diode output terminal voltages should be identical, and should read approximately 0.75 volt higher if a non-battery sensed regulator is being used. The input terminal from the alternator should be zero when the engine is off. Test with power off and batteries disconnected.

(1) Set the meter scale to ohms x1, and connect red positive probe to input terminal. Connect black negative probe to output terminals 1 or 2.

(2) If it is good the meter will indicate minimal or no resistance.

(3) Reverse the probes, and repeat the test. The reading should indicate high resistance, or over range.

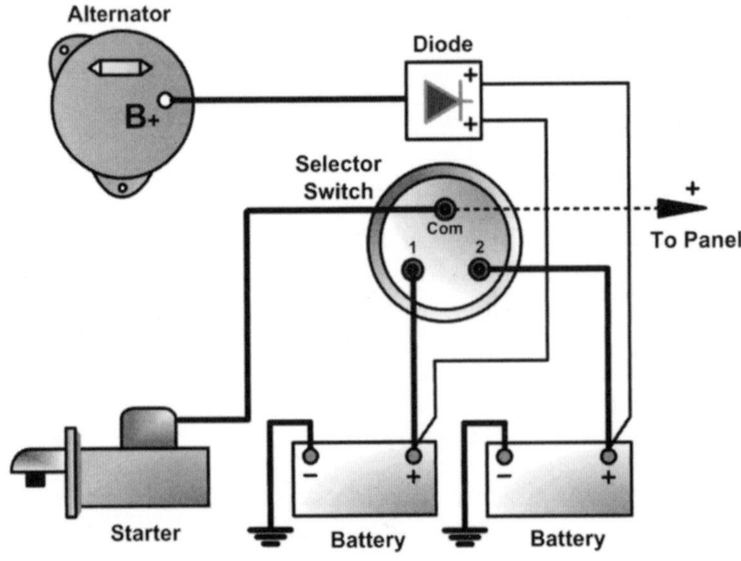

Figure 2-6 Single Engine Diode Charging System Configuration

2.6 Electronic Battery Switches. These are also known as charge distributors or integrators and characterized by the following systems:

(1) **NewMar Battery Bank Integrator (BBI).** When a charge voltage is detected that exceeds 13.3 VDC the unit switches on. The unit consists of a low contact resistance relay that closes to parallel the batteries for charging. When charging ceases and voltage falls to 12.7 VDC the relay opens isolating the batteries. The unit also incorporates a voltage comparator and time delay circuit. This prevents the unit cycling in the event of a voltage transient or load droop on the circuit dropping voltage below the cutout level.

(2) **PathMaker (Heart Interface).** These devices allow charging of two or three batteries from one alternator or battery charger. The units use a high current switch rated at 800 and 1600 amps for alternator ratings up to 250 amps. The unit has an LED status indicator.

(3) **Isolator Eliminator (Ample Power).** This is a multi-step regulator that controls charge to the second battery bank, typically used for engine starting. It is temperature compensated like an alternator control system.

2.7 Failure Analysis and System Redundancy. I am actively involved in large ship systems, and carrying out what is called a Failure Mode and Effects Analysis (FMEA), and the subsequent trials. This principle can also be adopted to sailing yachts. This requires to analyze the starting and charging systems, as both are critical to propulsion or power and to identify single point failures. In a single engine boat there is virtually no redundancy. It is necessary to first identify all of the points that upon failure will also fail the system, and then devise methods to improve redundancy. It is important when assessing risk to consider the following factors and statistics:

a. A failure in the battery charging system means no charging of batteries and therefore an eventual loss of all electrical power, and possibly propulsion starting.

b. A failure in the engine starting system means no propulsion and no charging of batteries, and eventual loss of all electrical power.

c. Approximately 80% of all electrical system circuit failures are due to faulty or failed connections and this is crucial to this exercise.

d. Approximately 70% of equipment and machinery failures are attributable to poor or improper maintenance.

2.8 Failure Analysis and Risk Assessment. A charging system must not be viewed as simply a collection of series connected components, but as a system. The typical charging system comprises a considerable number of elements:

a. **The Charging System.** It is a good idea to trace out each circuit on your boat, and draw in each component and mark each connection on it. As a minimum, you will have 4 main positive circuit connections, 4 main negative circuit connections, 4 control circuit connections, 2 changeover switch contacts, a meter shunt, the alternator, the regulator and the battery.

   (1) The alternator (which includes several components such as brushes, brush gear, slip-rings, bearings, diodes and windings).

   (2) The regulator, (which may be integral or separate).

   (3) The DC positive circuit, (which includes connections at alternator and battery, and the changeover switch).

   (4) The DC negative circuit, (which includes connections at alternator and battery, the cable back to the battery, and the meter shunt if fitted). In addition, the engine block also becomes part of the negative circuit, along with alternator bracket, holding bolts etc).

   (5) The battery.

b. **Charging System Failure Mode Analysis.** There is a total of 14 connection points plus the alternator, regulator and battery that can impact on the starting system. Each point represents a single point failure with subsequent total system failure, with no apparent redundancy. For this exercise wind, water and solar panels are considered extra or supplementary charge sources, as are generators with chargers. These however can be factored into redundancy provisions. The operational factors also must be considered. If a changeover switch is opened or fails during operation, the alternator diodes can be destroyed.

c. **Auxiliary Systems Failure Mode Analysis.** A similar analysis should be carried out on air start systems, fuel supply and filtering systems, the engine cooling water system, both salt and fresh water. The engine air system should be examined for failure modes, and that may include ventilation and fans. The final key area is the propulsion system, and that is the shaft, gearbox, stern tubing, propellers and related equipment.

2.9 Systems Redundancy. The key to minimizing failure or mitigating the effects of failure is the provision of redundancy. In the average single engine yacht, systems do not incorporate any redundancy on charging, power or starting systems. In most commercial shipping, this is a basic premise in all systems design. There are several methods for improving redundancy and the following are the easiest and most economical to carry out. Redundancy is the process of having a backup system. This includes carrying the appropriate spare parts.

2.10 Charging Systems. There are relatively simple modifications that can be carried out on the charging system to improve efficiency and reliability:

a. **Second Alternator.** Installing a second alternator on the engine will require adding a second pulley. The second alternator is for the house battery charging circuit, with the existing alternator being used for charging the start battery. Each alternator will have a separate positive circuit without any switches or other devices in it. This will eliminate changeover switch problems on alternators that commonly destroy the alternator rectifier diodes. This reduces connections to just 2. It also eliminates accidental (human error) switch operation under load, or switch contact failures, which are very common. Each alternator will have a separate negative circuit cable running back to the respective battery from the alternator. This provides separation from the starter motor to battery negative, with the main starter negative serving as a backup. This reduces connections to just 2. It also takes the engine block out of the circuit, and generally reduces voltage drop. There is anecdotal evidence that current flow through a bearing also results in reduced engine bearing life.

b. **Separate Charging System.** Separate the charging system from the starting circuit. Previous illustrations show how these various methods can be done, and in the long term, will considerably reduce problems and increase reliability. This process entails the deletion of battery selection changeover switches, and the installation of a separate charging circuit, which may include charge splitting diodes or relays. An emergency crossover switch between battery banks can be installed; however this does not affect the circuit during operations.

c. **Install Separate Negative Cables.** Install a separate negative conductor of at least 6 AWG (15mm^2) from each alternator case or negative terminal directly back to the corresponding battery negative. This bypasses the engine block and all the cumulative resistances of mountings and brackets. It offers a good low resistance path and reduces stray currents through the block, which can cause pitting of bearings. It also eliminates a single point failure of the main negative connection to the engine block.

d. **Replace Positive Cable.** Most installed positive cables are too small, especially if a fast charge device is installed. The cable size should be doubled. Ideally install a minimum of 6 AWG (15mm^2). A problem is that besides having a maximum current going through it with fast charge devices or when heavily discharged batteries are recharged, the heat of the engine compartment also de-rates the current capacity of the cable. In most cases a significant voltage drop develops across the cable under full output conditions.

2.11 Maintenance Factors. Perform the recommended maintenance on all critical equipment and systems.

a. **Alternators.** Alternators have a relatively low failure rate, as actual operating hours are relatively low. Failures are generally caused by diode failures, or overheating, in particular with fast charge regulators and oversized battery banks. Alternators should be cleaned and overhauled on a regular basis, ideally every two years. Consider a higher rated alternator to reduce overloading and heating.

b. **Batteries.** Batteries have the highest failure rate. This is generally due to either inadequate charging with resultant sulfation, lost capacity and failure, or flattening of the battery with subsequent damage. The second highest failure is inadequate inspection and topping up of electrolytes, with resultant plate damage. Consider different batteries such as AGM types with lower failure rates.

c. **Connections.** All connections on alternators, starters, engine blocks, and batteries should be check and tightened every six months. It is an easy task and results in fewer intermittent and complete failures.

d **Spares.** It is rare to see a boat with a spare starter motor, or alternator and these should be a prerequisite on an extended voyage. While some boats may carry spare bearings, diode plates, brushgear etc., it is easier and quicker to change out the entire alternator. Invest in a spare starter motor and alternator.

2.12 FMEA Results. There is now a significant reduction in exposure to single point failures. With two separate charging circuits there is full redundancy and there are now only a total of 4 connections in both the positive and negative circuits, the negative having a backup with the starter motor negative. In a typical system using a changeover switch, that is a reduction of up to 75% in possible failure points. There is a significant improvement in charging efficiency, with a gain of up to or exceeding 0.5 VDC due to lower circuit resistance in both positive and negative circuit. This reduces alternator loads, and can shorten charging time, reduce charge current and extend alternator life. The starting system is more efficient with the negative at the starter reducing voltage drops, lowering current, reducing run time, and improving starting times. Coupled with a spare starter motor/solenoid, there is a reasonable chance of being operational within an hour. Separation of start and charging systems eliminates the many problems of voltage surges and transients. There are now two redundant power systems, each one being capable of powering the vessel, and any single failure of one system will not affect the other. The alternator negatives provide some redundancy to the main starter negative. As critical equipment has been evaluated and appropriate maintenance strategies implemented there is increased Mean Time Between Failure (MTBF) rates.

2.13 Multiple Alternator Charging System Configurations. In twin engine vessels there is by default two charging systems. In single engine boats, the option of fitting a second alternator is a useful option where redundancy is required. There are a number of different system configurations for multiple alternator installations.

a. **Discrete Systems.** These systems usually have the original engine alternator charging the engine start battery only. The additional alternator, usually a higher rated type of 80 amps or greater charges the house batteries only. If there is more than one bank, this may be split through either a diode isolator or a switch. Ideally the start battery alternator should be used to charge a third battery bank as the alternator is under-utilized given that start batteries require very little charging.

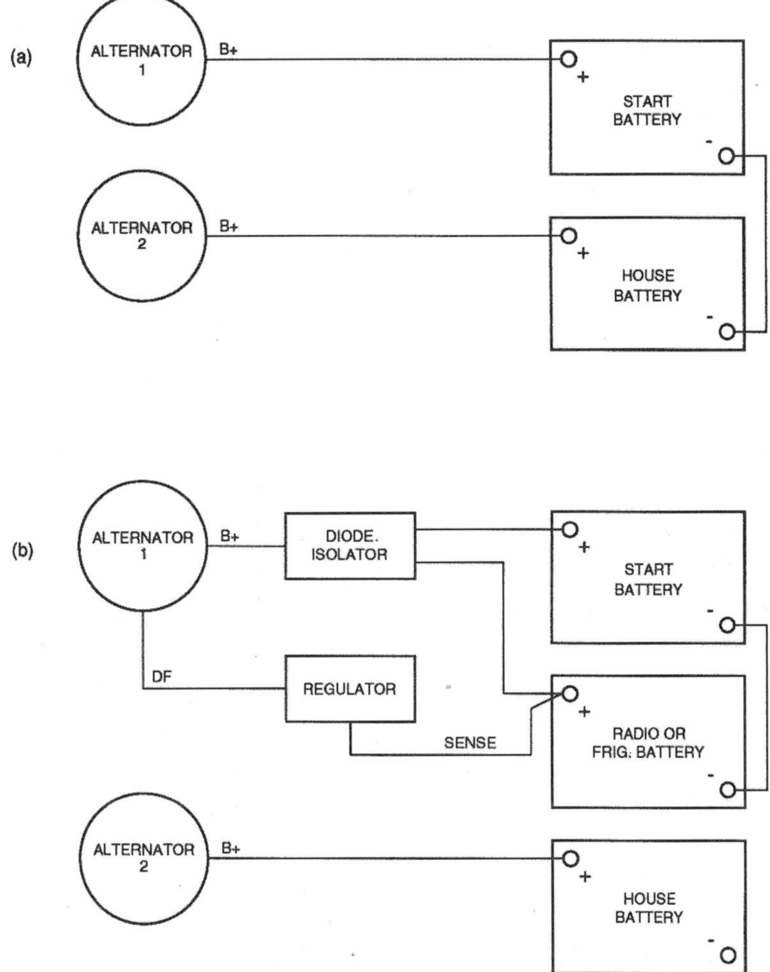

Figure 2-7 Two Alternator Charging Systems

b. **Cross Feed Systems.** These systems usually have each alternator charging a primary battery bank except that each alternator cross feeds to the other battery bank via a diode isolator. Ideally, a fast charge device should still be used. Although initially it looks complicated, it is in fact simple and the advantages of such a layout are:

(1) **Alternator Redundancy.** The arrangement allows charging of both battery banks even if one alternator should fail.

(2) **Load Balancing.** It is easier to balance loads between battery banks in order to achieve similar discharge levels of the same periods. This allows both batteries to be charged at a similar rate, which overall is faster, assuming that alternators and regulators are the same.

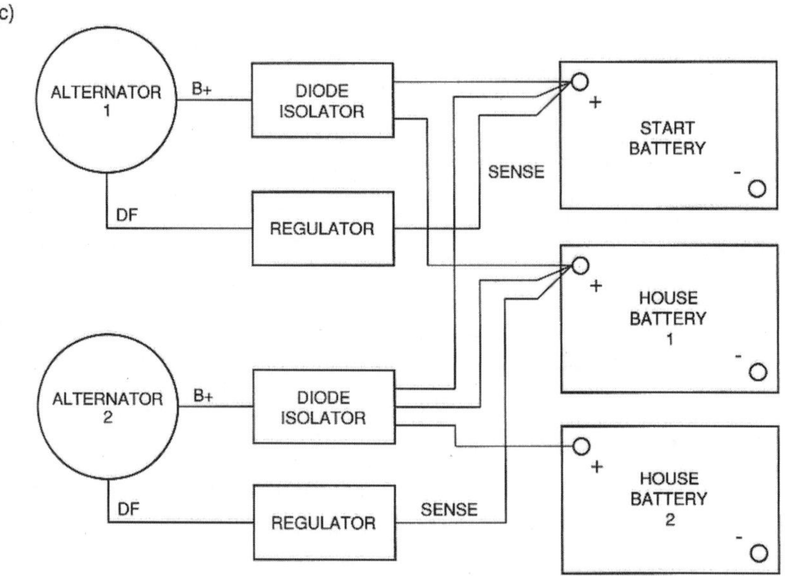

Figure 2-8 Two Alternator Diode Charging Systems

2.14 Battery Charging Recommendations.

The primary battery charging source should be calculated with a minimum output of 30% of the total installed battery capacity.

From the power analysis table we have calculated the maximum current consumption. Added to this is a 20% margin for battery losses giving a final charging value. A battery requires the replacement of 120% of the discharged current to restore it to full charge. This value is required to overcome losses within the battery due to battery internal resistances during charging. A popular benchmark is that alternator rating should be approximately 30% of battery capacity. An 80-amp alternator is recommended, which is approximately the largest rating possible without advancing to high priced or exotic high output alternators. I avoid where possible installing a battery bank in excess of 300 amp-hours and usually fit a bank of two six-volt cells rated at 230 amp-hours. With a suitable regulator system, this is usually adequate for most charging and load requirements. As a battery is effectively self-limiting in terms of charge acceptance levels, we cannot simply push in the discharged value and hope that it will recharge. The battery during charging is reversing the chemical reaction of discharge, and this can only occur at a finite rate. The alternator therefore must be selected if possible to recharge at the battery optimum charge rate as specified. Charging by necessity has a tapered characteristic, which is why start and finishing rates are specified. These ratings are largely impractical in marine installations. The required charging current is the sum of the charge rate plus anticipated loads during charging.

Alternative charging sources such as wind, solar and water should not be included within power calculations. These systems are to be classified as supplementary charging sources.

Other charging sources should be viewed as additional and not be used in the primary calculations, as they are reliant on weather and other conditions.

Where two alternators are installed, one alternator should be dedicated to the starting battery. The two alternators should not be connected in parallel to a single battery.

Where two alternators are installed to provide redundancy or improved charging capability, one should be dedicated to starting. Both alternators should not be charging in parallel to the same battery bank. In most cases, one will act as a slave and be very inefficient.

The positive cable from the alternator to the battery, or charge distribution device (diode, relay) should be rated at the maximum rated alternator current and for a maximum voltage drop of 5%.

All charge circuit cables must be rated for maximum current capacity of the alternator, with minimal voltage drop and allow for high ambient temperatures. Many installations are under-rated and overheat and fail.

A negative cable should be installed equivalent in size to the positive cable, from the negative terminal or case of the alternator to the battery. Where more than one alternator is installed the negative cables should be connected to the respective battery negatives.

To maintain system separation and minimize voltage drops in the charging circuit, which normally includes the engine block, a separate negative should be installed for each alternator. These should go the correct battery under charge, and crossovers are frequent.

All charging system cable terminations and connectors should be rated for the maximum alternator current. All charging system cable connectors should be crimped.

Many charging system terminations are under-rated for the current capacity of the cable. Ensure that crimp connections of the right capacity are used with rings of the correct size for the termination bolts on alternator and battery. Many are over-sized and make poor contact. Soldered connections frequently fail or are high resistance points in the circuit. After charging physically touch the alternator output terminal, and if it is very hot then the connection is probably undersized and therefore overheating causing charging system power losses.

No alternator output cable should have any isolation switch or fuse installed within the circuit so that opening of the circuit during operation could cause damage or failure of the alternator.

Switches and fuses should not be installed in any alternator output circuit. Alternator failures caused by inadvertent operation of changeover switches are common. When a switch is opened the spike as the field collapses normally destroys the alternator diodes.

Where a fast charging regulator or similar device is installed, it should not increase voltages to a level that can cause excess gas generation from the batteries. Maximum voltages should not exceed installed equipment ratings or be able to cause damage.

The majority of alternators have a fixed output of 14 volts, with some makes having the option of regulator adjustment up to around 14.8 volts for isolation diode voltage drop compensation. The regulator should not be able to cause high voltages that cause excessive gassing of batteries or in excess of normal equipment voltage input ranges. In the case of AGM or gel batteries this might cause catastrophic damage to the batteries and also the venting of dangerous gases.

Alternator-Charging Systems

3.0 Alternator Charging. The alternator is a robust and reliable piece of equipment. It is the principal charging source on most marine installations. Automotive alternators, or derivatives of them, are used in the majority of marine charging systems. Most alternators, however, are incorrectly rated for the installed battery capacity and therefore are unable to properly restore the discharged current. The alternator is generally sized by the engine manufacturer and is designed to recharge the engine start batteries. Yanmar is now offering a 80 amp and 115 amp alternator, and on one optional list 80A or 115A and optional additional 80 amp alternator. The standard alternator is just 55 amps. Universal/Westerbeke has an optional alternator from 32 HP engines and above, and this upgrades the standard 55 amp unit to a 72 amp alternator. The larger engines offer an 90 amp alternator and also dual output 135 amp, 165, and 190 amp units. The basic alternator consists of several components, which are described below. The typical automotive type alternator generates a 3-phase AC alternating current, which is where it derives its name. This is then rectified to produce a DC output for charging via the full wave bridge rectifier. There are variations in design, and some 12 pole units are rated to run at up to 20,000 rpm, with modified cooling arrangements. There are vehicle derived units that have no brushes or windings. The excitation winding is fixed and pre-excitation is not required. Although water cooled alternators are common in vehicles they are not on boats.

a. **Stator.** The stator is the fixed winding. It consists of a three phase winding that is connected in a “star” or a “delta” arrangement. The windings are formed onto a solid laminated core. They supply three phases of alternating current (AC) to the rectifier.

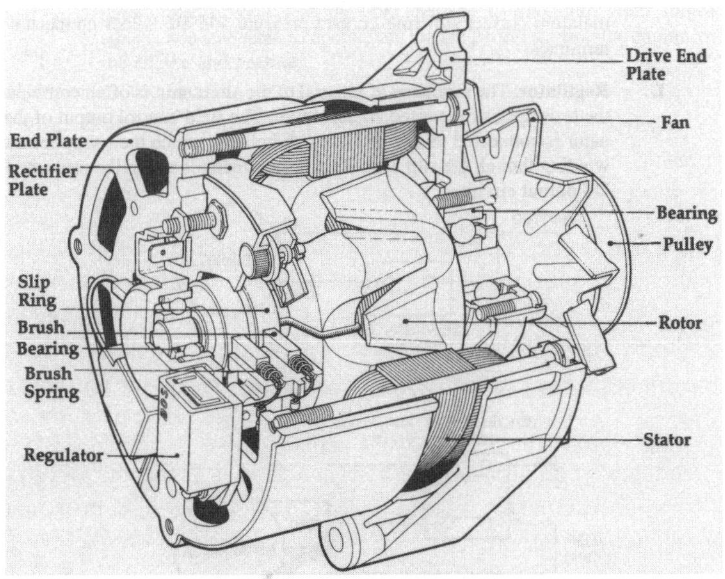

Figure 3-1 Bosch Alternator

b. **Rotor.** The rotor is the rotating part of the alternator. The shaft has the pole or claw shaped magnet poles attached, the excitation winding, the cooling fan at one end, the bearings, and the collector sliprings.

c. **Rectifier.** The rectifier consists of a network of six diodes, which are connected across the positive and negative plates. These plates also function as heat sinks to dissipate the heat from power generation. This rectifies the three generated AC phase voltages into the DC output for charging. Two diodes are used on each winding to provide full wave rectification. In some alternators Zener diodes are used to limit voltage peaks that arise during sudden load changes.

d. **Exciter Diodes.** The exciter (D+) or pre-excitation diodes consist of three low power diodes which independently rectify each AC phase and provide a single DC output for the warning light or auxiliary control functions. They are required as the residual magnetism (or remanence) in the iron core is insufficient at low speeds and starting to initiate the self-excitation required to build up the magnetic field. This only occurs when the alternator voltage is higher than the voltage drop across the two diodes. The warning lamp functions as a resistor and provides pre-excitation current, which generates a field in the rotor. In this respect the power or watts rating of the lamp is important and 2 watts is typical.

e. **Brushgear.** The brushes are normally made of copper graphite. The brushes are spring-loaded to maintain correct slipring contact pressure and are soldered to the terminals.

f. **Voltage Regulator.** The voltage regulator is usually combined with the brushgear or mounted adjacent to it. The field control output of the alternator is connected to one of the brush holders, which then supplies the rotor winding though the slipring. Regulator sensing is normally connected to the D+ output circuit. The regulator maintains a constant voltage output over the entire operating range of the alternator. Earlier electro-magnetic contact type regulators are relatively uncommon now with most being electronic types with no moving parts. The electronic regulator allows precise control with short field switching periods.

3.1 Field Circuits. The field circuit is used to vary the output of the alternator. It can be simply defined as the alternator "controller" because all alternator output is controlled by the field current level. On modern sailing boats alternators come with integral regulators and modification is a difficult task that requires a new external regulator. There are a number of variations in the connection of fields besides the normal regulator and these are as follows:

a. **Advanced Field Switching.** This method is comparatively rare in modern integral regulator alternators. The field is taken through the battery selector switch auxiliary contacts, so that the field circuit is broken, de-energizing the alternator immediately before the main output contacts break. This will prevent any accidental circuit interruption and subsequent diode destruction through generated surges.

b. **Oil Pressure Switch Control.** This method has two configurations. The first senses battery voltage through an oil pressure switch on the engine. The alternator does not commence generating until after engine oil pressure has built up. The second method takes the field directly through an oil pressure switch. It is not common except on older vessels.

c. **Field Isolation Switch.** This circuit is common in small engines or where small output auxiliary engines drive more than one piece of equipment. This enables the alternator to be switched off to reduce engine loadings so that other equipment such as refrigerators or watermakers can operate. To avoid circuit disturbances and possible damage to alternator from surges and spikes, it is advisable to operate the switch before starting the engine, or stop the engine and operate. It is not a very common arrangement.

3.2 Alternator Selection. Boat owners have a number of important factors to consider when selecting alternator output ratings. Along with regulators, the alternator is probably the most common item to fail onboard, therefore careful selection is required.

a. **Engine Run Times.** The engines in a majority of cruising vessels are run excessively in an attempt to recharge batteries. The maximum run time goal is one hour in the morning and one hour in the evening, which also coincides with mechanical refrigeration pull down times.

b. **Engine Loading.** Diesel engines should not be run with light loads because unloaded engines suffer from cylinder glazing. A high output alternator can provide loads of up to 1.5 horsepower at rated output.

c. **Engine Speeds.** Ideally, the engine should be able to charge at maximum rates at relatively low speeds. The preferred speed is generally a few hundred revs/min above idle speed. Alternator speed is dependent on the drive-pulley ratio and the alternator cut-in speed.

d. **Battery Capacity.** Nominal charging rates are specified by manufacturers, and they generally specify starting and finishing rates. A battery requires the replacement of 120% of the discharged current to restore it to full charge. This value is required to overcome internal resistances within the battery during charging.

e. **Charging Current.** As a battery is effectively self-limiting in terms of charge acceptance levels, we cannot simply push in the discharged value and hope that it will recharge. The battery during charging is reversing the chemical reaction of discharge, and this can only occur at a finite rate. If possible, therefore, the alternator must recharge the battery at the optimum charge rate specified. Charging by necessity tapers off as full charge is reached, which is why start and finishing rates are specified. These ratings are largely impractical in marine installations. Alternator output current is the sum of electrical loads on the system during the charging period, plus the actual battery charging current.

f. **Charge Voltage.** The majority of alternators have a fixed output of 14 volts, with some makes having the option of regulator adjustment up to around 14.8 volts for isolation diode voltage drop compensation. Charge voltage is probably the single most important factor in charging, as all other factors are related to it.

g. **Alternator Output Current Selection.** From the power analysis table, we have calculated the boat's maximum current consumption. Added to this is a 20% margin for battery losses, giving a final charging value. One popular opinion is that alternator ratings should be approximately 30% of battery capacity. In practice, this is at best optimistic and difficult to achieve.

I always specify and install an 80-amp alternator, which is about the highest rating possible without going into high-priced or exotic alternators. I avoid wherever possible installing a battery bank in excess of 300 amp-hours and usually fit a bank of two 6-volt cells rated at 230 amp-hours. With a suitable regulator system, I have never found this to be inadequate for charging and load requirements. You can go and fit large output units, but economic considerations weigh against that solution.

h. **Marine Alternators.** Marine alternators are essentially enclosed and ignition protected with a UL listing to prevent accidental ignition of hazardous vapors. Windings are also protected to a higher standard by epoxy impregnation. Marine units have a corrosion-resistant paint finish and are designed for higher ambient operating temperatures.

i. **Marinized Alternators.** An alternator can be marinized to a reasonable degree if you wish to marinize and improve your alternator. Bearings should be totally enclosed. Replace if they are not. Windings should be sprayed or encapsulated with a high-grade insulating spray. The back of the diode plate can also be sprayed with an insulating coating to prevent the ingress of moist, salt-laden air and dust which can short out diodes and connections.

j. **High-Output Alternators.** It is, regrettably, a fact of life that many so-called marine electrical people push high output alternators (typically 130 amps or more) as the first step toward solving battery-charging problems. These alternators are expensive, and in many cases mask the more common problems of poor circuit design, poor installation, and inadequate regulation. Be warned! This may be a typical automotive electrician's answer, but not a marine one; a high output alternator will not necessarily solve your charging problems. In most cases, an 80–90 amp alternator is all that's required. A considerably cheaper and more reliable solution may be to replace the regulator. If you choose to upgrade your alternator, install a quality unit such as Silver Bullet, Lestek, Balmar, Niehoff or Powerline. Beware of rewound standard alternators; they are notoriously unreliable.

k. **Outboard Motor Alternators.** On many multihull vessels up to 40 feet with outboard motors, charging problems are commonplace. Outboards have a flywheel driven alternator, and they generally have a low output, typically in the range 10–15 amps.

3.3 Overvoltage and Surge Protection. Some alternators are provided with separate surge protection units. Overvoltage protection comprises several methods.

a. **Zener Diodes.** As described earlier the rectifier diodes are Zener diodes that limit the high voltage spikes or peaks that arise below a safe value, which can damage the regulator. The typical limiting voltages of Zener diodes in use are 25–30 volts for 14-volt alternators and 50–55 volts for 28-volt alternators.

b. **Surge-proof Alternators.** Some alternators are equipped with high specification components. The components are rated up to 200 volts for 14-volt systems and 350 volts for 28-volt systems. This is supplemented by the installation of a capacitor across the alternator output and ground. Lucas/CAV alternators incorporate a surge protection avalanche diode within the alternator (ACR & A115/133 range). This protects the main output transistor in the regulator.

c. **Overvoltage Protection Devices.** Often these are only installed in 28-volt alternators. These electronic semi-conductor devices are connected across the alternator output. They operate by short-circuiting the alternator through the excitation winding when peaks rise over a set value. Some alternators use what is called a freewheeling diode, anti-surge or suppressor diode. This is connected in parallel with the excitation winding of the alternator.

d. **NewMar Filters.** The 80-A and 150-A are designed for installation in the alternator output lead adjacent to the alternator. They will reduce noise in the 70kHz to 100 MHz range that commonly affects GPS, and radios. They are relatively heavy and large so they will require careful fastening on the engine, or on an adjacent bulkhead if close.

e. **Additional Protection.** A Metal Oxide Varister (MOV) installed across the B+ and negative terminals will provide additional surge protection. Another good method is to solder a capacitor rated at 0.047F/250 V across each of the AC windings.

f. **Interference Suppression.** Alternator diode bridges create noise (RFI) that can be heard on communications or electronics equipment. Always install an interference suppression capacitor. As a standard, install a 1.0-microfarad suppressor. In some cases, a suppressor is required in the main output cable.

3.4 Alternator Installation. Optimum service life and reliability can only be achieved by correctly installing the alternator. The following factors must be considered during installation:

a. **Alignment.** It is essential that the alternator-drive pulley and the engine-drive pulley be correctly aligned. Pulley misalignment can impose twisting and friction on drive belts and additional side loading on bearings. Both can cause failure.

b. **Drive Pulleys.** Drive pulleys between the alternator and the engine must be of the same cross section. Differences will cause belt overheating and premature failure. Ideally, the split, automotive-type pulleys on some alternators should be replaced by solid pulleys of the correct ratio.

c. **Drive Belt Tension.** Belts must be correctly tensioned. Maximum deflection should not exceed 10 mm. When a new belt is fitted, the deflection should be re-adjusted after 1 hour of operation and again after 10 hours. Belts will stretch in during this period. Under-tensioning causes belt overheating and stretching, as well as slipping and subsequent undercharging. The excess heat generated also heats up pulleys and the high heat level conducts along the rotor shaft to the bearing, melting bearing grease and increasing the risk of premature bearing failure. Over-tensioning causes excessive bearing side loads, which leads to premature bearing failure. Signs of this condition will be characterized by sooty deposits around the belt area and wear on the edges of the belt.

d. **Drive Belts.** Belts must be of the correct cross section to match the pulleys. Notched or castellated belts are ideal in the engine area as they dissipate heat easily. If multiple belts are used, always renew all belts at the same time to avoid varying tensions between them. For any alternator over 80 amps, a dual-belt system should be used because a single belt will not be able to cope with the mechanical loads applied at higher outputs.

e. **Ventilation.** An alternator, similar to electrical cable, cannot achieve its rated output in high temperatures. Ideally, a cooling supply fan should be fitted to run when the engine is operating and its airstream should be directed to the alternator. Many alternator failures occur when boost charging systems are installed because such systems run at near maximum output for a period in high ambient temperatures. Always ensure when fitting an additional alternator that the alternator's fan is rotating in the correct direction.

f. **Mountings.** Yanmar engines have a dual foot mounting that has a 3" separation. Universal/Westerbeke uses a single 1" so replacement is always a problem. Alternator mountings are a constant source of problems. When tensioning the alternator, always adjust both the adjustment bolt and the pivot bolt. Failure to tighten the pivot bolt is common and causes alternator twisting and vibration. Vibration fatigues the bracket or mounting and may cause it to fracture. Additionally, this can cause undercharging and radio interference. Make sure that the slide adjustment arm is robust. Most marine engines have a vibration level that will fatigue the slide and break it. In my experience, Volvo engines are notorious for this problem.

g. **Warning Light.** The light circuit is not simply for indicating failure—the lamp excites the alternator. In many cases, an alternator will not operate if the lamp has failed because the remnant voltage or residual magnetism has dissipated. Ideally, a lamp should be in the range of 2–5 watts. Undersized lamps are often characterized by the need to rev the engine to get the alternator to kick in. This is often highly visible with alternator driven tachometers. Many newer engine panels have a printed circuit board type of alarm panel.

h. **Interference Suppression.** Alternator diode bridges create radio frequency interference (RFI) which can be heard on communications or electronic equipment. Always install an interference suppression capacitor. As a standard, install a 1.0 microfarad suppressor. In some cases, a suppressor is required in the main output cable.

3.5 Alternator Drive Pulley Selection. Ideally, maximum alternator output is required at a minimum possible engine speed. This is typically a few hundred revs/min above idle speed. Manufacturers install alternators and pulleys assuming that the engine is only run to propel the vessel, when in fact engines spend more time functioning as battery chargers, at low engine revolutions. Alternators have three speed levels that must be considered and the aim is to get full output at lower speeds.

a. **Cut-in Speed.** A voltage will be generated at this speed.

b. **Full Output Operating Speed.** This is the speed where full rated output can be achieved.

c. **Maximum Output Speed.** This is the maximum speed allowed for the alternator, otherwise destruction will occur.

d. **Pulley Selection.** An alternator is rated with a peak output at 2300 revs/min. At a typical engine speed of 900 revs/min and a minimum required alternator speed of 2300, a pulley ratio of approximately 2.5:1 is required. The alternator's maximum output speed is 10000 rev/min. Maximum engine speed is 2300, so 2300 multiplied by 2.5 = 4000 revs/min. This falls well within safe operating limits and is acceptable. A pulley with that ratio would suit the service required.

e. **Selection Table.** Table 3-1 gives varying pulley ratios with an alternator pulley diameter of 2.5 inches.

Table 3-1 Drive Pulley Selection Table

Engine Pulley	Pulley Ratio	Engine RPM	Alternator RPM
5 inch	2:1	2000	4000
6 inch	2.4:1	1660	4000
7 inch	2.8:1	1430	4000
8 inch	3.2:1	1250	4000

f. **Alternator Characteristics.** The graph below illustrates the relationship between output current, efficiency, torque, and horsepower against rotor revolutions. The optimum speed can be selected from these characteristics. The performance curves and characteristics illustrated are for a Lestek high output alternator, and for a 9135 series 135-amp alternator.

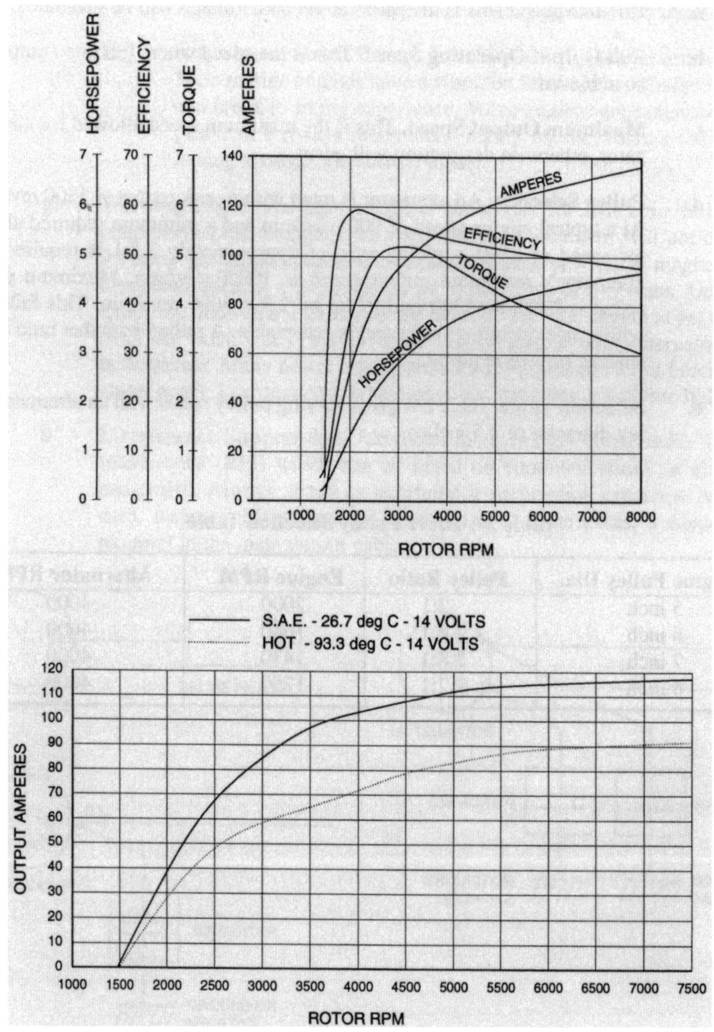

Figure 3-2 Alternator Output Characteristics

3.6 Alternator Maintenance. Many alternator failures can be avoided by performing basic maintenance tasks.

a. Drive Belts. Check monthly as follows:

(1) Check and adjust tension. Deflection: 10mm maximum.

(2) Examine for cuts, uneven wear or fatigue cracks.

(3) Ensure belts are clean, with no oil or grease.

b. Connections. Check monthly as follows:

(1) Clean and tighten all alternator terminals.

(2) Check cable and connectors for fatigue.

c. Vibration. Check monthly as follows:

(1) Check alternator for vibration when running.

(2) Examine mounts for fatigue cracks.

d. Bearings. Check every 1500 operating hours as follows:

(1) Remove alternator and turn rotor. Listen for any bearing noises.

(2) Renew every 3000 hours or at major overhaul.

e. Brushes. Check every 1500 operating hours as follows:

(1) Check brushes for excess or uneven wear.

(2) Check sliprings for scoring.

f. Cleaning. Clean yearly as follows:

(1) Wash sliprings, diode plate, and brushgear with electrical solvent. Do not use any abrasives on sliprings; they must be cleaned only to preserve a film that is essential for brush contact.

(2) Wash out windings and dry.

g. Pre-cruise. Take alternator to a quality marine or automotive electrical workshop. Request the following tests:

(1) Test alternator output for maximum current.

(2) Check diodes.

(3) Clean windings, sliprings and brushgear. Renew worn brushes.

(4) Renew bearings.

3.7 Alternator Faults and Failures. Failures in alternators are primarily due to the following causes, many of which can be prevented with routine maintenance.

a. **Diode Bridge Failures.** Diode failures are generally caused by:

(1) **Reverse Polarity Connections.** Reversing the positive and negative leads destroys the diodes. This is a common occurrence.

(2) **Short Circuiting Positive and Negative.** A short circuit will cause excess current to be drawn through the diodes and the subsequent failure of one or more diodes. The most common cause is reversing the battery connections.

(3) **Surge.** A high voltage surge is generated by the inductive effect of the field and stator windings. This occurs if the charge circuit is interrupted, most commonly when an electrical battery selector switch is accidentally opened.

(4) **Spikes.** Short-duration, transient voltages several times greater than the nominal voltage can be caused by high inductive loads when starting up, say, a pump. Most spikes, however, are caused by lightning strikes. Countermeasures are covered in the lightning protection chapter.

b. **Winding Failures.** Stator winding failures are usually due to the following causes:

(1) **Overheating.** Normally due to insufficient ventilation at sustained high outputs, causing insulation failure and intercoil short circuits.

(2) **Short Circuit.** Due to mechanical winding damage, overheating, or ingress of moisture.

(3) **Rotor Winding.** Short circuit or a ground fault due to overheating or over-voltage if the voltage regulator fails.

c. **Brushgear.** Brushgear failures are not that common in a properly maintained alternator, but they are generally due to:

(1) **Brushes.** Brushes worn and sparking, and characterized by fluctuating outputs and radio interference.

(2) **Sliprings.** Scoring and sparking due to build-ups of dust, also causing radio interference.

d. **Bearing Failure.** The first bearing to fail is normally the front pulley bearing. Rotating it by hand will usually indicate grating or noise.

3.8 Alternator Troubleshooting. Troubleshooting should be carried out in conjunction with charging system troubleshooting as described in Table 3-4.

a. **Check Output.** This initially depends on the lamp and the regulator. Using a voltmeter, check that the output across the main B+ terminal and negative rises to approximately 14 volts. No output indicates either total failure of alternator or regulator. Partial output indicates some diodes failed or a regulator fault.

b. **Regulator Check.** If there is no output either the alternator is faulty or the regulator is failing to excite the alternator. This is not difficult with external regulators but if an internal regulator is fitted, the alternator will need to be opened and a wire attached to the brush-holder. Switch off all electrical and electronic equipment at the switchboard circuit breaker before commencing test. *If in doubt, don't try it.* Check that the alternator gives full output by shorting the wire to negative in negative type machines or positive in positive types. If the alternator gives full output voltage the regulator is probably faulty.

c. **Alternator Test.** The other components are tested after confirming the function of the regulator. I recommend first that you remove the alternator, and take it to any good automotive electrician with a test bench if in port. This saves a considerable amount of time and effort. If you don't carry spares you can do little. To get home with partial diode failure, you can disconnect the regulator, and apply a full field voltage to get maximum output.

d. **Auxiliary Diode and Warning Light Tests.** On some occasions, the auxiliary diodes may fail. Put your multimeter on the 20-volt range and connect across 61/D+ and negative. If there is any reading, the diode may be faulty. Turn on the ignition key without starting. The reading should be around 1–2 volts. If lower the wiring may be faulty, if higher, the diode may be faulty; or there is excessive rotor resistance or a bad connection. Check that the warning light is operating and on when the ignition switch is turned on. If not the lamp may be faulty, or seating badly if a replaceable lamp, or there is a lamp connection fault. Check that the wire is not off the D+ terminal, or connection is loose.

e. **Rotor Testing.** If a regulator has failed, particularly in an overcharge condition, prior to replacing the regulator, the rotor should be checked for damage. The test is as follows:

(1) **Test Insulation Resistance.** Place one multimeter probe on a slipring, and the other on the rotor core. Resistance should be infinite or over-range.

(2) **Test Winding Resistance.** Place the multimeter probes on each slipring. Resistance should be around 4 ohms. If it is very high, an open circuit may exist, and if very low, a coil short circuit may exist.

3.9 Alternator Terminal Designations. Alternators have a variety of different terminal markings, which are listed in Table 3-2.

Table 3-2 Alternator Terminal Markings

Make	Output	Negative	Field	Auxiliary	Tachometer
Bosch	B +	D -	DF	D+/61	W
Ingram	B +	B -	F	IND/AL	W
Lucas	BAT	E	F	L	
Paris-Rhone	+	-	DF	61	W
Sev Marchal	B +	D -	DF	61	
Motorola	+	-	F	AUX	AC
CAV	D +	D -	F	IND	
AC Delco	BAT	GND	F		
Niehoff	BAT+	BAT -	F	D+	X
Valeo	B +	D -		D+	W
Mitsubishi	B +	E	F	L	
Nippon Denso	B +	B	F	L	
Prestolite	POS+	GND		IND LT	AC TAP
Silver Bullet	+	-	F		R

3.10. Alternator Remagnetization. After dismantling or stripping down an alternator, it is not uncommon to find it simply won't work at all. Before you hurriedly dismantle it again to find your mistake, perform the following checks:

a. **Field Disconnect.** Disconnect the regulator field connection (assuming you have installed a separate regulator or controller).

b. **Manual Field Activation.** With the engine running at idle speed and all electrical and electronic equipment off, temporarily touch the field connection to the following:

  (1) **Positive Control.** (Bosch, Paris-Rhone, Motorola, Sev-Marchal) If the field control is on the positive side, touch the lead to main alternator output terminal B+; or if a diode is fitted, then to the diode battery output terminal.

  (2) **Negative Control.** (Lucas, CAV, Hitachi) If the field control is on the negative side, touch the lead to the negative terminal or to the case.

c. **Output.** If the alternator is operational, it will immediately generate full output, you will hear the engine load up, and voltage will rise up to 16 volts. Only do this for a second or two. Reconnect the regulator back to normal. In many cases, this will restore magnetism to the alternator and it will operate normally. If there is little or no output after this test, it generally indicates a fault in the alternator. Normally this is caused by a faulty diode bridge or the brushes not seating on sliprings.

d. **Warning Light.** Make sure the light is operating and on when the ignition switch is turned on. If not, the following may be faulty:

(1) Lamp fault, or seating badly if a replaceable lamp.

(2) Lamp connection fault.

(3) Wire off D+ terminal, or a loose connection.

(4) Faulty alternator-excitation diodes.

3.11 Emergency Repairs and Getting Home. The following gives basic survival methods where an alternator or regulator has failed and you have neglected to carry spares. In some cases it may not work, but could get you home.

a. **Regulator Failure.** No output or full, uncontrolled high voltage output:

(1) **No Output.** To overcome this, apply full field voltage as described in the above paragraph. For sustained motor sailing in this condition, place a spare navigation lamp or bunk light in the field circuit to limit field current value.

(2) **High Voltage Output.** Run the engine for limited periods only, until the voltage rises across the battery. Disconnect all electronics to avoid damage. The internal regulator should also be disconnected and a lamp placed in the circuit if motoring for extended periods.

b. **Alternator Diode Failure.** This is indicated by low charge voltage. In many cases, only a few diodes may have failed. If you do not carry a spare diode plate, the following actions are required to get some charging capability:

(1) Identify any short-circuited diodes using a multimeter.

(2) Disconnect and remove the short-circuited diodes.

(3) Reduce battery capacity to one battery to prevent overloading the reduced diode bridge.

c. **Warning Light Failure.** In many cases, an alternator will not operate without a warning light. Place any small lamp in series with the lead off the auxiliary output (D+), and touch it to the battery's positive terminal. Excitation is usually immediate. Remove straight away.

3.12 Diode Isolator Testing. On rare occasions, a diode isolator may fail because of an external event such as a surge or spike. The following tests can be carried out to verify its operation:

a. **Engine Operating.** Output terminal voltages should be identical. The input terminal should read approximately 0.75 volt higher if a non-battery-sensed regulator is being used. The diode system should not be used in these installations.

b. **Engine Off.** Output terminal voltages should read the same as the service and starting batteries. The input terminal from the alternator should be zero.

c. **Ohmmeter Test.** Make sure all power is off before testing.

(1) Disconnect battery input and output cables.

(2) Set meter scale to x1.

(3) Connect red positive probe to input terminal.

(4) Connect black negative probe to output terminals 1 or 2.

(5) If the diode is good, the meter will indicate minimal resistance.

(6) Reverse the probes, and repeat the test. The reading should indicate high resistance, or over range.

3.13 Alternator Regulators. The regulator is the key to all alternator charging systems. The function of the regulator is to control the output of the alternator and to prevent the output from rising above a nominal set level, typically 14 volts. Higher voltages would damage the battery, alternator, and equipment.

a. **Principles.** An alternator produces electricity by the rotation of a coil through a magnetic field. Its output is controlled by varying the level of the field current. This is achieved by applying the field current through one brush and slipring to the rotor winding, and completing the circuit back through the other slipring and brush. Essentially, the regulator is a closed loop controller, constantly monitoring the alternator output voltage and varying the field current in response to output variations.

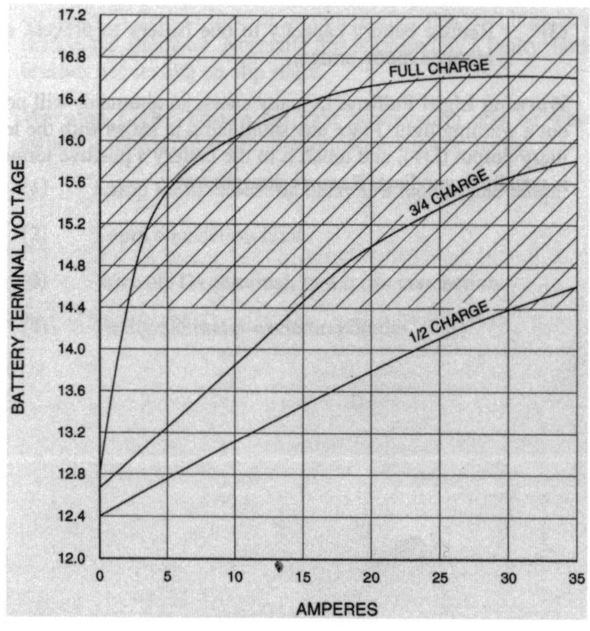

Figure 3-3 Regulator Operating Range

b. **Regulator Operating Range.** A regulator does not control the charging process significantly until the battery's charge level reaches approximately 50%. When the voltage of the battery rises to this threshold, the regulator starts limiting the voltage level. The charge current levels off as the voltage level rises; this is called the regulation zone.

c. **Standard Regulators.** The traditional automotive alternator is fitted with a regulator designed for automotive service. This requires the replacement of a relatively small amount of discharged power within a short time. The alternator then supplies the vehicle's electrical power as the engine runs. This arrangement is totally inadequate for marine applications.

To recharge a battery properly on a boat, the charging system must overcome the battery's counter voltage, which increases as charging levels increase. The typical scenario is one of a high charge at initial start-up and then a rapidly decreasing current reading on the ammeter. As a result, few boat batteries are ever charged much above 70% of capacity.

One of the many undesirable effects of standard regulators is that when a load is operating on the electrical system, charging current also decreases. As an example, I tested an alternator with a total output of 30 amps at 14 volts aboard a vessel with an electrical load of 24 amps. I found that only 6 amps was flowing into the battery with a terminal voltage of only 13.2 volts. The more load you apply on the system during charging, the less goes to charging the battery. It is better to have as much load switched off as possible.

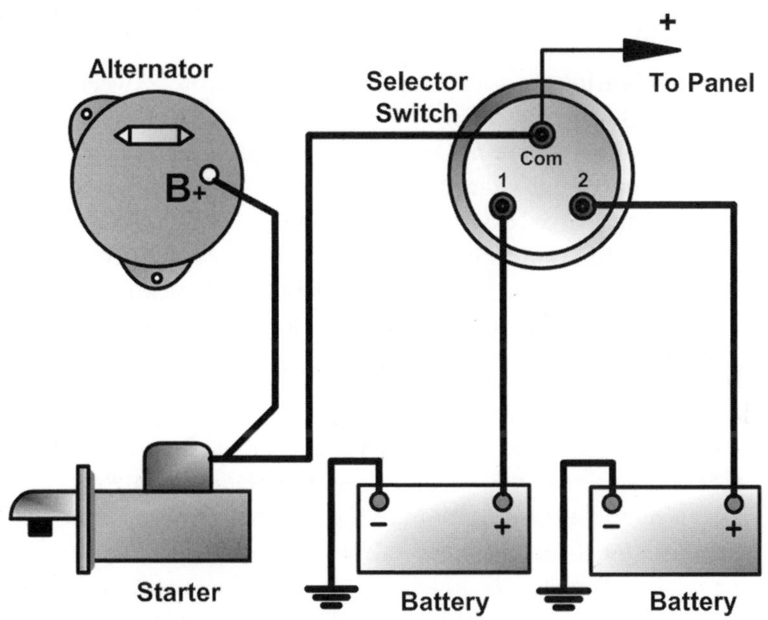

Figure 3-4 Standard Engine Charging Configuration

3.14. Alternator Regulator Sensing. With any type of charging system, there is a voltage drop between the alternator output terminal and the battery. With a nominal alternator output of 14 volts, it is not uncommon to have a totally inadequate 13 volts reach the battery. This voltage drop increases as current increases. Regulator sensing consists of the following configurations:

a. **Machine Sensed.** The machine-sensed unit simply monitors voltage at the output terminal and adjusts alternator output voltage to the nominal value, which is typically 14 volts.

   (1) **Charge Circuit Voltage Drops.** The machine-sensed regulator makes no compensation for charging circuit voltage drops. Voltage drops include inadequately rated terminals, cables, and negative path back through the engine block.

   (2) **Diode Isolators.** If a diode-isolator, charge-distribution system is installed, this also contributes a further drop, typically 0.75 volt.

b. **Battery Sensed.** The battery-sensed unit monitors the voltage at the battery terminals and adjusts the alternator output voltage to the nominal voltage. Always install battery sensing if possible.

   (1) **Charge Circuit Voltage Drops.** The battery-sensed regulator compensates for voltage drops across diodes and charge circuit cables. By sensing the battery terminal voltage, the regulator varies the output from the alternator until the correct voltage is monitored at the battery. Some alternator manufacturers such as Bosch, Lucas, Prestolite, and Sev-Marchal are introducing modifications so that regulators can be compensated with a separate sense connection that goes directly to the battery.

   (2) **Caution.** In some cases, the voltage drop between alternator terminals and battery may be considerable, and figures of 1.5 to 2 volts and above are not uncommon. With a multimeter, check the output and battery voltage to find out the drop, ideally at the full output current. An excessive voltage drop is a fire risk. Excessive current flow, along with high, ambient engine space temperatures, can literally melt and ignite the cable insulation, or typically first burn off the terminals. Check output terminal to see if it is hot.

c. **Temperature Compensation.** Very few alternator manufacturers incorporate temperature compensation, even though electrolyte is affected by temperature. In hot climates, charge voltage should be marginally decreased; in cold climates, it should be increased. Regulators with compensation usually have it sensed at the regulator, as the sensing element is part of the regulator circuit. In most vessels, however, batteries are not located near the engine, so the regulators reduce charging output when they sense high engine compartment temperatures. Compensation should be based on the ambient temperature of the batteries.

3.15 Alternator Regulator Types. It is extremely important to distinguish between a regulator and a controller. There are a number of new alternator-control devices which do not fit into the definition of a regulator.

a. **Regulator Function.** A regulator is a fully automatic device which ensures a stable output from the alternator. The primary function of a regulator is to prevent overcharging the battery and damaging the alternator. This crucial function is frequently forgotten—with disastrous results—when selecting a controller.

b. **Alternator Control Devices.** There are now five main categories of alternator control devices:

(1) **Standard Regulators.** These are factory fitted to alternators.

(2) **Cycle Regulators.** These devices use a cyclic regulator control principle that is microprocessor controlled.

(3) **Stepped Cycle Regulators.** These use a timed cycle of voltage steps.

(4) **Regulator Controllers.** These devices either parallel connect or override existing standard regulators.

(5) **Manual Controllers.** These devices have no regulator function and control alternator output manually by operator control.

3.16 Standard Regulators. Standard alternator regulators are simple and inexpensive voltage regulators with associated circuitry. They are normally an integral part of the alternator, are incorporated with the brush gear as a removable module, or are located externally on the engine or an adjacent bulkhead. The best arrangement is to have a separate regulator mounted on an adjacent bulkhead to minimize engine heat and vibration damage.

3.17 Regulator Polarity. Regulators and field windings have two possible field polarities. It is important to know the difference when installing different regulators or testing regulator function. The two types are as follows:

a. **Positive Polarity.** The positive regulator controls a positive excitation voltage. Inside the alternator, one end of the field is connected to the negative polarity. Alternators with this configuration include Bosch, Motorola, Ingram, Sev-Marchal (older models), Silver Bullet, Lestek, Balmar:

(1) **Polarity Test.** To test, use a multimeter on the ohms x 1 range and connect across the field connection to an unpainted part of the alternator case or negative output terminal.

(2) **Meter Reading.** The reading should be in the range of 3 to 8 ohms.

b. Negative Polarity. The negative regulator controls a negative excitation voltage. Inside the alternator, one end of the field is connected to the positive polarity. Alternators with this configuration include Hitachi, Lucas A127, ACR 17-25, and AC5, CAV, Paris-Rhone, new model Sev-Marchal, and Valeo, AC Delco, Mitsubishi:

(1) **Polarity Test.** To test, use a multimeter on the ohms x 1 range and connect across the field connection to the alternator's positive terminal.

(2) **Meter Reading.** The reading should be in the range of 3 to 8 ohms.

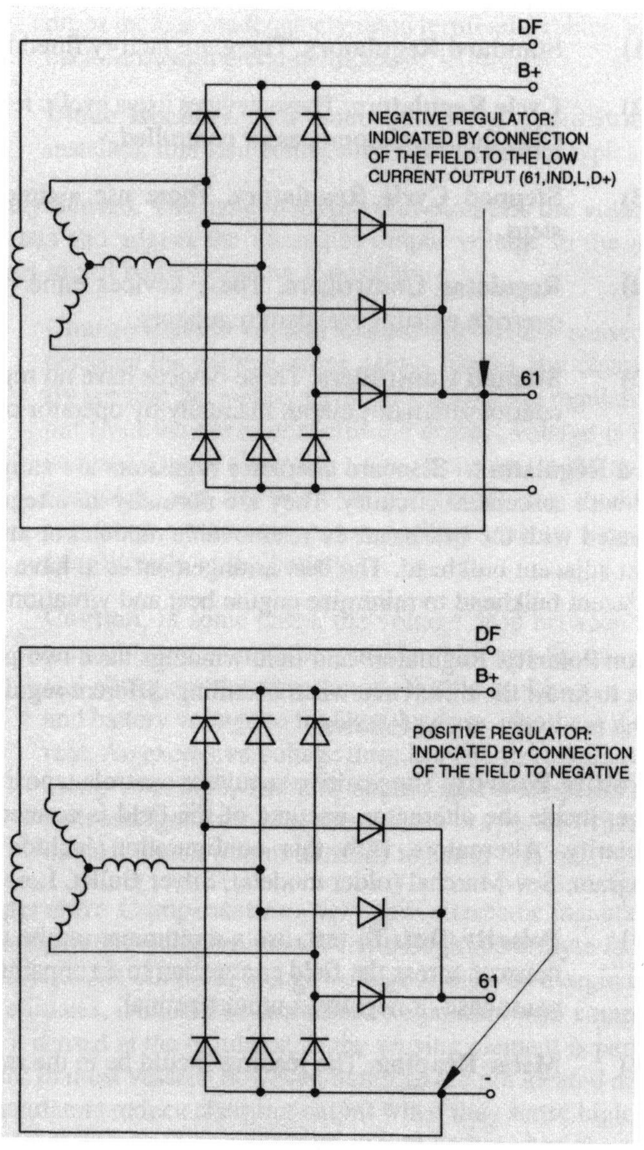

Figure 3-5 Alternator Regulator Field Polarity

3.18 Regulator Removal. If a regulator must be removed or checked, certain procedures should be used to avoid damage. The following diagrams illustrate various alternators and disassembly procedures. Mounting a separate regulator on the engine bulkhead makes replacement simple and inexpensive, and makes testing easier.

a. **Bosch (K1/N1 Series).** Dismantle as follows:

(1) Unscrew the two screws retaining the regulator.

(2) Carefully lift the regulator up and out. Be careful not to damage the brushes.

(3) Disconnect the (D+) lead from the back of the regulator.

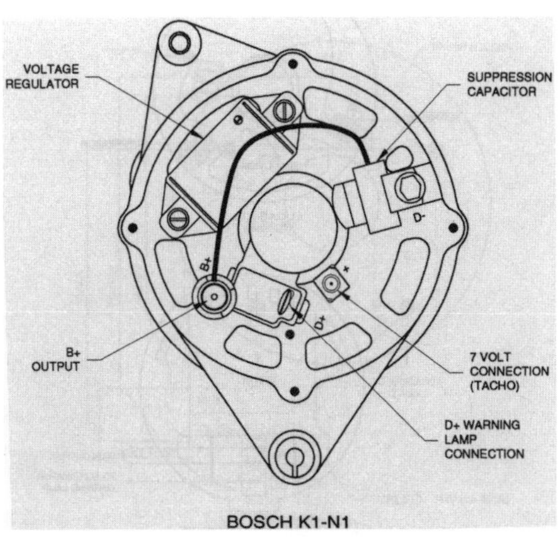

Figure 3-6 Bosch Series K1/N1 Alternator

b. **Lucas.** This is the standard type fitted to Perkins engines. There are a large number of Lucas alternators around and all are different. This procedure covers both removal and conversion to an external regulator:

(1) Unscrew and remove the two screws securing the integral regulator and brushgear to the alternator housing.

(2) Carefully lever open the two halves of the regulator, which are held together with an adhesive.

(3) Cut and disconnect the three joining links from the brushes.

(4) Solder a new wire to the spring-loaded connector immediately below the inner brush-holder. You must use a special solder to do this because normal solders will not work. (RS Stock Number 555-099) Run it out through the cover for connection to the new regulator. This is the field control connection.

(5) Place the two regulator halves together and refit into the alternator.

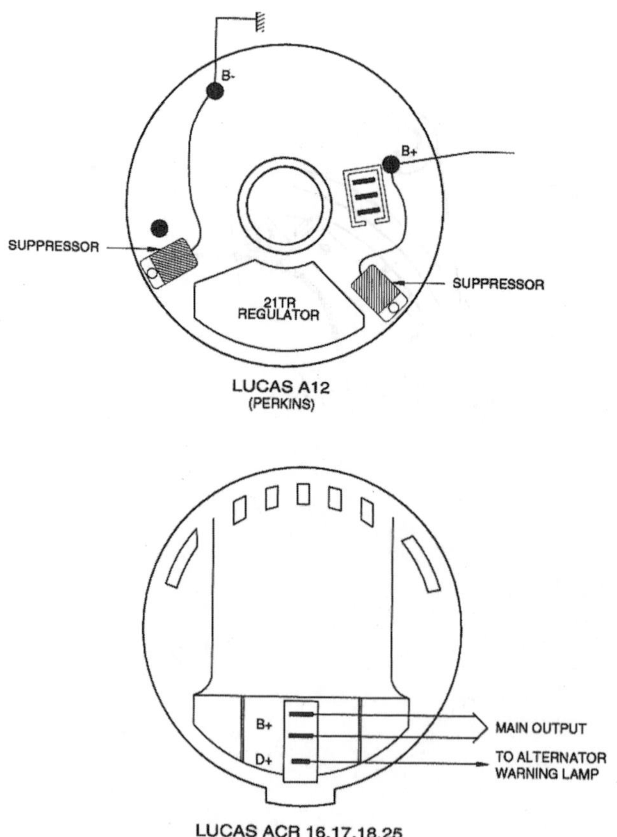

Figure 3-7 Lucas Alternator

c. **Paris-Rhone/Valeo.** Usually a standard type fitted to Volvo engines, Paris-Rhone and Valeo are now all the same as Valeo alternators, though there are some differences in the design. Use the following procedure to disconnect and install a new, external regulator system, or to replace the existing one:

(1) Unscrew and remove the 4 screws securing the regulator to the casing.

(2) There are 4 cables leading from the regulator (5 on the new Valeo). If you are replacing the regulator with an external type, cut the cables off at the regulator. The regulator and housing act as a spark arrestor cover for the brushgear.

(3) Remove the negative cable to the regulator entirely.

(4) Take the cable running internally under the plastic cover to terminal 61 and solder it to one of the brush-holder connections. This cable was initially connected to the regulator until cut off.

(5) Solder a wire to the remaining brush-holder and run it out through the cover for connection to the new regulator. This is the field control connection.

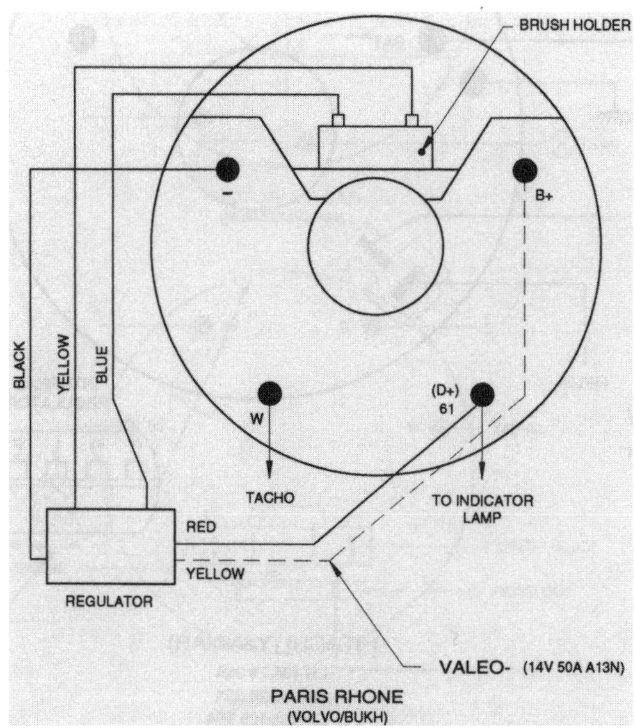

Figure 3-8 Paris-Rhone/Valeo Alternator

d. **Hitachi.** This is the standard alternator fitted to Yanmar engines (Models LR 135-74 35A, LR 155-20 55A and LR 135-105 35A). Use the following procedure:

(1) Remove the rear casing from the alternator. The screws are generally torqued extremely tight, so use the correct screwdriver size.

(2) Carefully cut off the connections to the existing regulator. There are 5 in total.

(3) Solder a 1.5 mm bridging wire between the R and F terminals, as illustrated.

(4) Replace the rear casing.

(5) Connect the external field connection to the socket at the rear of the alternator. A cable and plug are normally fitted and can be removed.

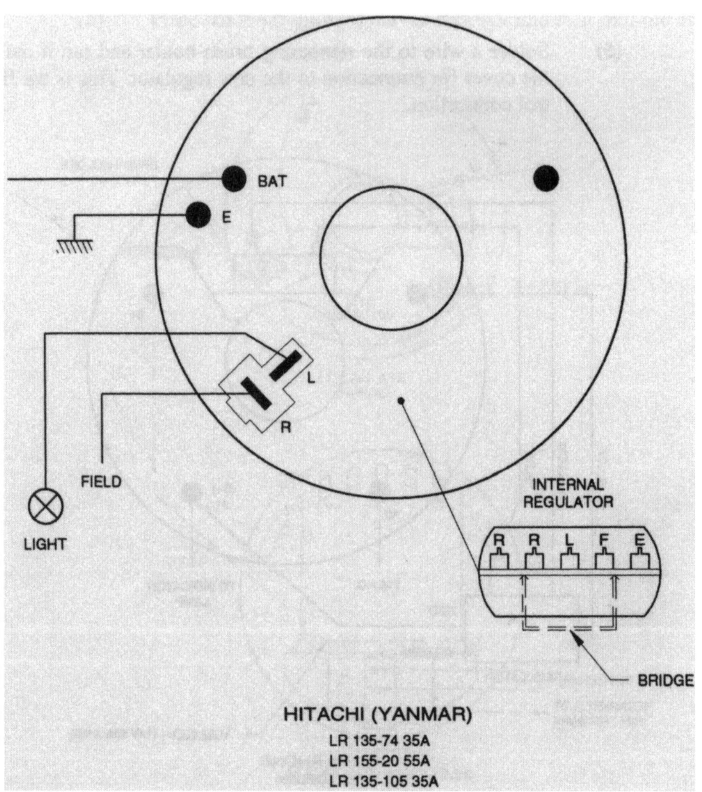

Figure 3-9 Hitachi Alternator

e. **Motorola.** Model 9AR is usually fitted to Nanni and Universal engines. Remove as follows:

(1) Unscrew and remove the two retaining screws holding in the existing regulator.

(2) Either cut or remove the two cables connecting the regulator to the alternator.

(3) Fit a new wire to the vacated lower terminal and run it out through the cover for connection to the new regulator. This is the field control connection.

(4) Refit the old regulator and housing, which acts as a spark arrestor cover for the brushgear.

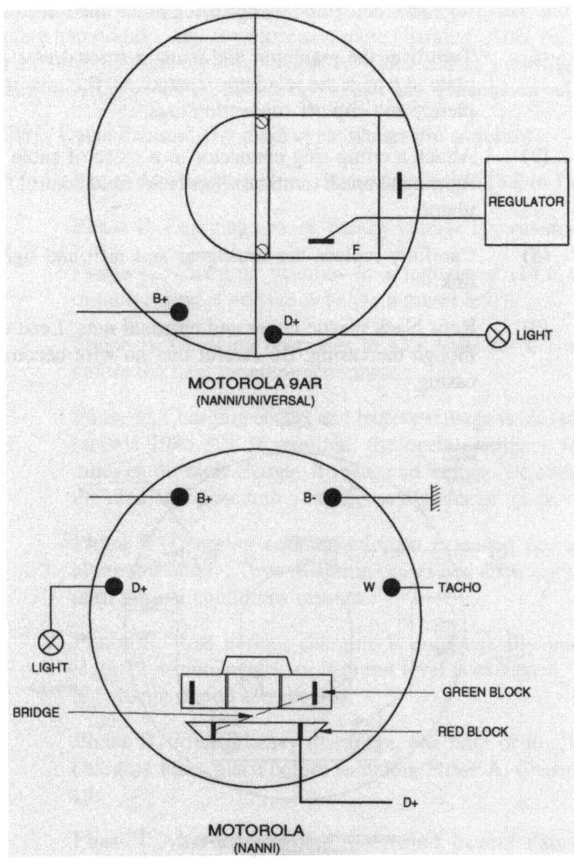

Figure 3-10 Motorola Alternator

f. **Prestolite.** Model 8EM2017KA, 51 Amp. is the standard alternator now often fitted to Universal and Westerbeke engines. Remove and modify as follows:

(1) Disconnect cables and make a note of installation points.

(2) Unscrew and remove terminal nuts and washers.

(3) Remove the two hex head bolts in the center of the rear casing and remove the black plastic cover.

(4) Remove the nut securing the bridging link from the brush terminal to the indicator light terminal.

(5) Carefully unscrew the two hex head bolts securing the brushgear and regulator. Slide up brushgear/regulator and remove.

(6) Turn over the regulator, and using a screwdriver, pry off the metal plate to which the regulator is attached. Remove the regulator completely and clip off connection tags.

(7) Attach a crimp ring connector to a piece of cable and fasten to the right-hand brush terminal. This is for field control from external regulator.

(8) Carefully replace the brushgear and refit and tighten the bridging link.

(9) Refit black plastic cover and terminal nuts. Lead out the field wire though the casing. Be careful that no wire becomes trapped under casing.

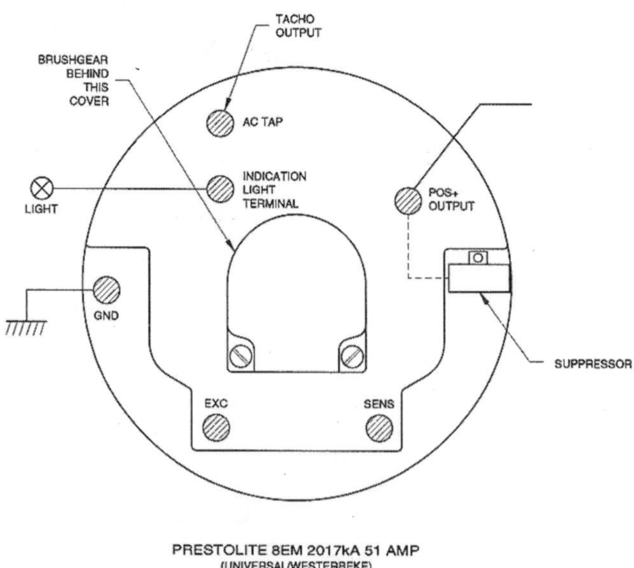

Figure 3-11 Prestolite Alternator

3.19 Adverc Cycle Regulator. A cycle program is the basis of the charging system. The regulator is also designed for parallel connection to the existing regulator, giving some redundancy should failure ever occur. Temperature compensation takes place; the Adverc has a linear one. The Adverc system has a light warning system, with indication given for low and high voltage conditions or a loss of sensing leads. Adverc regulators are designed with value engineering principles, and components are over rated by 400%. This of course is important where robust construction is a key consideration. Latest versions have had circuit enhancements that attenuate RFI and smoother control outputs. I have an Adverc installed and the system has proven itself.

a. **Cycle Period.** The cycle periods on the Adverc system consist of four 20-minute intervals followed by a one-hour rest period. Voltage levels used within the charging cycle are a normal charge rate of 14.0 volts, and a high level of 14.5 volts. To check operation always connect a digital meter across the battery and observe the cycles.

b. **Adverc Regulator Color Codes.** The following are color codes for Adverc (and TWC) regulators:

(1) **Green.** Connects to the field connection of the alternator.

(2) **Brown.** Connects to the auxiliary D+ output terminal.

(3) **Yellow.** Connects to alternator warning light, taken off the D+ terminal of the alternator at installation. Splice them together.

(4) **Black.** Connects to the alternator negative or case.

(5) **Blue.** This is a sense wire, and connects to the alternator main output B+, but note that where a diode isolator is used, it must be connected to the house battery side of the diode isolator. This may entail lengthening the blue wire. The lead also has a 0.5 "clamp" applied to the circuit protect against high output voltages.

(6) **Red.** This is the sense wire that connects to the house battery or changeover switch common terminal.

3.20 Balmar Regulator. Balmar have a range of regulators (www.balmar.net):

a. **Max Charge MC-612 Regulator.** This is a microprocessor-controlled unit with several user selectable multi-voltage variable-charge time programs for six battery types. The settings are via dipswitches. The principle is the use of an automatic absorption time program, alarm outputs, LED status and alarm indicators. The amp manager function has a remote controlled power reducer if required. The unit has a data output port, and the option of a soft start and ramp up function, this is to save belt wear. There is an optional battery compensation sensor and alternator temperature sensor for over-temp protection. As a comment, battery compensation should have been standard.

b. **Max Charge MC-412 Regulator.** This is a microprocessor-controlled unit with several user selectable multi-voltage variable-charge time programs for four battery types. It also has an LED display for program mode indication, and self-diagnostics. It has connections for warning light and electrical tachometer output. The amp manager function is included along with a data port. There is an optional battery compensation sensor and alternator temperature sensor for over-temp protection.

c. **BRS-2 Regulator.** This single stage regulator has a nominal setting of 13.7 volts, and has a range of settings for various battery types. It has connections for warning light and electrical tachometer output.

d. **ARS-4 Regulator.** This is a 3-step controller with user settings for Deep Cycle, Gel, AGM and Optima batteries, and uses bulk, absorption and float charge principle. It also has an LED display for program mode indication, and self-diagnostics. It has connections for warning light and electrical tachometer output.

3.21 Heart Interface Alpha (InCharge) 3-Stage Charge Regulator. This is a 3-stage regulator that has user definable settings for Accept, Float and Time. The unit operates on what is called the I-Uo-Uo characteristic. The first step is the bulk charge phase, where voltage rises steadily up to approximately 14.2–14.4 volts, and maximum current output occurs up until approximately 80% charge level. The second step is the absorption or acceptance phase where the voltage is maintained constant and the current slowly reduces. The third step is the float phase where voltage reduces to approximately 13.8 volts and maintains a float charge to the battery.

3.22 PowerTap Regulator. PowerTap has a range of regulators, which are as follows:

a. **Smart Regulator SAR-V3.** The regulator uses a microprocessor controlled cycle type program. It has no operator adjustable functions with respect to the charging cycle, and operates based on 12 programmed charging cycles. Battery temperature compensation is incorporated and it is for use with P type alternators only. An alarm function uses a coded flash system. An over-voltage runaway circuit detects over-voltage conditions that occur when regulator output has a short circuit and runaway. This is indicated via the alarm lamp circuit. Current limiting is via a user adjustable function that requires connecting an externally operated switch. The switch will reduce output to a relatively low level to avoid overheating of an alternator or to remove load off a smaller engine. Equalization function is a user adjustable feature that requires connection of an externally operated switch. The function enables an equalization current to be applied until battery voltage reaches 16.2 volts. The regulator has some very commendable features. The field output driver is short protected to prevent damage in the event of a field circuit failure. Additionally all inputs are voltage transient protected, although normal precautions should still be installed. The lamp circuit is also over-rated to provide alarm buzzer load capability as well. In addition, a voltage limit function enables charge voltage to be held at 13.8 volts to prevent halogen light damage during long night motoring passages.

b. **3-Step Deep Cycle Regulator.** The 3-Step device uses a step type program, that is fully automatic, and operates based on the charging cycles of absorption and float. The unit consists of a timer circuit rather than an intelligent program chip, and has simple battery and ignition inputs. Users are able to manually alter absorption and float voltage settings which is useful in applications such as NiCad cells that require different charging voltage levels. The manufacturer states that due to full alternator output requirement in step 1, many alternators may not be able to cope, and may suffer failure. This is generally due to windings overheating and diode failure. The regulator is suitable for P type alternators only (i.e. Bosch, Prestolite, Motorola, Valeo/Paris-Rhone etc). The regulator has the following control steps:

(1) The alternator is controlled to give full output until the absorption set point is reached. The time required to reach this level depends on the initial battery level and output speed of the alternator.

(2) The absorption set point is maintained for a period of 45 minutes (14.5 volts).

(3) The charge level reduces to the float voltage set point (13.8 volts).

c. **Next Step Regulator.** The Next Step deep cycle regulator is an improved version of the 3-step unit. The unit is a microprocessor-controlled unit and incorporates temperature compensation. Due to full alternator output requirement in step 1, many alternators may not be able to cope, and may suffer failure. This is generally due to windings overheating and diode failure. Users are able to manually alter both absorption voltage and time as well as float voltage settings. The regulator has the following control steps:

(1) The alternator is controlled to give full output until the absorption set point is reached. The time required to reach this level depends on the initial battery level and output speed of the alternator.

(2) The absorption set point is maintained for a period of 45 minutes (14.5 volts).

(3) The charge level reduces to the float voltage set point (13.8 volts).

3.23 Ideal Regulator. This regulator is different in that current is a factor in the charging process, and not just a voltage dependent device. I have not seen this in any other regulator types I have come across. The regulator is used in conjunction with a digital circuit monitor. The regulator has the following control steps.

(1) **Delay Period.** A 20 second delay period after voltage is applied from ignition to allow engine speed to rise to normal running speed.

(2) **Ramping Up Period.** This allows a controlled increase of alternator output over a 10 second period until the default current limiting value is reached. This reduces shock loadings, allows belts to warm up, and reduces power line surges that occur when full outputs are applied.

(3) **Charge Cycle.** The charge cycle allows full alternator output until the battery voltage reaches 14.3 volts.

(4) **Acceptance Cycle.** Charging continues at 14.3 volts until charge current decreases to a default value of 2% of capacity. Once the 2% level is reached the acceptance hold cycle begins.

(5) **Acceptance Hold Cycle.** Charging is held at 14.3 volts and the charging current is monitored and continues for a minimum of 10 minutes. A maximum of 20 minutes is imposed on this cycle.

(6) **Float Ramp Cycle.** This is a transition phase between charged and float cycles. Voltage is reducing to the float setting of 13.3 volts during the cycle.

(7) **Float Cycle.** The voltage is held constant at 13.3 volts.

(8) **Condition Cycle.** This is a manually activated function. Current is held at 4% of battery capacity, until a maximum voltage of 16 volts is attained. Once voltage reaches 16 volts, it is maintained until charge current falls to charge current % setting. The cycle then automatically terminates and reverts to the float ramp cycle to bring the voltage down.

3.24. Battery Charge Settings. Different battery types require different charge voltages and regulator systems should be adjusted to suit the installed batteries.

Table 3-3 Battery Regulator Charge Levels

Temp	Flooded Hi/Float	Gel Hi/Float	AGM Hi/Float
90°F	14/13.1	14.0/13.6	14.4/13.9
80°F	14.3/13.3	14.0/13.7	14.5/14.0
70°F	14.4/13.5	14.1/13.8	14.6/14.1
60°F	14.6/13.7	14.3/13.9	14.7/14.2
50°F	14.8/13.9	14.2/14.0	14.8/14.3

3.25 Alternator Manual Control Devices. Manual devices are those that require total operator control of the alternator output without regulation. Some handbooks give information on how to make your own controllers. From personal experience, I can say that once these homegrown controllers and circuits are installed, the charging system, batteries and alternator will be burned out not too far into the future. There is no such thing as a cheap solution, and if you really care about your power system, don't risk it. There is no sense in having and relying on electronics worth thousands only to balk at paying relatively small sums to improve charging. The following control methods are used at your own risk. While there are many around who boast how reliable and cheap they are, I do not mind making a very nice living off the majority who have subsequent problems. The savings initially achieved on these methods are more than negated by one mishap which often shortens battery life through overcharging and battery plate damage.

a. **Field Switches.** A typical manual method is to connect the switch directly to the field connection. It simply puts on a full field voltage resulting in maximum alternator output. The results can be quite spectacular and very damaging to both battery and alternator. Once while crossing a dangerous bar on a his vessel, a friend casually flicked a switch, which was followed by sparks and smoke curling out of the engine compartment. After investigation, I found this same set-up, which was potentially a disaster at the time.

b. **Field Rheostats.** The most common type of control is the rheostat. A rheostat is simply a variable resistance rated for the field current. The term rheostat is still in common usage and low value variable resistances are generally termed potentiometers. Operation is very reliant on operator control, with no safety cutouts or regulation. As a general alternator charging control it is not recommended, as both alternator and battery are easily and commonly damaged.

3.26 Alternator Controllers. Controllers are devices that require the yacht owner to manually select or partially over-ride the existing regulator to fast charge. It is important to remember the basic phases of charging a battery, i.e. bulk, absorption, float and equalization and that at no stage does battery voltage exceed gassing level. In most cases, controllers do not adhere to these basic charging principles.

a. **Operating Principles.** Controllers are either direct regulator replacement units or are connected in parallel to the existing regulator. Some units have an ammeter to monitor output and require continual adjustment of field current to maintain required charge current level, but they do not monitor or take into account the high and damaging system voltages that are imposed while maintaining the initial high charging currents.

b. **Precautions.** All controllers will have some beneficial outcome, and can improve the charging process to varying degrees. There are however serious risks that must be considered to avoid damage:

(1) **Power System Disturbances.** If you apply excessive voltages or full alternator outputs, spikes and surges can arise on the system that will damage regulators and electronics equipment.

(2) **Battery Damage.** Forcing current into batteries above the natural ability to accept charge will simply damage plates, heat the battery up, and generate potentially explosive gases. Failure of automatic cutouts, or forgetting about the regulator may cause all of the mentioned problems.

c. **Performance and Efficiency.** There are some important factors to consider before purchasing controllers.

(1) **Efficiency.** At best, these types of units can offer a 10–15% improvement that brings charge levels up to approximately 85% of nominal capacity.

(2) **Performance.** It is interesting to note that virtually none of the controller manufacturers can offer any verifiable proof or independent testing to support claims that they in fact improve charging.

d. **Controller Types.** Some of the more common controllers on the market include the AutoMAC. This controller is parallel connected to the existing regulator. A potentiometer is used to adjust alternator current in conjunction with an ammeter. When a predetermined voltage is reached the unit automatically cuts off and existing regulator takes over.

3.27 Diesel DC Charging Systems. An alternative or addition to main propulsion energy charging systems is a dedicated engine powering an alternator. Systems are as follows:

a. **Balmar.** Balmar in the United States has a unit driven by a FW cooled 13 hp Yanmar diesel. It is installed with a 310 amp brushless alternator, Max Charge regulator, and small start battery alternator. Optional equipment is a 20 or 40 gph watermaker. It is an efficient charging solution and consumes around 0.25 gph, which is economical.

b. **Ample Power Genie.** This unit uses a seawater cooled Kubota diesel fitted with a 120-amp alternator and Smart regulator system.

c. **Stirling Cycle Machine.** This new system is known as the WhisperGen and follows a principle developed in 1816, which uses a continuous combustion process, with no noise as the motor/generator is hermetically sealed, requires no oil lubrication as there are no moving parts, and no exhaust fumes. Maintenance requirements are also low with burners requiring cleaning every 2000 hours and a low fuel consumption of 0.7l/hr. The DC output is from a permanent magnet DC generator. Ratings are in the 4-6 kW range.

3.28 Regulator Troubleshooting. There is a simple test to check whether your regulator or controller is working properly. This is not difficult with external regulators, but if an internal regulator is fitted, the alternator will need to be opened and a wire attached to a brush-holder. Switch off all electrical and electronic equipment at the switchboard circuit breaker before starting this test. *If in doubt, don't try it.*

a. **Alternator Test.** Check that the alternator gives full output. If the alternator operates after testing, then the regulator is suspect.

b. **Rotor Testing.** If a regulator has failed, particularly in an overcharge condition, checked the rotor for damage before replacing the regulator. The test is as follows and is illustrated below:

(1) **Test Insulation Resistance.** Place one multimeter probe on a slipring, and the other on the rotor core. Resistance should be infinite or over-range.

(2) **Test Winding Resistance.** Place the multimeter probes on each slipring. Resistance should be around 4 ohms. If it is very high, an open circuit may exist; if very low, a coil short circuit may exist.

c. **Auxiliary Diode Test.** On some occasions, the auxiliary diodes may fail. Put your multimeter on the 20-volt range and connect across 61/D+ and negative. If there is any reading, the diode may be faulty. Turn on the ignition key without starting. The reading should be around 1–2 volts. If less, the wiring may be faulty; if higher, the diode may be faulty, there is excessive rotor resistance or there are bad connections.

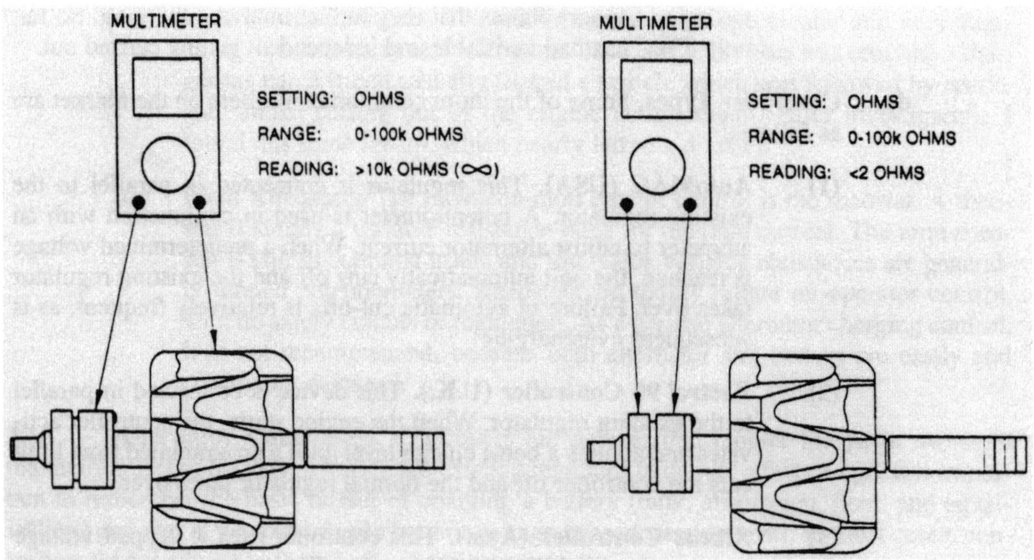

Figure 3-12 Rotor Testing

Table 3-4 Charging System Troubleshooting

Symptom	Probable Fault	Corrective Action
Reduced charging	Drive belt loose	Adjust to 10 mm
	Oil on belt	Clean belt
	Loose alternator connection	Repair connection
	Partial diode failure	Repair alternator
	Suppressor breaking down	Replace suppressor
	Regulator fault	Replace regulator
	Diode isolator fault	Replace diode
	Negative connection fault	Repair connection
	Solder connection fault	Re-solder connection
	Under-rated cables	Uprate cables
	In-line ammeter fault	Repair connections
	In-line ammeter fault	Replace ammeter
	Ammeter shunt fault	Repair connections
Over charging	Regulator fault	Replace regulator
	Sense wire off	Replace wire
No charging	Drive belt loose	Re-tension belt
	Drive belt broken	Replace belt
	Warning lamp failure	Replace lamp
	Auxiliary diode failure	Repair alternator
	Regulator fault	Replace regulator
	Diode bridge failure	Repair alternator
	Jammed brushes	Clean brush gear
	Stator winding failure	Repair alternator
	Rotor winding failure	Repair alternator
	Output connection off	Repair connection
	Negative connection off	Repair connection
Fluctuating ammeter	Alternator brushes sticking	Repair alternator
	Regulator fault	Replace regulator
	Loose cable connections	Repair connections
High initial start current, low charge current	Ammeter fault/overcurrent	Replace ammeter
	Batteries sulfated	Replace batteries
	Battery cell failure	Replace batteries
	Battery charge very low	Recharge extended time

Alternative Energy Systems

4.0 Alternative Energy Charging Systems. More misconceptions exist about the capabilities of alternative energy systems on cruising yachts than virtually any other equipment. In most cases, expectations are wildy optimistic, and the realities are at best disappointing. Some absolute truths must be recognized before embarking on projects that entail large expenditures, and often a lot of engineering. They must be faced in spite of the philosophical and environmental arguments. The important factors are outlined below for consideration in that decision-making process.

a. **Secondary Power Sources.** Alternative energy sources at the prevailing technology levels can only be considered as auxiliary charging sources. They should be integrated into the power system as a secondary power source where no further charging capacity can be derived from the engine alternator. In most cases, alternative energy sources significantly reduce dependence on engine-based systems. A battery on a poorly maintained vessel can lose as much as 14% of its charge per month, so an alternative energy charging system would be ideal in this situation.

b. **Primary Power Sources.** Many people, for a variety of reasons, choose to rely solely on renewable power sources to supply electrical power. Yet I have observed in a large number of cases a complete lack of understanding of basic electrical design.

(1) **Design Considerations.** Alternative energy systems require considerably more stringent design criteria, a sailing philosophy that excludes a large number of electrical and electronics equipment, and a very disciplined lifestyle while cruising. If you want all the comforts and technologies of home, you are going to require a very large number of solar panels, wind generators, and probably water-powered ones as well. Regrettably, the natural forces that control alternative sources are far from predictable, which is why many have had to adjust their cruising behavior to one dominated by the search for ways to conserve battery power, and recharge batteries.

(2) **Output Data.** You must realize that the quoted output data is almost always in absolutely ideal laboratory conditions. In practice, you will require a large safety factor to get a reasonable result.

(3) **The Downward Spiral.** In practice, battery charge levels tend to slowly spiral downward. The real trick in getting the most out of alternative energy systems is to fully charge the battery to 100% with a good, fast-charge engine system before the batteries sink low enough to be damaged. That allows the alternative systems to keep up.

4.1 Solar Systems. Many cruising yachts use solar panels and wind generators to trickle charge batteries when anchored or moored, as well as under sail. Solar energy concepts are not new, and date back to 1839 when the French scientist Becquerel discovered the photovoltaic phenomenon. Solar systems are the most commonly used alternative energy sources and offer a renewable and nearly maintenance free energy source. The fundamental process of a solar cell is that when light falls on to a thin slice of silicon P & N substrate, a voltage is generated. This is called the photovoltaic principle. Cells consist of two layers, one positive, and one negative. When light energy photons enter the cell, the silicon atoms absorb some photons. This frees electrons in the negative layer, which then flow through the external circuit (the battery) and back to the positive layer. When manufactured, the cells are electronically matched and connected into an array by connecting in series to form complete solar panels with typical peak power outputs of 16 volts. There are a number of solar cell types and this is based on the cell material or structure used:

a. **Mono-crystalline.** Pure, defect-free silicon slices from a single grown crystal are used for these structures. The cell atomic structure is rigid and ordered and unlike amorphous cells cannot be easily bent. The cells are approximately 12%–15% efficient. The thin pure silicon wafers are etched within a caustic solution to create a textured surface. This textured surface consists of millions of four sided pyramids, which act as efficient light traps reducing reflection losses. Panels are made by interconnecting and encapsulating 34–36 wafers onto a glass back.

b. **Polycrystalline.** These cell types use high purity silicon 0.2mm wafers from a single block, and are high power output cells. The wafers are bonded to an aluminium substrate. Solarex cells are covered with a tempered iron glass, and a titanium dioxide anti-reflective coating to improve light absorption. The polycrystalline cell has better low light angle output levels and is now the most commonly used.

c. **Amorphous Silicon.** These cells are formed from several layers applied to a substrate. They have a characteristic black appearance and Solarex cells have a tin oxide coating to improve conductivity and light absorption. Unlike crystalline cells, these thin film panels have a loosely arranged atomic structure and are much less efficient. They do have the advantage that the cells can be applied to flexible plastic surfaces and as such flexible panels are made. Additionally they are capable of generating under low light conditions. Crystalline cells won't do this. The big disadvantage is that power outputs are nearly a quarter of crystalline cells of the same size.

d. **Construction.** Cell arrays are normally laminated under ethylene vinyl acetate (EVA). Anti-reflection coatings with titanium dioxide are used, and some are characterized by a blue coloring. This also increases the gathering of light at the blue end of the light spectrum. Panels are constructed to be moisture and ultraviolet resistant. Glass surfaces are tempered and sometimes textured to reduce reflection, increase surface area, and improve light gathering at low lighting angles. Solar arrays often utilize front and rear connections to improve faulty cell redundancy.

4.2. Solar Ratings, Efficiency and Regulation. Efficiency is at an optimum when a solar panel is angled directly towards the sun and manufacturers rate panels at specific test standards. The most effective panels are rigid units while the flexible units have significantly lower outputs. Output ratings are normally quoted to a standard, typically 1000W/m^2 at 25°C cell temperature, and the level of irradiance is measured in watts per square meter. The irradiance value is multiplied by time duration to give watt-hours per square meter per day. Location and seasonal factors affect the amount of energy available. Cells are approximately 15% efficient and start producing a voltage as low as 5% of full sunlight value. Solar angles are important to the efficiency of panels. With the sun at 90° overhead, panels give 100% output. When angled at 75°, the output falls to approximately 95%, at 50° output falls to 75%, and a lower light angle of 30° gives a reduction to 50%. Many panels now will give some output on dull days, The table shows typical seasonal hours and yearly averages based on solar array tilted towards the sun at an angle equal to latitude of the location +15°.

Table 4-1 Peak Solar Level Table

Location	Winter Hours	Summer Hours	Average
California	4.0	5.0	4.5
Miami	3.6	6.2	4.9
Central Pacific	4.5	6.0	5.3
Caribbean	5.5	5.5	5.5
Azores	2.2	6.0	4.1
Northern Europe	1.5	4.0	2.7
Southern England	0.6	5.0	2.8
South France	2.5	7.5	5.0
Greece	2.4	7.4	4.9
SE Asia	4.0	5.5	4.7
Capetown	4.0	5.0	4.5
Red Sea	6.0	6.5	6.3
Indian Ocean	5.0	5.5	5.3
Eastern Australia	4.5	5.5	5.0

a. **Panel Regulation.** In any panel larger than a small 12–15 watt unit, a regulator is required to restrict the voltage to a safe level. It is not uncommon to have solar panel output rise to 15–16 volts and boil batteries dry over an extended, unsupervised period. There are solar control devices in use which must not be confused. One simply limits voltage to safe levels and the other device, called a linear current booster, increases power for certain conditions:

(1) **Regulators.** The regulator serves to limit panel output to a safe level and prevent damage to a battery. Some units simply limit voltage to 13.8 volts, the maximum float level, and dissipate heat through a heat sink. More sophisticated regulators get more from the panel. These units incorporate an automatic boost level of 14.2 volts and a float setting of 13.8 volts. The regulator float charges the battery until a lower limit of approximately 12.5 volts is reached before switching to boost. The units normally eliminate the need for an additional blocking diode. Check the manufacturer's data sheet first. Some regulators also have temperature compensation and must be installed adjacent to the batteries.

(2) **Linear Current Booster.** These electronic devices boost current from the solar module. They are designed to prevent permanent magnet motors from stalling, but effectively they are constant current devices. Such units are used primarily in applications where panels directly supply a load. They are not useful on boats where the panel is used to charge a battery.

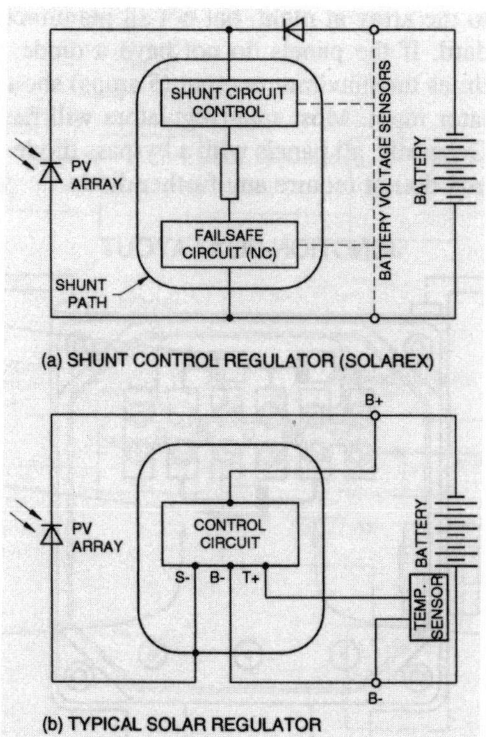

Figure 4-1 Typical Solar Regulator Systems

b. **Diodes.** Most panels have diodes installed. There is a rather flawed argument that the use of a diode reduces charging voltage. This is true, as a diode reduces voltage by approximately 0.75 volt. But if you are installing a couple of 3-amp panels, which is typical, you will need a regulator to reduce the voltage to avoid overcharging and damaging your batteries. If the regulator is a good unit, the control will float between 14.5 and 13.8 volts, so the small voltage drop will not be a problem. If the regulator has the appropriate reverse-current protection diode, then the panel-installed diode can be removed to increase the input voltage to the regulator, which gives a marginally higher output. If you do not regulate the solar supply, failing to install or removing the diode will result in a dead battery overnight. There are two functional uses of diodes:

(1) **By-pass Diodes.** By-pass diodes, normally installed at the factory in solar module junction boxes, reduce power losses that might occur if a module within the array is partially shaded. For 12-volt systems, these offer sufficient circuit protection without the use of a blocking diode. A 24-volt array requires two 12-volt panels in series. An array for larger current outputs requires the parallel connection of these series arrangements. If one module of a parallel array is shaded, reverse current flow may occur.

(2) **Blocking Diodes.** Blocking diodes are often connected in series with the solar panel output to prevent the battery from discharging back to the array at night, but not all manufacturers install them as standard. If the panels do not have a diode, then a diode rated to 1.5 times the maximum output (5 amps) should be installed at the regulator input. Most solar regulators will have the diode incorporated. Generally, all panels with a by-pass diode installed in the connection box do not require any further diode.

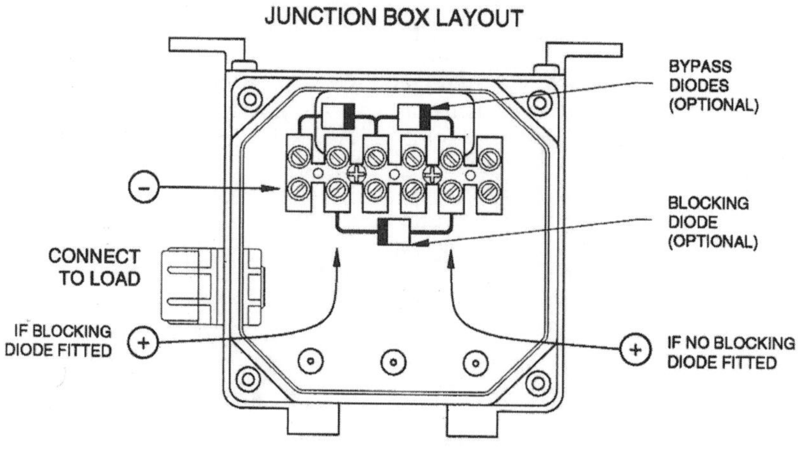

Figure 4-2 Typical BP Solar-Diode Junction Box

c. **Charging System Interaction.** There is often an interaction between solar panels and alternator charging regulators during engine charging. If the solar panels are not regulated, it is quite common to see a voltage of up to 16 volts or more across the battery. When an alternator regulator senses this high voltage level, it simply registers it as a fully charged battery, and as a result the alternator does not charge the battery, or does so at a minimal rate. When installing panels and regulators, consider the following features:

(1) **Isolation Switch.** Install an isolation switch on the incoming line to the panel so that it can be switched out of circuit.

(2) **Regulator By-pass Switch.** Always install a switch that can by-pass the regulator and apply full-panel output to the battery. This will make periodic equalization easier and charging a dead battery more efficient.

(3) **Engine Interlock.** This circuit automatically disconnects the solar panel via a relay so that the solar panel does not impress a higher voltage and "confuse" the alternator's regulator.

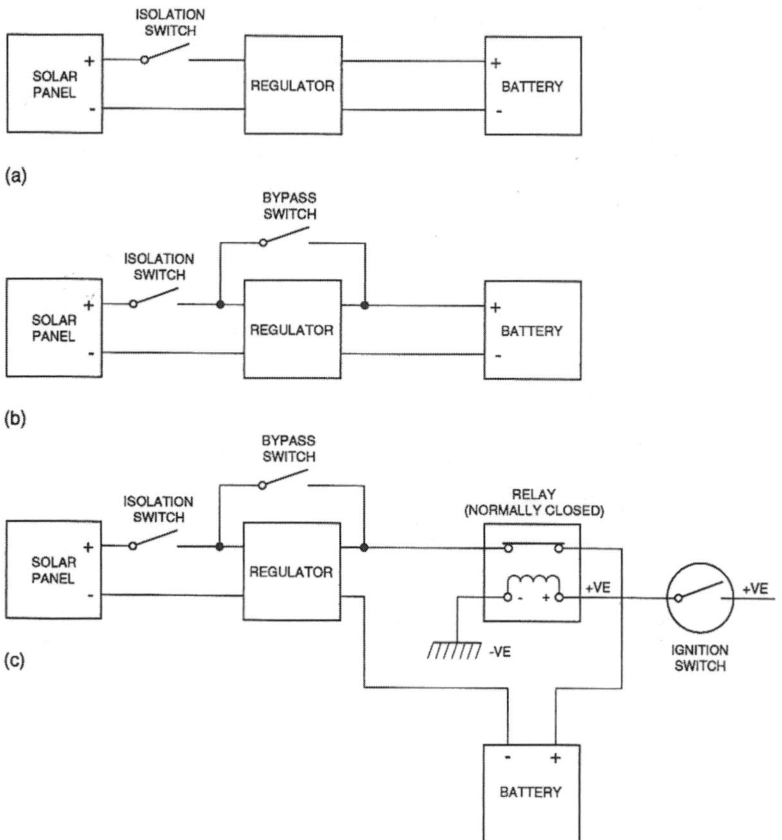

Figure 4-3 Typical Solar Interlock Systems

d. **Site Selection.** Solar-panel siting is largely dependent on the physical space available for installation. The following options are the most common and most efficient. In all cases, it is essential to ensure that panels are not shadowed by sails, spars, or any other equipment. Ideally, panels should be angled towards the sun if at all possible, but on a cruising yacht this is not always practical. Generally, flat-mounted panels offer the best compromise, which is why the stern arch configuration is becoming so popular:

(1) **Coach House.** Panels can be mounted on coach-house tops, but one panel will often be shaded and the other illuminated, depending on the tack you are on.

(2) **Stern Mounting.** This arrangement is really only suitable for a single panel, and is usually designed to allow the panel angle to be adjusted. The panel support brackets are welded to the stern pulpit rail.

(3) **Stern Arches.** This is becoming the most popular method as it allows the easy installation of at least two clear and unobstructed 3-amp panels.

(4) **Stern Pulpit (Pushpit) Rails.** This arrangement uses two panels mounted on swing-up brackets on each side of the vessel, normally close to and on the pulpit rails. Depending on tack, or direction of sun, the panels can be put into service, and folded down if not used.

(5) **Multihulls.** The greater deck area of a multihull and its nearly flat sailing attitude make site selection much easier, and offer increased efficiency. In most cases, a large coach house can be utilized, and on trimarans, arrays can be mounted on the outer hulls well clear of shadows.

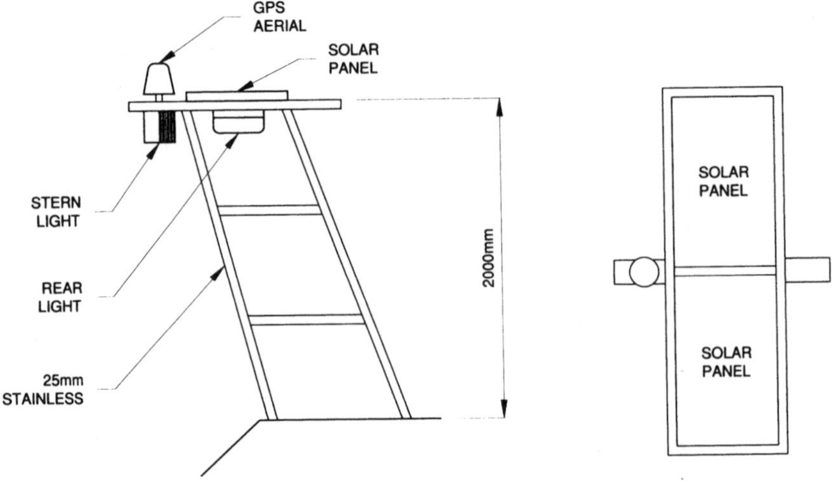

Figure 4-4 Stern-Arch Arrangement

e. **Installation.** Solar panels are manufactured in either rigid or flexible form. Cabling should be properly rated to avoid voltage drop. To cope with two 65-watt panels, 2.5 mm2 (15-amp) cable is the minimum size. Use only tinned copper marine cable. Most panels have weatherproof connection boxes and connections can be simply twisted and terminated in terminals. Do not use connectors or solder the wire ends. Manufacturers also specify grounding the array's or the module's metallic frames. I have heard concerns about corrosion aboard vessels with automatic bilge pumps and a solar panel charging the battery. If the pump cable develops a fault, once the battery is flat, serious electrolytic corrosion may corrode through-hull (skin) fittings and hull as a voltage is being applied directly to them. Although theoretically possible, I have never heard of this occurring and it would be extremely rare. If it is a concern, operate the bilge pump off the non-charging battery:

(1) **Panel Safety.** Cover solar panels to prevent voltage from being generated during installation or removal so that accidental short circuiting of terminals or cables cannot occur.

(2) **Mounting.** Each panel should be securely mounted and able to withstand mechanical loads. Ideally, they should be oriented to provide unrestricted sunlight from 9 to 3 PM solar time.

(3) **Stand-offs.** Allow sufficient ventilation under the panel. Excessive heat levels will reduce output and damage cells. Most panels in frames have sufficient clearance incorporated into them.

f. **Maintenance.** Maintenance requirements for solar panels are minimal:

(1) **Cleaning.** Panels should be cleaned periodically to remove salt deposits, dirt, and seagull droppings. Use water and a soft cloth or sponge. Mild, nonabrasive cleaners may be used; do not use scouring powders or similar materials.

(2) **Connections.** Make sure the terminal box connections are secure and dry. Fill the box with silicon compound (sealer).

g. **Troubleshooting.** Faults are normally the result of catastrophic mechanical damage. A single cell failure will not seriously reduce performance as multiple cell interconnections provide some redundancy. Reliability is very high and manufacturers give 10-year warranties to support this. Faults can be virtually eliminated by proper mounting and regular maintenance. As with all electrical systems, the most common faults are cable connections. The following checks should be carried out if charging is not occurring:

(1) Check regulator output for rated voltage, typically 13.2 V DC.

(2) Check regulator input; voltage will typically be 14+volts. Disconnected from battery, it can be up to 17–18 volts.

(3) Check panel junction boxes for moisture or corroded connections.

4.3 Wind Charging Systems. Wind generators are the second most used alternative energy source. As with all charging systems, there are important factors to consider when deciding whether to install a unit as part of a balanced power system. The following chapter outlines the various factors to consider.

a. **Cruising Patterns.** Wind generators are more effective in some areas than others. In the Caribbean, they are very effective. In the Mediterranean, solar power is considered more efficient. If you sail downwind following the trades, wind generators are not effective as the apparent wind speed is reduced, along with charging capability. If your cruising takes you primarily to sheltered anchorages, they may not be an economical or practical proposition. It is at anchorages, however, that wind generators can be the most useful and give 24-hour charging.

b. **Generator Types.** Essentially a wind generator is either a DC generator or an alternator driven by a propeller. In the U.S. the trend is normally for large two or three bladed DC generator units. The UK/European trend is for smaller diameter multi-bladed AC alternator units. These units incorporate a heavy hub that acts as a flywheel to maintain blade inertia. Many units have a permanent magnet rotor, with up to 12 poles. A three phase alternating current is generated and rectified to DC similar to engine driven alternators. The aerodynamically shaped AeroMarine is a three bladed unit and has a brushless permanent magnet alternator with internal regulator. From many observations, in protected marinas and bays the 6 bladed Rutland appears more efficient than the 3 bladed AeroMarine.

c. **Generator Operation.** Many units have a permanent magnet rotor, with up to 12 poles. A 3-phase alternating current is generated and rectified to DC, similar to engine-driven alternators. The Rutland 910 unit circuit is illustrated below.

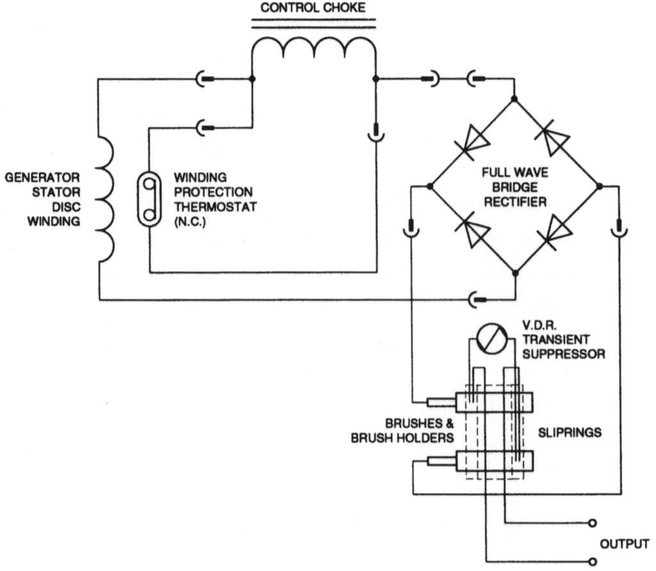

Figure 4-5 Rutland Wind Generator Circuit

d. **Ratings.** The average wind generator produces anything from 1 to 10 amps maximum. Ratings curves are always a function of wind speed and are quoted at rated output voltages.

Table 4-2 Wind-Generator Output Table

Make and Model	Output Current	Wind Speed
Aerogen 3	1 Amp	10 knots
	2	13
	3	18
	6	25
	10	35
Rutland 910	2 Amps	15
	3	18
	5	27
	6	35
	8	45
LVM 50	0.5 Amps	7
	1	11
	2	24
	7	34
Ampair 100	1 Amps	10
	3	15
	5.5	25
Fourwind III	4 Amps	10
	7	15
	12	25
Windbugger	4 Amps	10
	8	15
	13	25

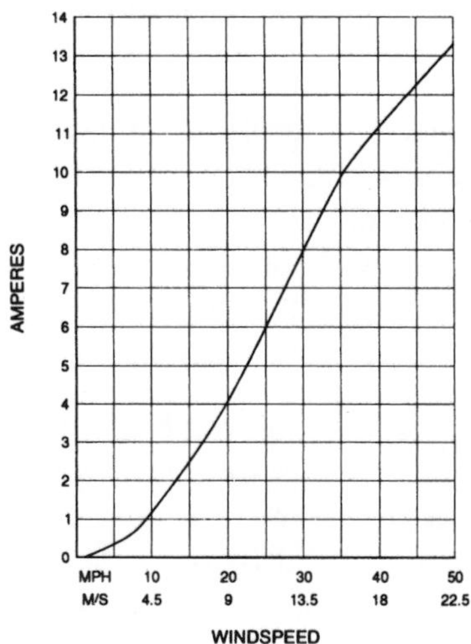

Figure 4-6 Wind Generator Performance Curve

e. **Charging Regulation.** There are a number of features incorporated into wind generators to protect batteries and generators. These are as follows:

(1) **Regulators.** A regulator is required to limit normal charging voltages to a safe level (14.5 volts) and to limit output at high wind speeds. Normally, a shunt regulator is preferred over a normal solar panel regulator as it is more suited to constant loads. Shunt regulators divert excess current to a resistor which functions as a heater and dissipates heat through a heat sink. If series regulators are used, a power zener diode should be installed to provide some load when the battery is fully charged. Twelve-volt systems should use an 18-volt zener diode. The zener must be rated for at least half the generator's rated output.

(2) **Regulator Interaction.** Like solar panel installations, interaction may occur with alternator charging systems. The charging should be either switched out of circuit or diverted to a battery other than the sensed one (e.g., the start battery).

(3) **Chokes.** Some units incorporate a choke to limit the charge produced at high wind speeds.

(4) **Winding Thermostats.** A number of generators incorporate a winding embedded thermostat which opens in overload conditions when the winding overheats.

(5) **Transient Suppressors.** These suppressors are installed to minimize the effects of intermittent spikes being impressed on the charging system. These could damage the rectifier and onboard electronics. The suppressor is usually a voltage dependent resistor (VDR).

f. **Installation.** Selection depends largely on available mounting locations. Arrangements vary. Some are mounted on the front of the mast. Others are hoistable.

(1) **Stern Posts.** The ideal mounting arrangement is on a stern post, which keeps the blades clear of crew and feeds it air coming off the mainsail. One of the major complaints is that under load wind generators create vibration. It is essential that the post section be as thick as possible and well supported. Usually this extra support is on the stern pulpit (pushpit); some install stainless steel wire stays.

(2) **Mountings.** Mountings can be cushioned with rubber blocks or similar material to reduce the transmission of vibrations. Rutland chargers have a tie-bar modification that strengthens the blades and prevents excessive blade deformation under load and the increased vibration that occurs. Newer models have improved blade design and strength that prevents blade breakage.

g. **Troubleshooting.** Always secure the turbine blades when installing, servicing, or troubleshooting a wind generator. The following performance tests should be carried out:

(1) If no ammeter is installed on the main switchboard, install an ammeter in line and check the charging current level. If there is no output, check the system according to the manufacturer's instructions.

(2) If there is no output and the generator has brushes, check that they are not stuck and are free to move. Instead of brushes and commutators, many generators have a set of sliprings installed with brushes to transfer power from the rotating generator down through the post to the battery circuit. They can jam, and on rare occasions cause loss of power.

(3) Some generators have a winding embedded thermostat. Check with an ohmmeter that it is not permanently open circuited. If it is open circuited, the generator will not charge. The thermostat opens in high wind charging conditions. If the thermostat has not closed after these conditions and the generator case is cold, the thermostat is defective. Regrettably, it cannot be repaired unless a new winding is installed. To get the generator back into service, connect a bridge across the thermostat terminals. Remember that there will be no protection in high wind and heavy-charging conditions, so the winding may burn out.

(4) Excess vibration may be caused by bearing wear. If the unit is a few years old, renew the bearings. Vibration can also be caused by damage to one or more blades, and these should be carefully examined for damage that may cause imbalances.

(5) Check the rectifier to be sure that it is not open or short circuited.

(6) If the generator output is correct, check for a malfunctioning regulator. The voltage input should be in the range 14–18 volts, and the output approximately 13–14 volts.

(7) Ensure that all electrical connections are secure and in good condition.

4.4 Prop Shaft Charging Systems. Prop shaft generator systems are either traditional alternators with prop shaft gearing to achieve rated output, or alternators wound to generate outputs at low speed. These systems can be used as an extra energy source while under power (not an economic proposition), or to take advantage of a free-wheeling propeller under sail. The following points must be considered:

a. **Cruising Patterns.** The viability of these units depends on your cruising pattern. Consider that only about one quarter to one third of your time is spent passage making, so the shaft alternator is used for a limited period.

b. **Drag.** Under any load, the alternator will brake the shaft by slowing shaft rotation, causing drag and a reduction in vessel speed. On a lightweight vessel, this can be as high as half a knot. On steel or other heavy displacement vessels, the inertia of the vessel will generally minimize the drag effect. For such cruising yachts, prop systems are a useful proposition. With an increasing number of yachts opting for two- and three-bladed folding props, shaft alternators may rarely be used.

c. **Output.** The maximum output will generally be in the region of 5–10 amps. The Lucas unit has a maximum output of 12 amps, with an approximate output of 1 amp per knot. Cut-in speed is 600 rev/min and requires a shaft-pulley ratio of 5:1. One major fear has been gearbox damage due to improper lubrication while freewheeling, but many major gearbox manufacturers have dispelled this fear.

4.5 Water Charging Systems. Water-based charging systems come in two configurations:

a. **Towed Turbine Generator.** The towed turbine water generator is essentially a slow speed alternator with the drive shaft mechanically connected to a braided rope and turbine assembly. When streamed off the stern, the turbine turns and rotates the alternator. Typical output is approximately 6 amps. The trail rope is typically around 30 meters long:

(1) **Drag.** Typical drag speed reduction is around half a knot. The trailing generator, like the old-fashioned trailing log, is reliable, and hungry ocean denizens rarely eat the turbine.

(2) **Turbine Skipping.** One problem is that the turbine tends to skip out of the water at speeds over 6 knots. There are a variety of methods to reduce skipping, which include adding sinker weights to the turbine, increasing the towline length, and increasing the towline diameter. The Ampair units have two turbine types, one for speeds up to 7 knots, and another coarse pitch turbine for higher speeds.

b. Submerged Generator. These units comprise a forward facing, three-bladed propeller that drives a permanent magnet alternator. The propeller is mounted at the end of a tubular arm at a depth of approximately one meter. As a water-driven power source, they are a good option, being easy to lift and service. Maximum output is approximately 8 amps.

(1) **Drag.** The drag on a submerged generator is approximately double that of a towed generator.

(2) **Physical Characteristics.** As the electrical alternator is underwater, the generator housing has double seals, as do the cable glands. The alternator body is filled with hydraulic fluid to equalize external pressures when fully immersed. A reservoir is fitted to allow for oil expansion and contraction.

(3) **Mounting Locations.** Generators can be mounted directly on the transom or on the taffrails.

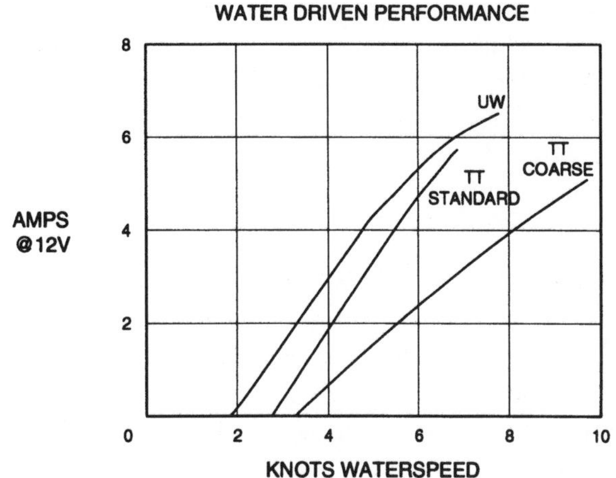

UW = UNDERWATER GENERATOR
TT STANDARD = TOWED TURBINE - STANDARD PITCH
TT COARSE = TOWED TURBINE - COARSE PITCH

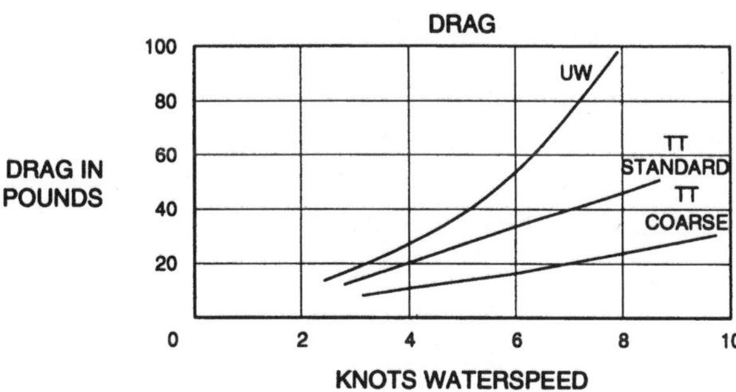

Figure 4-7 Ampair Water Generation System Characteristics

4.6 Fuel Cells. Fuel cells, an alternative power charging source of the future, are now available for sailboats. The fuel cell is classified as an electrochemical energy device. The cell converts the chemical energy of a fuel such as hydrogen, natural gas, methanol etc. and an oxidant such as air or oxygen into water and outputs electricity for battery charging. In principle, the fuel cell operates similarly to other electrochemical devices such as the battery. The main difference is that where the battery requires recharging, the fuel cell does not discharge or have to be recharged from a charging source. A fuel cell will produce electrical and heat output, as long as the fuel and an oxidizer are available in the required quantities. The battery similarities are that both are electrochemical devices. Both have a positively charged anode and a negatively charged cathode. Also they both have an ion-conducting material that is also termed an electrolyte. Fuel cells come in several varieties. Each type utilizes different chemistry and the classification is based on the electrolyte material. The proton exchange membrane fuel cell (PEMFC) device is the most common application in powering motor vehicles and boat charging systems.

4.7 Construction and Operation. The basic construction consists of a fuel electrode called the anode; an oxidant electrode which is termed the cathode. The anode and cathode are separated by an ion-conducting membrane. Oxygen is continuously passed over one electrode, and hydrogen is continuously passed over the other. This generates electricity, water and heat. What occurs in fuel cells is that they chemically combine the molecules of a fuel and oxidizer without any combustion or burning. This means there is no pollution and no toxic or high temperature exhaust emission. The by-products of a fuel cell are a very small quantity of carbon dioxide, pure water and minimal heat. The carbon dioxide level is approximately the same as a person's breath, and some basic ventilation is required.

a. **The Anode.** The anode is the negative part of the fuel cell. The function of the anode is to conduct electrons that are released from the hydrogen molecules, and these electrons can then be used in an external electrical circuit. The anode has a series of channels etched into it. This is to aid in the even dispersal of the hydrogen gas across the catalyst surface.

b. **The Cathode.** The cathode is the positive part of the fuel cell. It also has a series of channels etched into it, to aid in the even distribution of oxygen across the surface of the catalyst. The cathode also conducts the electrons in the external circuit from the catalyst. They then recombine with the hydrogen ions and oxygen to release water.

c. **The Electrolyte.** The electrolyte is the proton exchange membrane (PEM). This consists of a material with a special treatment. It resembles a kitchen plastic wrap except that it will also conduct the positively charged ions. The membrane also will block the passage of electrons.

d. **The Catalyst.** The catalyst is manufactured from a special material that triggers, or is the catalyst for the reaction between the oxygen and hydrogen. The catalyst comprises a thin coating of platinum powder layered onto a substrate of either carbon paper or cloth. The catalyst is also fairly rough and very porous. This is designed so that a maximum surface area of the platinum material is exposed to the hydrogen or oxygen, and this ensures a maximum reaction. The catalyst is always oriented towards the PEM.

e. **The Reformer.** Hydrogen used to be a major component of basic fuels cells, however is not readily available. To overcome this, a device known as a reformer is used. The reformer converts hydrocarbon or alcohol fuels into hydrogen, and this is then fed to the fuel cell. Reformers generate heat and produce other gases in addition to hydrogen and this tends to lower the efficiency of the fuel cell. Methanol is a liquid fuel that has many similar properties to gasoline and it is a farmed biofuel distilled from sugar cane.

4.8 Fuel Cell Eficiency. The ideal fuel cell is powered with pure hydrogen, and these cells can be up to 80-percent efficient. When a reformer is used to convert more easily available methanol to hydrogen, this efficiency falls dramatically to approximately 30 to 40 percent. When the electrical energy is converted into mechanical work using either an electric motor and an inverter the overall efficiency drops to around 24 to 32 percent. This makes the fuel cell considerably more efficient than solar, wind or generators. The chemical reaction in a single fuel cell produces a fairly low 0.7 volt. Similarly to a 12-volt battery that comprises 6 cells, the cell stacks must be also built to form a fuel-cell stack of the required voltage. To increase power source availability the fuel cells can be connected in parallel as you do with batteries.

4.9 Fuel Cell Cartridges. The Maxpower fuel cell requires a 4.4 kg (10lb) fuel cartridge that will last around 3 days on full power, and electrically produce 340 Ah at 12V/DC power. The MaxPower unit is made by the Navimo group (Plastimo) and uses the Navimo distribution network to distribute the fuel cartridges through ship-chandlers. Fuel cells are not service free, but since there are no moving parts this is significantly reduced. The fuel cell stack will degrade with a slow decrease in performance. The stack life will be a minimum of around 1500 hours and as high as 5000 hours. The stack replacement task is relatively simple and fast. The basic unit weighs in at just 16 lbs (7 kg) so this makes it reasonably easy to ship to the nearest service center.

Battery Chargers

5.0 Battery Chargers. Battery chargers are generally used as the primary charging source in large vessels with AC generators in continual service. Many vessels have had batteries ruined by poor quality chargers due to a marginal overcharge voltage level. In reality, battery chargers are not a principal charging source on a cruising yacht, and a relatively small output automatic charger of approximately 10–15 amps will meet the normal requirements while in port. The basic principles of most battery chargers are as follows:

a. **Transformation.** The AC voltage, either 220/230 or 110 volts AC, is applied to a transformer. The transformer steps down the voltage to a low level, typically around 15/30 volts depending on the output level.

b. **Rectification.** The low level AC voltage is then rectified by a full-wave bridge rectifier similar to that in an alternator. The rectifier outputs a voltage of around 13.8/27.6 volts, which is the normal float voltage level.

c. **Regulation.** Many basic chargers do not have any output regulation. Chargers that do have regulation are normally those using control systems to control output voltage levels. These sensing circuits automatically limit charge voltages to nominal levels and reduce to float values when the predetermined full-charge condition is reached.

d. **Protection.** Battery chargers have a range of protective devices, from a simple AC input fuse to the many features that are described as follows:

   (1) **Thermal Overload.** This device is normally mounted on the transformer, or rectifier. When a predetermined high temperature is reached, the device opens and prevents further charging until the components cool down.

   (2) **Input Protection.** This is either a circuit breaker or fuse that protects the AC input against overload and short circuit on the primary side of the transformer.

   (3) **Reverse Polarity Fuse.** A fuse is incorporated to protect circuits against accidental polarity reversal of output leads.

   (4) **Current Limiting.** Limiting circuits are used to prevent excessive current outputs, or to maintain current levels at a specific level.

   (5) **Short Circuit Protection.** This is usually a fuse that protects output circuits against high current short circuit damage.

e. **Interference Suppression.** Most chargers have an output-voltage ripple superimposed on the DC. This is overcome by the use of chokes and capacitors across the output. This ripple can affect electronics and cause data corruption on navigation equipment.

5.1 **Battery Charger Types.** There are a number of charger types and techniques in use:

a. **Constant Potential Chargers.** Chargers operate at a fixed voltage. The charge current decreases as the battery voltage reaches the preset charging voltage. Unsupervised charging can damage batteries if electrolytes evaporate and gas forms. Additionally, such chargers are susceptible to input voltage variations. If left unattended, the voltage setting must be below 13.5 volts or batteries will be ruined through overcharging.

b. **Ferro-Resonant Chargers.** These chargers use a ferro-resonant transformer which has two secondary windings. One of the windings is connected to a capacitor, and they resonate at a specific frequency. Variations in the input voltage cause an imbalance, and the transformer corrects this to maintain a stable output. These chargers have a tapered charge characteristic. As the battery terminal voltage rises, the charge current decreases. Control of these chargers is usually through a sensing circuit that switches the charger off when the nominal voltage level is reached, typically around 15% to 20% of the charger's nominal rating.

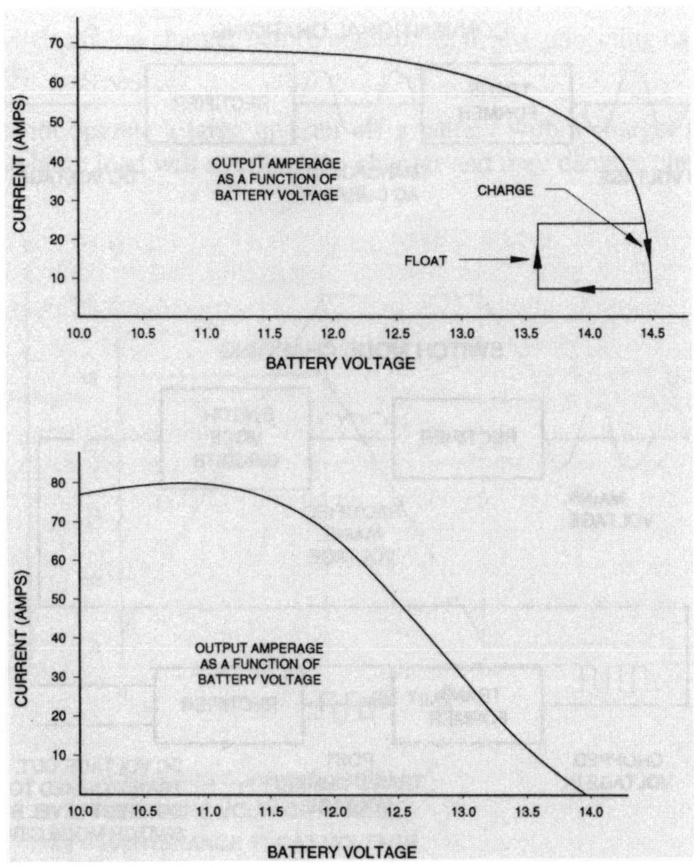

Figure 5-1 Newmar Battery Charger Characteristics

c. **Switch Mode Chargers.** Compact switch mode chargers are becoming increasingly popular due to their compact size and low weights. These charger types convert the input line frequency from 50 to 150,000 hertz, which reduces the size of transformers and chokes used in conventional chargers. An advantage of these chargers is that line input and output are effectively isolated, eliminating the effects of surges and spikes. These chargers are my favorites, and the units from LEAB of Sweden are technically very advanced. The chargers are battery-sensed, temperature-compensated, have integral digital voltmeters and ammeters, and are compact.

d. **Automatic Chargers.** This term covers a wide range of electronically controlled charging systems. These include chargers that have SCR or Triac control, a combination of current and voltage settings with appropriate sensing systems and control systems, as well as overvoltage and overcurrent protection. The ideal charger characteristic is one that can deliver the boost charge required and then automatically drop to float charge levels so that overcharging does not occur.

5.2 Multiple Battery Charging. Most marina based sailing yachts have a battery charger connected permanently to charge a single house battery bank although many yachts have multiple house banks and twin engines with separate batteries. A separate battery charger is required or a method of splitting the charge to each battery. Remember that gel cells or AGM batteries may have different requirements and this should be checked prior to using any system, as batteries when fully charged can loose water rapidly if charging is imprecise.

a. **Multiple Output Chargers.** Install a battery charger with multiple outputs, such as those from Newmar, where each battery bank has its own isolated charging outputs. This prevents any interaction and is an efficient way of having two or more separate chargers.

b. **Diodes.** A diode isolator can be used to split the charge between the two or three battery banks. For three battery banks use 2 diode isolators, and link the diode isolator inputs. There are problems of voltage drop across the diode that have to be considered, and battery chargers with battery sensing are required to compensate for this. The typical voltage drop is around 0.7 volt so the charger outputs without sensing will require adjustment of output voltage up an additional level equivalent to the drop.

c. **Relay/Solenoid.** A relay or solenoid can be used to direct the charge current to each battery bank. This is activated either with the monitored charging voltage or via a manually operated switch. The configuration effectively parallels all of the batteries to form a single battery bank. Possible systems include the NewMar Battery Bank Integrator (BBI). When a charge voltage is detected that exceeds 13.3 VDC the unit switches on. The unit consists of a low contact resistance relay that closes to parallel the batteries for charging. When charging ceases and the voltage falls to 12.7 VDC the relay opens isolating the batteries. The unit also incorporates a voltage comparator and time delay circuit, which prevents the unit cycling in the event of a voltage transient or load droop on the circuit dropping voltage below the cut-out level. Another similar device is the PathMaker from Heart Interface. These devices allow charging of two or three batteries from one alternator or battery charger. The units use a high current switch rated at 800 and 1600 amps for alternator and charger ratings up to 250 amps.

d. **Smart Devices.** These are intelligent charge distribution devices such as the Ample Power Isolator Eliminator. This is a multi-step regulator that controls charge to the second battery bank, typically the one used for engine starting. It is temperature compensated like an alternator control system and is effectively a secondary charger. Another system is the AutoSwitch from Ample Power, which is a smart solenoid system. An electronic sensing circuit will enable the setting of the different modes. One mode is a timed function that terminates the charging to the second paralleled start battery once the period expires. There is also a voltage mode, which disconnects the second battery after the preset voltage is reached. Other devices such as Charge-Link and Echo-charge perform a similar function. These smart devices reduce the chances of overcharging secondary batteries such as the start or generator battery.

5.3 Battery Charger Installation. Chargers should be mounted in a dry and well ventilated area. The following precautions should be undertaken when using chargers:

a. Always switch off battery charger during engine starting if connected to the starting battery.

b. The AC connection should be an industrial grade outlet in engine areas or normal outlet in dry areas.

c. The metal case of any charger must be properly grounded to the AC ground.

d. To prevent cables from moving, clips or permanent fasteners should be used on cables if the charger is permanently installed.

e. Switch off the charger before connecting or disconnecting cables from battery.

f. Do not operate a large inverter off a battery with a charger still operating. The large load will overload the charger and may damage circuitry.

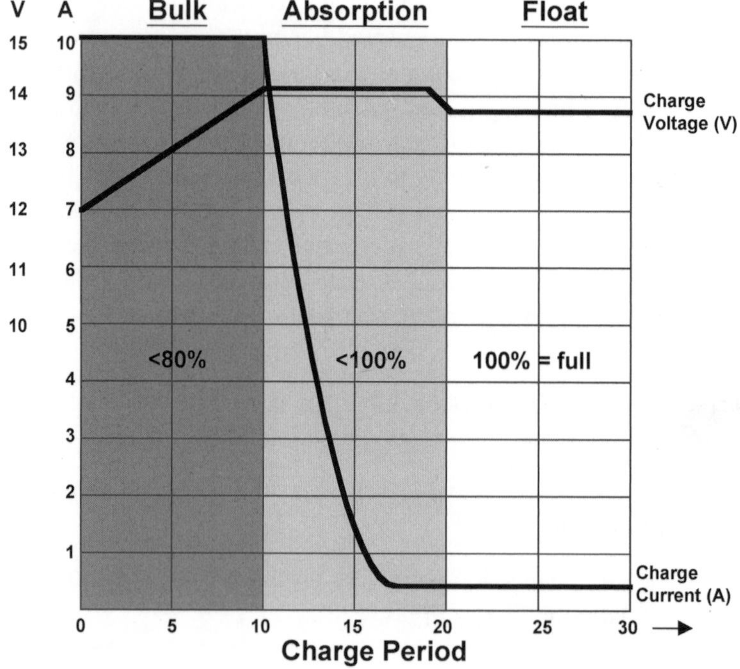

Figure 5-2 Automatic Charging Characteristic

DC Systems, Installation and Wiring

6.0 DC Systems. A significant proportion of yacht electrical system failures can be attributed to improper connections and terminations or incorrectly installed cables. Using accepted practices could eliminate these failures. Unfortunately the common attitude is to still treat vessel low voltage systems like automotive installations, and the high failure rates on boats reflect this attitude.

6.1 Electrical Standards. Electrical systems should be installed to comply with one of the principal standards or recommendations in use, and most standards are similar. The following are what I consider the most recognizable. Many people do not buy copies of the recommendations or rules, as they are relatively expensive, and many are intimidated by the complexities and are unable to interpret them, as they are not written in plain language but consist of technical jargon. To make readers aware of typical requirements I am using a set of recommendations that I developed and which encompass or exceed many of the provisions of the various standards listed below. They will not cover all of the provisions but they use best practice and will assist in getting your installation to a similar level. Where you are required to use standards, a copy of the relevant standard is a prerequisite.

a. **American Boat and Yacht Council (ABYC).** *Standards and Recommended Practices for Small Craft.* (www.abycinc.org). These are voluntary standards and recommendations that are widely used by many US boat builders and are the de facto standard within the US.

b. **International Standard ISO 10133.** *Small Craft – Electrical systems – Extra-low-voltage DC installations, 1994.* The standard is prepared and ratified by a large group of nations including the US, France, and the UK. Standards are also made in conjunction with the International Electrotechnical Commission (IEC).

c. **NFPA 302,** *Fire Protection Standard for Pleasure and Commercial Motor Craft, 1994 Edition.* This standard is approved by the American National Standards Institute and is also relevant to yacht installations. The technical committee includes representatives from ABYC, USCG, Underwriters Laboratories (UL), and others such as the National Association of Marine Surveyors.

d. **Lloyd's Register of Shipping.** *Rules and Regulations for the Classification of Yachts and Small Craft.* Normally used when a vessel is to be built to class, it serves as a very high benchmark. Commercial vessels, trawlers and large super yachts may fall under other large ship rules.

e. **The International Recommendations for Boat Electrical Systems (© *IRBES*).** These generic recommendations include many of the provisions in the various other recommendations and standards, and incorporate best installation practice. They were developed over several years by myself in conjunction with several marine electrical engineers. Where a vessel must comply to Class (Lloyd's, DNV, Bureau Veritas, ABS), survey, or other requirements such as the USCG in the United States; the USL (Uniform Shipping Law) in Australia; the MCA in the United Kingdom, Transport Canada or others, the rules should be obtained and referenced for the particular installation as required.

6.2 DC System Voltages. It is quite common to see vessels having both 12- and 24-volt systems in use, and 42 volt will also have the same factors. They should be treated as two entirely separate entities. In polarized ground systems the negatives will be at the same potential. This will mean two alternators and two battery banks. The merits of 24 volts for heavy current consumption equipment such as inverters and windlasses are obvious, because the cables are half the size and weight of 12-volt systems. In many cases electronics will be able to operate on 24 volts without modification.

a. **12 (14) Volt Systems.** The 12-volt system is the most common system. This is because of automotive influences, which have led to a large range of equipment being available. 12 and 24 volts are used to power most boat electrical and electronics equipment. It is also possible to purchase virtually any appliance rated for 12 volts.

b. **24 (28) Volt Systems.** This system is prevalent in most commercial applications. It has the advantage of lower physical equipment sizes, cabling, and control gear. Additionally voltage drops are not as critical. Because much equipment is commonly 12 volts a DC-DC converter must be used to step down to 12-volt equipment. Although complicating the system, this does isolate sensitive electronics equipment from the surge and spike-prone power system.

c. **32 Volt Systems.** This is an old system voltage still found on some vessels.

d. **36 (42V) Volt Systems.** A new automotive standard voltage which may eventually transfer to boats.

e. **48 Volt Systems.** This system is now starting to become more prevalent, in particular for powering thrusters.

6.3 DC Voltage Conversion. In many vessels, a mix of voltages requires the use of DC converters to step down from 24 to 12 volts. If implemented on boats the same will be required for 42 volt systems. There are a number of technical points that must be considered when selecting converters.

a. **Power Input.** Converters may be either galvanically isolated or only isolated in the positive conversion circuit. Galvanically isolated units will totally isolate input and output providing protection to connected loads, and these are preferable. Good quality converters have a stabilized output of around 13.6 (27.2) volts. Stability is typically about 1% between line and load at rated output voltage. Typical power consumption of a converter without a load connected is approximately 40 to 50 milliamps, so there will always be a battery drain. The converter should ideally have an isolation switch on the input side. Most converters are installed with automatic thermal shutdown, short circuit fuse protection, current limiting and reverse polarity protection.

b. **Power Output.** Converters are able to withstand short surge current. Normally a 50% over current can be applied for intermittent surges, and approximately 70% for a very short duration of up to 30 seconds for peak loads. Some high power units can withstand peak overloads of 200% for up to 30 seconds. Output ratings vary but I usually install one rated at approximately 15 amps continuous. Duty cycle ratings are also applicable to converters. Intermittent overloads can only be sustained on a cycle of 20 minutes every hour, and peaks for a 30 to 60 seconds per hour. Failure to observe these duty cycles will result in a burnt-out converter. Converters in common with most electrical equipment are designed to provide an output at a specific temperature range, typically 0–40°C. At 50°C, converters should be de-rated to 50%.

c. **Installation.** Good ventilation is essential. Converters should be mounted vertically so that fins are also vertical to facilitate convection cooling. Sufficient clearance must be allowed between top and bottom.

6.4 How to Wire and Rewire Your Boat. The average yacht has many systems installed. Equipment is often purchased before the actual impact on the system is considered. Planning the installation requires a carefully considered systems approach. In the majority of cases, systems are over complicated, follow no accepted electrical practice, and have inherent problems that are only overcome with costly total rewires. Do it once, and do it right!

Each yacht should have a complete wiring diagram showing all the wiring and systems installed. The diagram should include Equipment Identification, Equipment Current Rating, Cable Sizes, Circuit Breaker and Fuse Ratings, and Circuit Identification.

Perform the following wiring planning tasks.

(1) Make a plan of your vessel and locate every item of equipment on it. Write down the equipment identification name.

(2) Write down the current draw for each item of equipment. Also enter these into the battery calculation table. This will allow calculations to be made on required battery capacity and charging requirements.

(3) Draw in the proposed cable route for each item of equipment, showing bulkheads, decks or other obstructions. Where the cable will be routed within bilge areas, or be exposed to mechanical damage, use an alternative route.

(4) Determine the cable size by using the current draw and calculating the voltage drop within the circuit. It is best to standardize on 15–16 amp cables for most applications, as it is economical to buy cable by the roll. It also eliminates the selection exercise based on lowest cable size to achieve specific volt drop values for most, but not all circuits.

(5) Enter the circuit breaker or fuse rating for the circuit and assign a circuit number. Use a logical sequence; such as switchboard left hand vertical row is No. 1 downward and so on.

6.5 **Wiring Considerations.** There are a number of important considerations.

a. **Hull Material.** The hull material has important implications with respect to wiring systems as well as grounding and corrosion. This is important if you are building a boat, as you can purchase an engine with a fully isolated electrical system for a steel or alloy boat.

b. **Boat Size.** This affects the length of cable runs; with consequentially greater cable weights and voltage drop problems. This affects voltages, and large boats have a real case for selecting 24 volts as voltage drop problems are reduced, the battery weight and sizes for a given capacity are less, and the weight and size of equipment is generally reduced. Larger multihulls and yachts have a greater level of accommodation, and therefore more people are often aboard, (the parties are longer!) putting greater demands on batteries through lighting, electric refrigeration, with consequential increased requirements on charging. Boat builders largely overlook this simple lifestyle factor, but it is significant.

6.6 **Wiring Configurations.** The two preferred systems for distribution are as follows:

The two wire insulated system. This system is preferred for all steel and alloy vessels.

All steel and alloy vessels should be of the two-wire insulated return. This configuration has no part of the circuit, in particular the negative, connected to any ground or equipment. The system is totally isolated and is floating above hull, and this includes engine sensors, starter motors and alternators.

In two-wire, insulated systems each outgoing circuit positive and negative supply circuit should have a double pole short circuit protection and an isolation device installed. This may be incorporated within a single trip free circuit breaker. The isolator should be rated for the maximum current of the circuit.

In this configuration, a short circuit between positive and ground will not cause a short circuit or systems failure. A short circuit between negative and ground will have no affect. A short between positive and negative will cause maximum short circuit current to flow.

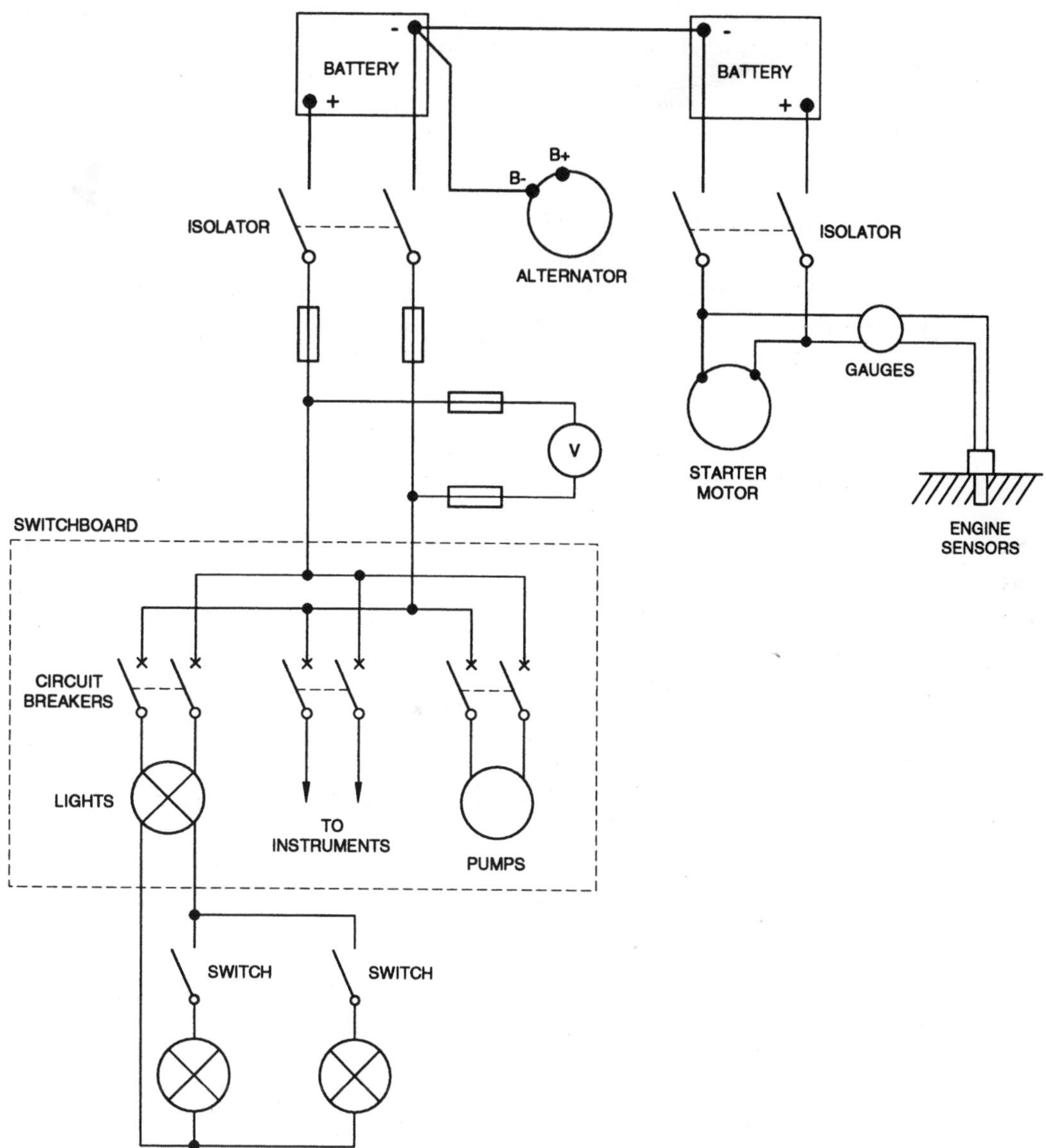

Figure 6-1 Insulated Two Wire Return Systems

The two-wire with one pole grounded system. This system is preferred for fiberglass and wooden boats.

This is also called a polarized system. It is the most common configuration, and holds the negative at ground potential by connection of the battery negative to the mass of the engine block.

The main negative cable is considered to be the grounded negative conductor in two-wire grounded circuit arrangements.

In most installations, the main negative to the engine polarizes the system, as the engine mass and connected parts such as shaft provides the ground plane. There should only be one ground conductor.

In a two-wire, one-pole grounded system, each outgoing circuit positive supply circuit should have a short circuit protection and an isolation device installed. This may be incorporated within a single trip free circuit breaker. The earthed pole should not have any protective device installed.

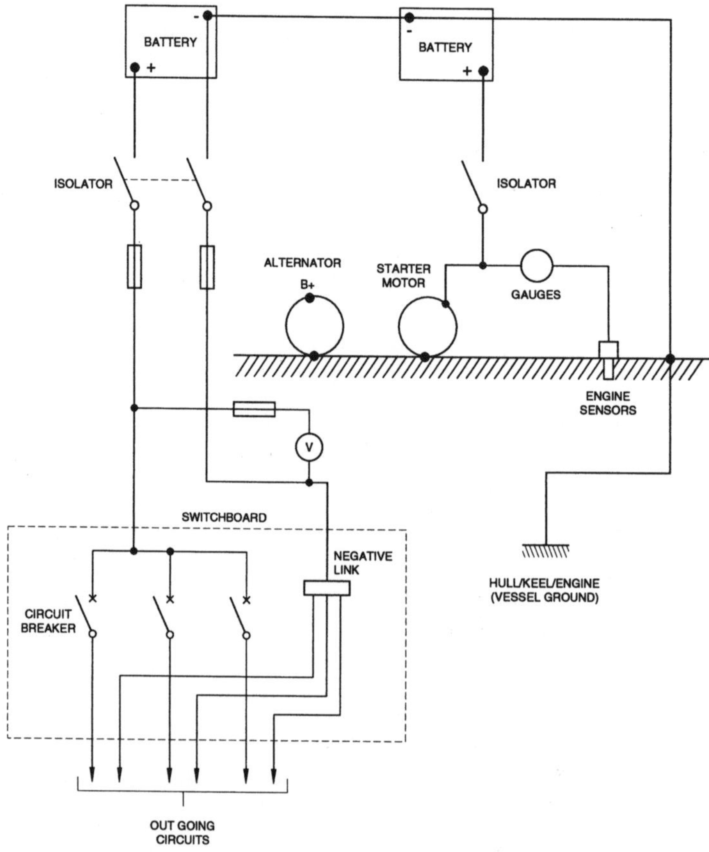

Figure 6-2 Insulated One Pole Grounded Systems

In this configuration, a short circuit between positive and ground will cause maximum short circuit current. A short circuit between negative and ground will have no affect. A short between positive and negative will cause maximum short circuit current to flow. The single pole circuit breaker will break positive polarity only.

6.7 Distributed Systems. These systems are typically broken down into a system of sub-panels, and this is increasingly preferable on larger yachts and multihulls. There are a number of significant advantages over a centralized system, including the separation of potentially interactive equipment such as pumps and electronics. Other options include intelligent controls systems that have remote control of circuits, and systems that have touch screens to switch circuits. The circuit control boxes have 16 circuits with ratings of up to 70 amps.

a. Separation enables a reduction in the number of cables radiating throughout the vessel from the main panel to areas of equipment concentration. This is a cause of RFI and requires a greater quantity of cable. Most distributed systems run all the sub-circuits from the central panel, with each circuit having a circuit breaker to protect it.

b. The illustration below shows the preferable breakdown of sub-circuits and panels, and is based on the successful implementation on a number of vessels. In the case below, only essential services are kept along with metering on the main panel. Lighting panels can be located anywhere practicable, once the circuits are on, lights are switched locally.

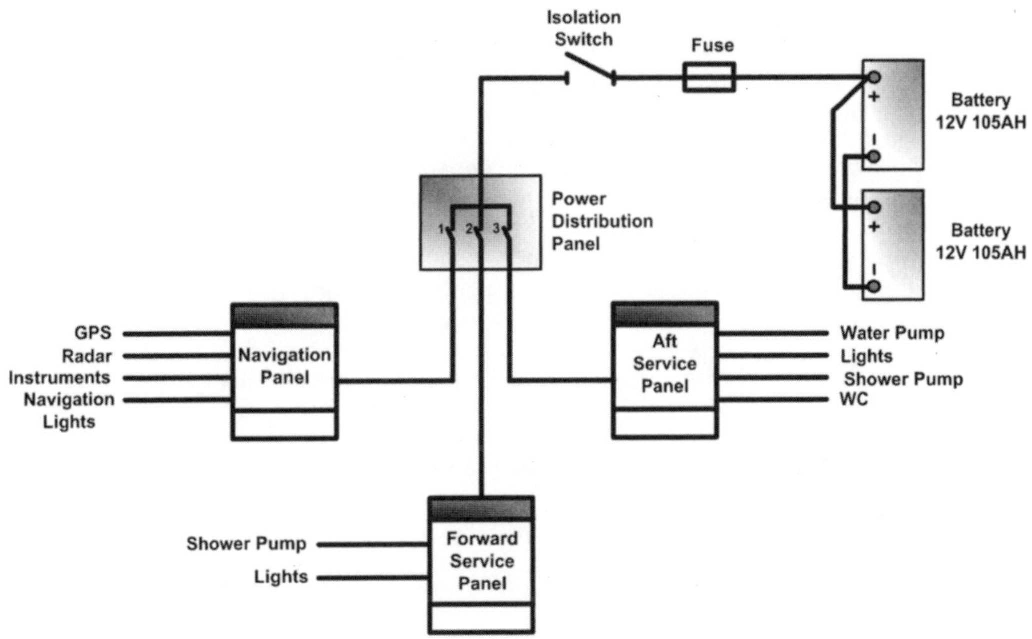

Figure 6-3 Distributed Power Systems

6.8 42 Volt Power Systems. When and if this new automotive voltage system migrates to boats it will require two systems in a distributed configuration.

a. **High Current Consumers.** The 42-volt system will be supplied by a high output water-cooled alternator rated at up to 4kW, which is around 100 amps. The output will supply a 36-volt battery bank, and a high current consumer panel. Typically this will supply the thrusters, hydraulic power unit (HPU), anchor windlass, winches, and DC electric propulsion. As these consumers are normally powered up when an engine is running the full 42 volts are available. A power management system will be also required.

b. **Low Current Consumers.** Low power consumers will be powered from a separate 12 or ideally a 24-volt battery system. These consumers will include the lighting loads, electronics, and auxiliary power circuits.

6.9 Circuit Control and Protection. The heart of all electrical systems is the switchboard or panel, which allows control, switching and protection of circuits. The purpose of protection systems is to prevent overload currents arising in excess of the cable rating. They are also to protect the cables and equipment from excessive currents that arise during short circuit conditions. Circuit protection is not normally rated to the connected loads, although this is commonly done on loads that are considerably less than the cable rating, such as VHF radios or instrument systems. The two most common circuit protective devices are the fuse and the circuit breaker.

a. **Short Circuit Current.** A short circuit is where two points of different electrical potential are connected, that is positive to negative.

b. **Overload Current.** An overload condition is where the circuit current carrying capacity is exceeded by the connection of excessive load. Excessive load can come from too many devices or equipment such as pumps with higher than normal load.

The switchboard or panel should be constructed of non-hygroscopic and fireproof material. The panel should be rated to a minimum of IP44.

Ideally panels should be non conductive; however many are also made of etched aluminum. They should be rated to meet either an IP or NEMA standard against the ingress of water.

The switchboard interior should be fireproof or incapable of supporting combustion.

Survey authorities specify that the internal part of the switchboard should be lined with a fire resistant lining. Line all interior walls with appropriate sheeting, and this will help in containing any fire that may arise in severe fault conditions.

The switchboard should be located in a position to minimize exposure to spray or water.

All switchboards should be installed in a location that is protected from seawater, spray or moisture. Where occasional spray is possible, some protection is recommended, which may be a clear PVC cover or similar measure.

DC systems should not be located or installed adjacent to AC systems. Where DC and AC circuits share the same switchboard, they should be physically segregated and partitioned to prevent accidental contact with the AC section. The AC section must be clearly marked with danger labels.

The DC switchboard should not be integrated with the AC system. Where possible the AC panel should be located in a different location. This eliminates the chances of accidental contact with live circuits, or confusion between wiring systems. Where systems are integrated, physical separation should be used to prevent contact. The barriers should be well marked warning of the danger.

A voltmeter should be installed to monitor the voltage level of the start battery.

A good quality voltmeter is essential for properly monitoring battery condition. A voltmeter will also tell you if the battery is charging at the correct voltage level. As a battery has a range of approximately one volt from full charge to discharge condition, accuracy is crucial. Analog voltmeters are the most common. The sense cable should go directly back to the battery, although on service battery connections most connect directly to the switchboard busbar. Direct connection gives greater accuracy and less influence from local loads. Voltmeters should be of the moving iron type and have a fuse installed on the positive input cable. Half a volt error is quite common. Switch off the meter after checking.

A voltmeter should be installed to monitor the voltage level of the service battery. A switch may be installed to enable monitoring of the service and start batteries from the same meter.

The same provisions apply as for start battery voltage monitoring. In practice more attention is given to house battery monitoring. Some switchboards also use LED voltage level indicators, and these devices are often used as a voltmeter substitute. They are not recommended, as they do not give the precise readings required. Digital voltmeters are relatively common and are far more accurate. They are susceptible to voltage spikes and damage; and many have maximum supply voltage ranges of 15 volts. There are a number of types, and these include Liquid Crystal Displays (LCD) and Light Emitting Diodes (LED). LED types consume power, an LCD meter consumes much less power and is more practical. Where one voltmeter is used to monitor two or more batteries, switching between batteries to voltmeter is through a double pole, center off toggle switch or a multiple battery rotary switch.

Voltmeters should have fuses installed within the meter circuit to provide short circuit protection.

A voltmeter is connected across the supply, that is positive and negative, and protection against short circuit is required.

An ammeter should be installed to monitor the discharge current rate from the service battery. An ammeter is not required for the starting battery. The installation of an ammeter in the primary charging circuit to monitor the charging current is recommended.

Ammeters are essential on the switchboard input positive to monitor service battery discharge levels. Analog ammeters should be selected for the calculated operating range. Shunt ammeters are also used in these applications. Although useful, an ammeter on the charging system can indicate that current is flowing. You do not know what should be flowing into the battery, and peace of mind is the greatest benefit. Cheaper ammeters are of the series type with the cable under measurement passing through the meter. The major failing of these is that often very long cable runs are required with resultant voltage drops, and if the meter malfunctions damage can occur. Preference should be given to a shunt ammeter. A shunt simply allows the main current to flow while monitoring and displaying a millivolt value in proportion to the current flowing. The advantage is that only two low current cables are required to connect the ammeter to the shunt, and the risk of damage is reduced. Do not run the main charging cables to where the meter is and connect it, as this can insert excessive voltage drop into the charging circuit. Install a shunt in the line wherever practical and run sense wires back to the panel mounted meter. The digital ammeter often uses a different sensing system. Instead of a shunt the digital ammeter has what is called a Hall Effect sensor on the cable under measurement. The Hall Effect transducer generates a voltage proportional to the intensity of the magnetic field it is exposed to. For vessel applications a 0–10 volt transducer output corresponds to a 0–200 amp current flow. Sensitivity is increased, and range reduced by increasing the number of coils through the transducer core.

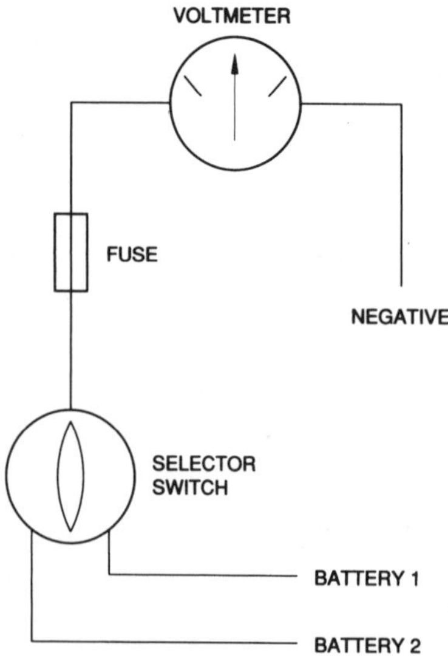

Figure 6-4 Voltmeter Connection

An integrated monitoring device that measures and displays all values is an acceptable alternative to installing separate meters.

Unlike starting batteries, house battery charge levels cycle up and down, and power level information is critical in determining charging periods. Typical of integrated monitors is the E-meter (Link 10). I have installed one of these on my own boat. These are "intelligent" devices in that they monitor current consumption, charging current, and a range of monitoring functions that also include voltage, high and low voltage alarms, amp-hours used and amp-hours remaining and allowing the battery net charge deficit to be displayed. The system also maintains accuracy by taking into account charging efficiency. The charging efficiency factor (CEF) is nominally set at 87%, with the factor being automatically adjusted after each recharge cycle. A falling CEF is indicative of battery degradation. In addition the E-meter also contains the *'n'* algorithm for calculation of Peukerts Coefficient. These meters offer a simpler diagnosis of battery power status without trying to guesstimate the actual level based on voltages.

a. **Shunt.** A meter shunt (500A/500mV) is installed in the negative load line. This is connected by twisted pair wires to prevent noise from induced voltages being picked up and carried into the meter, corrupting data.

b. **Protection.** The battery sense lead and DC meter power supply have fuses installed, and these should be checked if the meter fails to function.

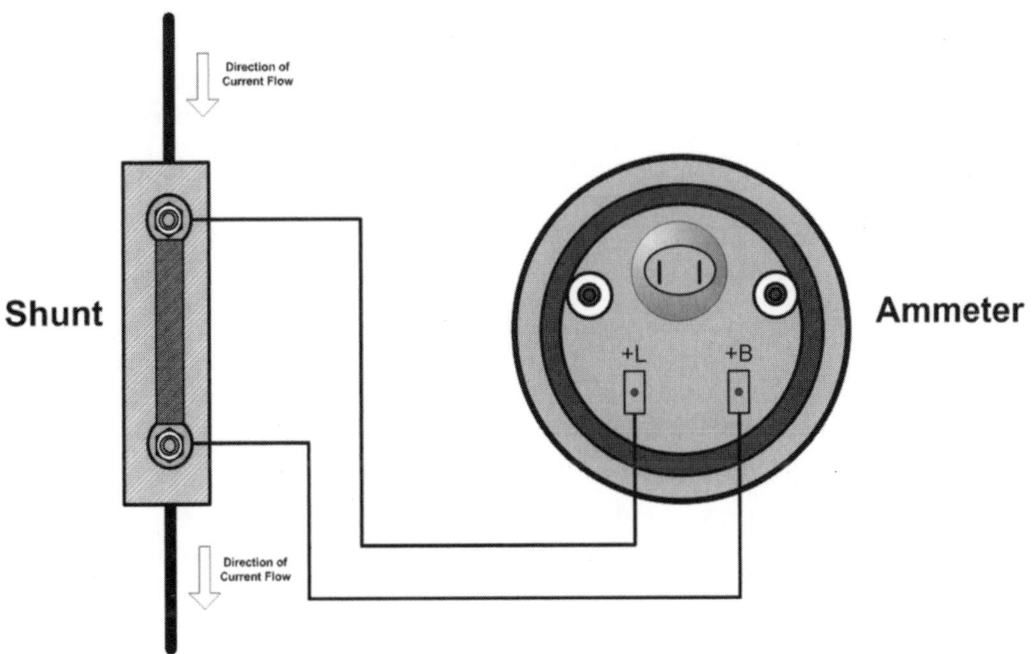

Figure 6-5 Shunt Ammeter

All protection and switchboard devices should be clearly marked to properly identify them.

All circuits should be properly labeled to allow easy identification. This should also include the circuit number if practicable.

All circuit isolation and protection devices should have visual status indication.

Circuit breaker status indicators consist normally of LED lights, filament lights, or backlit nameplates. Generally Green indicates off and Red is on. An LED requires a resistor in series and this is typically valued at 560 ohms for 12-volt systems. Red filament lamps are also commonly used. The one disadvantage of these is that they consume power, typically around 40 mA. If there are twenty circuits on this adds up to a reasonable load on the system, and a needless current drain. If you have a very large switchboard, allow for the current drain. In many cases people assume they have a current leakage problem when in fact it is the switchboard indicators causing the drain.

All cables, and cable looms to switchboard panels should permit opening of the panel without placing strain on connections or cables.

In many installations, cable looms are too short to allow easy opening of panels for inspection or installation. It is common to have the connectors pulled off the rear of circuit breakers due to the strain on conductors or wiring looms. Looms should be neatly tied in 2 or 3 separate looms. They should have sufficient length to allow complete opening of the switchboard, and the circuit cables should be secured so as to prevent undue stress on the connectors.

All protection and isolation devices should have an assigned DC fault rating and be approved by a relevant national or international standard. Such standards may include NEMA, UL, CSA, Lloyd's Register and others.

Install only circuit breakers that are approved by UL, CSA or Lloyd's. Approvals for small vessel breakers categorize them as supplementary protectors. I normally use ETA, Ancor and Carling circuit breakers. They must be approved for DC operation and be marked with the rating.

The power supply to the switchboard should have short circuit protection and circuit isolation installed as close as practicable to the battery in both the positive and negative conductors. The isolator should be accessible. This may be incorporated within a single trip free circuit breaker. The isolator should be rated for the maximum current of the starting circuit.

A circuit breaker rated for the cable should be installed as close as possible to the battery, and be accessible. Fuses can be used, but it is better to combine isolation and protection within one device that can be easily reset. They should also be mounted as high as practicable above potential bilge and flooding levels.

The power supply to auxiliary equipment connected directly to the battery should have short circuit protection and circuit isolation installed as close as practicable to the battery in both the positive and negative conductors. This may be incorporated within a single trip free circuit breaker.

These auxiliary supplies generally include high current equipment such as thrusters, electric windlasses, winches, toilets etc connected directly to a battery. A circuit breaker rated for the cable should be installed as close as possible to the battery, and be accessible. They should also be mounted as high as practicable above potential bilge and flooding levels.

Each navigation light circuit should have a short circuit protection device installed.

A circuit breaker supplying all navigation lights has the risk of a single fault tripping the breaker and all lights become unavailable until the fault is cleared. This may not be possible in adverse weather conditions. Where possible, separate circuit breakers should be installed. Alternatively where a single breaker is used, each circuit should have a replaceable fuse installed. This may be a multi-circuit fuse block or the rear of the switchboard, carrying fuses or circuit breakers for all circuits.

Circuits for power and lighting should be separate.

Circuits should not have mixed consumers, such as power to outlets or motors also connected to lighting equipment.

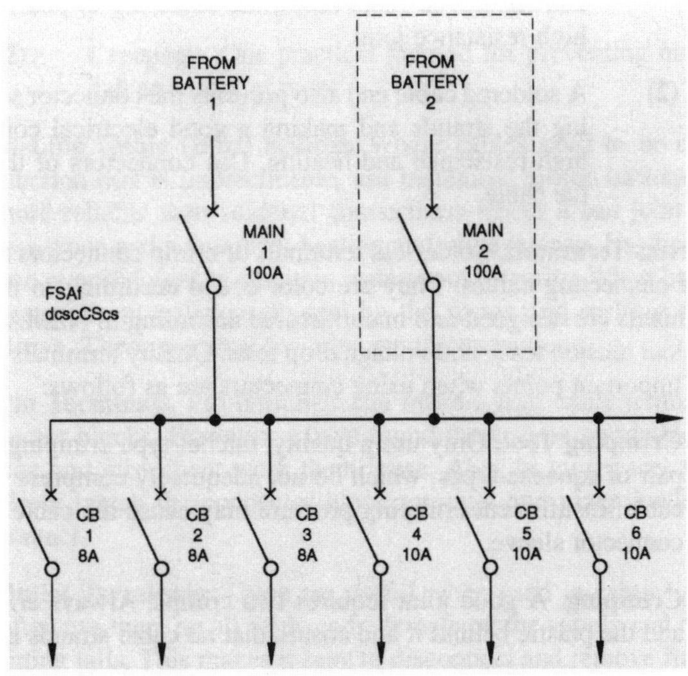

Figure 6-6 Supply Circuit Isolation

All fuses, distribution busbars, and terminals should be covered.

Covers should be fitted over all positive and negative busbars, distribution busbars and fuse holders such as slow blow ones used with anchor windlasses. This is also a requirement of ABYC and protects against accidental contact and water.

6.10 Fuses. Fuses are still widely used, and although cheaper than circuits breakers, they have many disadvantages. Control and DC circuit fuses are of the ceramic or glass type. There are either simple fuse holders or a combination fuse switches. Fuses are either fast acting or dual-element time-delay types. Fast blowing current-limiting fuses, also known as high rupturing capacity (HRC) fuses are in many AC machinery installations. The advantages are a lower initial capital cost. The disadvantage is that when you're in trouble you can't find a spare fuse.

a. **Rating Variations.** The typical glass fuse is not always accurate and can rupture as much as 10–50% above or below nominal current rating.

b. **Service Fatigue.** Fuse elements fatigue in service with the fuse element properties altering. Vibration also commonly causes failure.

c. **Voltage Drop.** There is added contact resistance in the fuse holder between each contact and the fuse ends which commonly cause voltage drops, intermittent supply and heating, and increases with corrosion.

d. **Troubleshooting.** Problems are amplified when a circuit has a fault and you go through a box of fuses on a trial and error troubleshooting exercise. A circuit breaker allows simple resetting.

6.11 Circuit Breakers. Circuit breakers are the most reliable and practical method of circuit protection. They are manufactured in press button aircraft types, toggle type, or rocker switch. Ideally they are used for circuit isolation and protection, combining both functions, which saves switchboard space, costs and installation time as well as improving reliability. Single-pole circuit breakers are normally fitted to most boats, however classification societies only allow these in grounded pole installations. This is because a fault arising on the circuit will provide a good ground loop and the large current flow will ensure proper breaker interruption. Double-pole breakers are recommended for all circuits, as they will totally isolate equipment and circuits. This is a requirement of many classification or survey authorities.

a. **Circuit Breaker Selection.** Circuit breakers must be selected for the cable size that they protect. The rating must not exceed the maximum rated current of the conductor. The cable sizes in Table 6-1 give recommended ratings for single cables installed in well-ventilated spaces. Bunching of cables and high ambient temperatures require de-rating factors. Ratings are given according to IEC Standard 157.

Table 6-1 Circuit Breaker Selection

Wire mm^2	AWG	Circ Mils	Current	CB Rating
1.5 mm^2	15	3260	7.9 - 15.9 A	8 amps
2.5 mm^2	13	5184	15.9 - 22.0	16 amps
4.0 mm^2	11	8226	22.0 - 30.0	20 amps
6.0 mm^2	9	13087	30 0 - 39.0	30 amps
10.0 mm^2	7	20822	39.0 - 54.0	40 amps
16.0 mm^2	5	33088	54.0 - 72.0	60 amps
25.0 mm^2	3	52624	72.0 - 93.0	80 amps
35.0 mm^2	2	66358	93.0 -117.0	100 amps
50.0 mm^2	0	105625	117.0 -147.0	120 amps

b. **Discrimination.** The principle of discrimination in both DC and AC circuits is extremely important and is rarely considered on boat electrical systems. A circuit normally should have two or more over-current protective devices, such as the main and auxiliary circuit breakers installed between the battery and the load. The devices must operate selectively so that the protective device closest to the fault operates first. If the device does not operate, the second device will operate protecting the circuit against over-current damage and possibly fire.

(1) Use circuit breakers with different current ratings. This effectively means that at a point on the time delay curve the first breaker will trip. If it does not and the current value increases the next will. A point is reached called the limit of discrimination. At this point the curves intersect and both breakers will trip simultaneously.

(2) Use circuit breakers with different time delay curves to achieve the same result.

(3) Use circuit breakers with different time delay curves, current ratings and different breaker types. This enables using all of the above to ensure discrimination.

c. **Tripping Characteristics.** The tripping characteristics are normally given by the manufacturer of the breaker in a curve of current against time.

(1) The greater the current value over the nominal tripping value the quicker the circuit breaker will trip. In cases of short circuit tripping is rapid due to the high current values.

(2) Slower tripping characteristics are seen where a small overload exists and tripping occurs some seconds or even minutes after switch on. This happens as the current levels gradually increase.

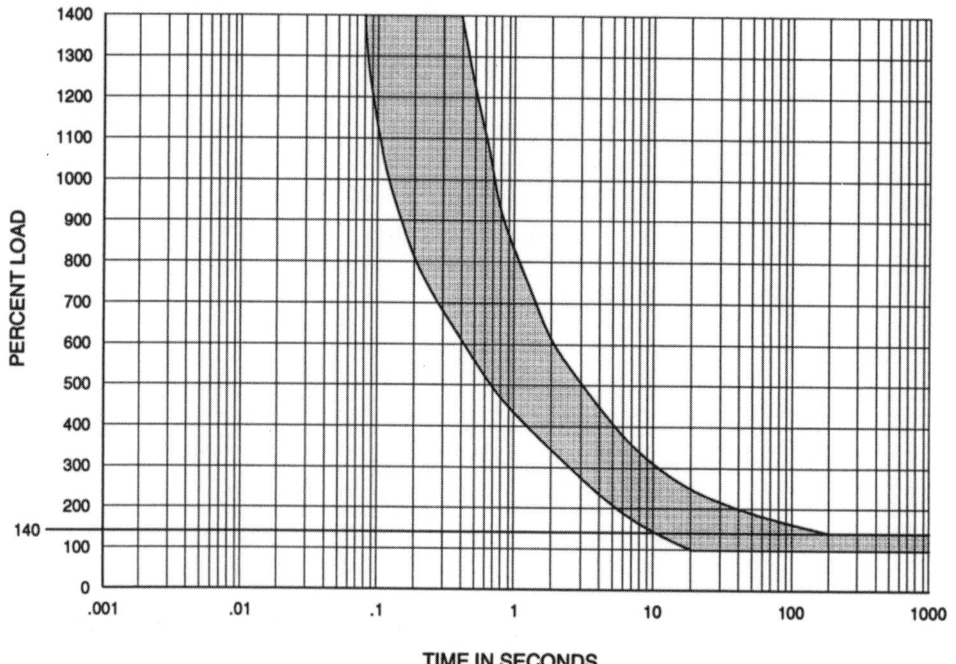

Figure 6-7 Circuit Breaker Time Delay Curve

6.12 Switchboard Troubleshooting. There are a number of faults that routinely occur on switchboards and their protective devices. The following faults and probable causes should be checked first. It is assumed that power is on at the switchboard.

a. **Circuit Breaker Trips Immediately at Switch On.** The ammeter shows in most cases an off-the-meter full-scale deflection that indicates a high fault current:

(1) **Load Short Circuit.** Check out the appropriate connected load and disconnect the faulty item before resetting.

(2) **Connection Short Circuit.** If after disconnection of the load the fault still exists, check out any cable connections for short circuit, or in some cases cable insulation damage.

b. **Circuit Breaker Trips Several Seconds After Switch On.** The ammeter shows a gradual increase in current to a high value before tripping off, and is typically an overload condition.

(1) **Motor Seizure.** This fault may arise if the electric motor has seized, or more probably the bearings have seized.

(2) **Load Stalling.** This fault is usually due to a seized pump.

(3) **Insulation Leakage.** This fault is usually due to a gradual breakdown in insulation, such as wet bilge area pump connection.

c. **There is No Power After Circuit Breaker Switch On.** If after checking that power is absent at the equipment connection terminals, check the following:

(1) **Circuit Connection.** Check that the circuit connection has not come off the back of the circuit breaker. Also check the cable connection to the crimp connection terminal.

(2) **Circuit Breaker Connection.** On many switchboards, the busbar is soldered to one side of all distribution circuit breakers. Check that the solder joint has not come away. In some cases breakers have a busbar that is held under breaker screw terminals. Check that the screws and connection are tight.

(3) **Circuit Breaker.** Operate the breaker several times. In some cases the mechanism does not make proper electrical contact and several operations usually solve the problem by wiping the contacts.

(4) **Circuit Negative.** If all tests verify that the positive supply is present, check that the circuit negative wire is secure in the negative link.

d. **Circuit Power On But No Indication Light.** The LED may have failed, and in some cases the resistor. Also check the soldered connection to the circuit breaker terminal.

6.13 Conductor Selection. Selection of the right conductors is the foundation for a reliable boat electrical system.

Conductors should be selected based on the maximum current demand of the circuit. Ambient temperatures exceeding the rated temperature of the cable should be de-rated by a factor of 0.05 for each 5 °C above.

All cables have nominal cross-sectional areas and current carrying capacities. The *ISO-10133* standard specifies nominal capacities for a range of cross-sectional areas and temperature ranges. Temperature reference is typically 25ºC. Table 6-2 illustrates typical current ratings for equivalent cable sizes. Standard cable sizes are used which is cheaper and simpler to calculate. All cable current carrying capacities are subject to de-rating factors. In any installation where the temperature exceeds the nominal value, the continuous current carrying capacity of the cable is reduced. This is important in engine spaces. Where the temperature exceeds 50ºC the de-rated capacity is the nominal capacity multiplied by 0.75. Consult actual cable manufacturer's ratings for accuracy.

Table 6-2 Typical DC Cable Nominal Ratings

Size AWG	Size mm^2	PVC Insulation (Heat Resisting)	Butyl Rubber (Lloyds 100A1)	Resistance OHMS/100m
17	1.0	11 amps	12 amps	1.884
15	1.5	14	16	1.257
	1.8	15		1.050
14	2.5	20	22	0.754
12	4.0	27	30	0.471
10	6.0	35	38	0.314
8	10.0	49	53	0.182
6	16.0	64	71	0.1152
4	25.0	86	93	0.0762
2	35.0	105	119	0.0537
1	45.0	127	140	0.0381
0	50.0	150	160	0.0295
2/0	70.0	161	183	0.0252

In the US the more common method of calculating cable current ratings or ampacity is the use of tables and the following formula.

CM = K x I x L/E

Where CM = Circular Mil area of the conductors

K = 10.75 (a copper resistance constant per mil-foot)

I = Current in Amps

L= Conductor Length in Feet

E = Voltage Drop at the load in Volts

Conductor size should be selected with a maximum allowable voltage drop of 5% for all circuits. The voltage drop can be calculated using the formula in ISO Standard 10133, Annex A.2.

Voltage drop must always be a consideration when installing electrical circuits. Unfortunately, many voltage drop problems are created by the poor practice of trying to install the smallest cables and wiring sizes possible. The maximum acceptable voltage drop in 12-volt systems is 5% or 0.6 volt. The voltage drop problem is prevalent in starting and charging systems, in thrusters, windlasses, and in long runs to equipment. The following formula is that specified in ISO Standard 10133, Annex A.2.

$$\text{Voltage Drop at Load (volts)} = \frac{0.0164 \times I \times L}{S}$$

S - is conductor cross-sectional area, in square millimeters

I - is load current in amperes

L - is cable length in meters, positive to load and back to negative.

Example: Anchor Windlass. These cable sizes must be calculated at working and peak loads. As the calculations show, a larger cable size ensures less voltage drop and fewer line losses. Working-load current = 85 amps, cable run = 12 meters, CSA = 35 mm^2, rating = 125 amps.

$$\text{Drop at 85 amps} = \frac{0.0164 \times 85 \times 24}{35}$$

$$= 0.96\text{ V } (35\text{mm}^2)\ 0.67\text{ V } (50\text{ mm}^2)$$

$$\text{Drop at 125 amps} = \frac{0.0164 \times 125 \times 24}{35}$$

$$= 1.41\text{V } (35\text{ mm}^2)\ 0.98\text{ V } (50\text{ mm}^2)$$

Conductors should be insulated and sheathed (double insulated).

Double insulated cables should be used on all circuits to ensure insulation integrity. Additionally insulation is temperature rated which has important implications with respect to ratings. In most vessels PVC insulated and PVC sheathed cables rated at 75°C are used. For classification societies, ship wiring cables that use Butyl Rubber, CSP, EPR or other insulating materials are specified. They have higher temperature ratings and subsequently also have higher current carrying capacities.

Conductors should be identified as red for the positive conductor, and black for the negative conductor. Numbered cores are an acceptable alternative; the numeral 1 should be positive and numeral 2 negative.

This is the common system worldwide. Many years ago AC systems moved to IEC standards to avoid confusion, leaving black and red for DC, and brown and light blue for AC. This did not take place in the US and recently this was further complicated with the ABYC nominating yellow as a negative polarity color. Yellow is also a primary AC phase, switching or control circuit color so be careful. Some ship wiring cables often have white cores only with numbering imprinted on the insulation. Make sure AC and DC are not in proximity to avoid any possible confusion.

Conductors should be of stranded and tinned copper.

When untinned copper is exposed to saltwater spray or moisture it will very quickly degrade and fail. The argument used against the installation of tinned copper is cost. The price differential is typically 30% greater and the reliability (and vessel resale increase) advantages far outweigh the lower priced plain copper conductor.

All conductors should have a minimum cross-sectional area of 16 AWG (1.0mm^2).

The minimum conductor size to be used should be 1.0mm^2 (16 AWG). It is recommended that conductor sizes be standardized to 2.5 mm^2 (13 AWG).

Where cables are bunched, the cable ratings should be de-rated.

When several (6–8) cables are bunched in a large loom, the current capacity of the cable is reduced. The factor is typically around the nominal rating multiplied by 0.85. This may become an issue in very large boats only.

Where cables carry large currents for short time durations, they should be used subject to duty cycles.

Heavy current carrying cables such as those used on windlasses, winches, thrusters and starter motors are in fact only used for short durations. As there is a time factor in the heating of a cable, smaller cables can be used. Table 6-3 shows battery cable ratings that are rated at 60% duty.

Table 6-3 Battery Cable Ratings

Size AWG	Size B & S	Size mm^2	Current Rating (60% Duty)
8	8	8	90 amps
6	6	15	150 amps
4	3	26	200 amps
2	2	32	245 amps
1	0	50	320 amps
00	00	66	390 amps

6.14 Conductor Installation. Good cable installation is essential if electrical problems are to be avoided. Follow these installation guidelines.

Cable runs should be installed as straight as practicable. Cable bend radii should be a minimum of 4 x cable diameter.

Cables should be neatly installed in as straight a run as practicable. Tight bends should be avoided to reduce unnecessary strain on conductors and insulation. The minimum cable bend radii is applicable to all cables but particular care should be taken with larger and more inflexible cables, and x 6 is a better target radius.

Cables should be accessible for inspection and maintenance.

The emphasis must be on accessibility, both for initial installation, maintenance and for the addition of circuits. Under no circumstances should you fiberglass in cables. All cables, in particular those entering transits, should be capable of routine inspection.

Cables should be protected from mechanical damage, either where exposed, or where they are within compartments.

All cables should be installed to prevent any accidental damage to the insulation, or cutting of the conductors, or place undue strain on the cable. Even though cables are routed through lockers, machinery spaces and cupboards, the cables require protection. In many cases, faults are traced to what are considered safe areas. Objects and equipment are thrown into the space, and sharp edges damage the cable or insulation. In machinery or engine spaces, cables are often damaged during engine repairs.

Cables passing through bulkheads or decks should be protected from damage using a suitable non-corrosive gland or bushing. Cables transiting decks or watertight bulkheads should maintain the watertight integrity.

Cable glands are designed to prevent cable damage and ensure a waterproof transit through a bulkhead or deck. A significant number of problems are experienced with the ingress of water through deck fittings and I have seen a variety of methods used. In addition running cables through fiberglass with some sealant invariably results in chafing and cable failure. Use circular multi-core cables if possible to ensure proper gland sealing is possible. The purpose designed Index (Thrudex) types are recommended. The structural material of a deck has to be considered before selecting glands. A steel deck requires a different gland type to a foam sandwich boat.

Cables should be supported at maximum intervals of 8 inches (200mm). Supports and saddles are to be of a non-corrosive material. Where used in engine compartments or machinery spaces, these should be metallic and coated to prevent chafe to the cable insulation. Cable saddles should fit neatly, without excessive force onto the cables, or cable looms, and not deform the insulation.

Cables can be neatly loomed together and secured with PVC or stainless saddles to prevent cable loom sagging and movement during service. While the recommendation is 450mm apart in *ISO 7.3* the closer support distances secure the cables more efficiently. I prefer standard electrical PVC-conduit saddles, which come in a variety of sizes. It is important to have a neat fit only, and not force saddles over cables or looms so that insulation is deformed. In machinery spaces metal saddles are often used, but they should have a plastic sleeve placed on them to prevent the sharp edges chafing the cable insulation.

The PVC cable tie or tie-wrap is universal in application, and should be used where looms must be kept together, or where any cable can be securely fastened to a suitable support. Do not use cable ties to suspend cables from isolated points; this invariably causes excessive stress and cable fatigue. For internal cable ties, you only require the white ones; any external cable ties should be the black UV-resistant type.

PVC spiral wrapping is an extremely useful method for consolidating cables into a neat loom. If a number of cables are lying loose, consolidate them into some spiral wrap, and then fasten the loom using cable ties.

A hot glue gun is often used to fasten small or single cables above headliners, or in corners behind trim and carpet finishes. It is useful where there is no risk of cables coming loose. Do not use on exposed cable runs.

Cables should be as far as practicable separated into power, signal and data, and heavy current carrying groups. Instrument and data cables should be installed as far as practicable from power cables, and communications aerial cables. A minimum distance of 12" (300mm) is recommended. AC cables should not be run within DC system cable looms, and should be kept separate.

Cables should be separated into signal or instrument cables, DC power supply cables, and where space allows heavy current carrying cables such as windlass or thrusters. This is to minimize induced interference between cables, in particular on long, straight runs. All data and instrument cables should be routed as far as practicable away from power cables. Aerial cables should also be routed well away from power cables. AC and DC cables must be kept separated.

Where cables may be exposed to heat, they should be installed within conduits or otherwise protected from the heat source.

Cables installed with machinery or engine spaces should be rated for the maximum heat of the space. In addition, where cables may be exposed to heat sources, such as exhaust manifolds or piping, they should be protected.

Where cables are installed within conduits, they should be supported within 3 inches (75mm) of both entry and exit points. Conduit ends should be treated, or otherwise protected to remove sharp edges and prevent chafe to cable insulation.

Conduits are often installed during the construction phase, and this allows cables to be easily pulled in, replaced, or added. Conduits offer good mechanical protection to cables and in many cases, single-insulated cables are run in conduit back to the switchboard. As they are single insulated, they are exposed where they enter or exit the conduits, and should be supported by saddle or clamp to prevent excessive movement. Try to avoid installing large bunches of cables in flexible conduits as they tend to move around and chafe. PVC conduits should not be used in machinery spaces. Where cables exit conduits, the exit should be bushed to prevent chafing. During installations when pulling in cables, insulation is frequently damaged as insulation rubs against sharp edges.

All externally installed cables should be protected against the effects of ultraviolet (UV) light.

Continued exposure to UV on external equipment cables will result in insulation degradation and failure. Small cracks in the insulation allow water to penetrate the conductor and subsequently degrade the copper. This is common on navigation lights, GPS aerial cables, radio aerial cables and other equipment. All exposed cables should be covered in black UV resistant spiral wrapping to prevent rapid degradation of insulation. Cable ties should also be of the black UV type. Use tinned copper conductors on all external wiring to navigation lights, spotlights, and cockpit lights etc.

Connections should be minimized within any circuit between the power supply and the equipment.

Connections and joins in cables should be avoided. Any connection adds resistance to a circuit and introduces another potential failure point.

Equipment grounds should be made to the same point as the battery negative ground point.

Equipment grounds, such as pump casings etc are usually connected to the boat ground. In many cases, a ground terminal block is installed close to instruments and a large ground conductor taken to the same point as the battery negative connection point. This is not the battery but actual termination point.

Electrical equipment and cables should not be installed within any compartment or space that may contain equipment or systems liable to emit explosive gases. This may include spark ignition engine fuel systems, LPG installations or flooded cell battery installations. Where any equipment or fittings are to be installed, they should be ignition-protected in accordance with the appropriate national standards.

No cables, connections or equipment should be installed in any space subject to gas or fumes. Equipment or fittings must be classified as ignition proof by an appropriate organization, such as UL.

6.15 Instrument and Data Cable Installation. Electrical and instrument cables generally are installed in close proximity. If precautions are not taken, this will result in electrical interference on electronics systems.

Instrument and data cables should be installed as far as practicable from power cables, and communications aerial cables. Long parallel runs close to power cables should be avoided.

Cables should be separated as far as practicable from power supply cables, and heavy current carrying cables such as windlass or thrusters. This is to minimize induced interference between cables, in particular on long, straight runs. Cables should also be routed well away from aerial feed cables.

Where instrument and data cables cross power cables, this should be done as close to an angle of 90° as practicable.

This is to prevent induced interference with right angle crossovers.

Navigation, autopilot and position fixing equipment should be located as far as practicable from radar, satellite communications equipment, VHF, HF, HAM and cellular telephone equipment, tuners and control units, cables, aerials, antennas and related components. A minimum clearance distance of 1m (39") is recommended.

Where practicable, electronics equipment, control modules, processors etc should be located clear of cable looms, aerial cables to prevent interference. This should include all satellite communications and television systems, and cellular telephones. Autopilots are prone to interference causing major uncontrolled course alterations.

Screens should be grounded at one end only, or in accordance with specific manufacturer's recommendations.

Conductor screens should be grounded as recommended by the equipment manufacturer. The termination is normally at the equipment end.

Electronic equipment grounds should be made to the same point as the battery ground, or in accordance with specific manufacturer's recommendations.

Equipment grounds are usually connected to the boat ground. In many cases, a ground terminal block is installed close to instruments and a large ground conductor taken to the same point as the battery negative connection point. This is not the battery but actual termination point.

6.16 Grounding Systems. The following chapters on lightning, corrosion, AC power systems, radio systems etc will all make reference to grounding systems. It is crucial to understand what the various grounds are and their importance within respective circuits as well as to each other. Confusion of the functions is a principal cause to system problems.

a. **DC Negative.** The DC negative is not a ground. It is a current carrying conductor that carries the same current that flows within the positive conductor. In a single circuit wiring configuration it may be bonded to a grounded point, usually the mass of the engine. The engine is connected to an immersed item such as the steel hull or prop shaft. This is used to polarize the system and doesn't actually carry current.

b. **Lightning Ground.** A lightning ground is also a point at ground potential that is immersed in seawater. It only carries current in the event of a lightning strike and the primary purpose is to ground the strike energy. It is not a functional part of any other electrical system, and should not be interconnected.

c. **Cathodic Protection System Ground.** The cathodic protection system ground is effectively the sacrificial anode that is connected to protected underwater items via bonding wires.

d. **AC Ground (or Earth).** The AC ground is a point at ground potential that is immersed in seawater. Under normal operating conditions, it carries no voltage or current. The primary purpose is that under fault conditions it will carry fault current to ground and hold all connected metal to ground potential, ensure operation of protective equipment, and protect against electric shock from exposed metal parts. In most cases where it is connected, it is to the same point as the DC negative. The main stated justification is that where DC negative and AC ground are not connected and a short circuit condition develops between the AC hot conductor and a DC negative, or bonding system this would result in the AC protection not tripping. This may cause energization of these circuits up to rated voltage, creating a risk to persons in contact. There are other stated risks of potentially fatal shock risks to swimmers. A separate ground plate is a good option to ensure total isolation from other systems, as is the proper separation of DC systems. The practice of bonding DC and AC has created much controversy due to well-documented cases of corrosion.

e. **Radio Frequency Ground.** The radio frequency ground is an integral part of the aerial system and is sometimes termed the counterpoise. The ground only carries RF energy and is not a current carrying conductor. It is not connected to any other ground or negative.

f. **Instrument Ground.** The instrument ground, which most GPS and radar sets have, is nominally vessel ground. In many cases, a complete separate ground terminal link is installed behind the switchboard, and to which the screens and ground wires are connected. A separate large low resistance cable is then taken to the same ground point as other grounds. Do not simply interconnect the DC negative to the link as equipment may be subject to interference.

6.17 Conductor Terminations. Conductor terminations are the single greatest cause of failure and they should be made properly if you wish to maintain reliability.

All conductors should be terminated where practicable using crimped connectors. Where cables are terminated within terminal blocks, they should be secured to prevent contact with adjacent terminals.

The most practical and common method of cable connection are the tinned-copper, crimp terminals or connectors. These are color coded according to the cable capacity that can be accommodated. Terminals are usually designed and manufactured according to NEMA standards, which cover wire pullout tension tests, and voltage drop tests. Where possible select double crimp types, which should be used in high vibration applications.

When crimping, always use a quality ratchet-type crimping tool. Do not use a cheap pair of squeeze types, which do not adequately compress and capture the cable. This subsequently causes failure and the cable pulls out of the connector sleeve. A good joint requires two crimps. Always crimp both the joint and the plastic behind it. Ensure that no cable strands are hanging out. Poor crimping is a major failure cause. A crimp joint can be improved by lightly soldering the wire end to the crimp connector. Avoid excessive heat. After crimping, give the connector a firm tug to ensure that the crimp is sound.

Table 6-4 Standard Cable Connectors Table

Color	AWG	Cables Sizes	Current Rating
Yellow	12-10	3.0 to 6.0 mm^2	30 Amps
Blue	16-14	1.5 to 2.5 mm^2	15 Amps
Red	22-18	0.5 to 1.5 mm^2	10 Amps

Quick-disconnect (spade) connectors are commonly used, particularly on switchboards. When using always select the correct quick-disconnect (spade) terminals for the intended cable size. Female connectors are easily dislodged, and there is a tendency for these types to slip off the back of circuit breaker male terminals, so ensure they are tight to push on. Ensure that the terminal actually goes on the CB terminal, and not in between insulation sleeve and the connector. For heavy duty, look at using heat-shrink fully insulated types. It is important not to apply too much strain on the cables.

Ring terminals are used on all equipment where screw, stud, bolt and nut are used. They should also be used on any equipment subject to vibration, or where accidental dislodgement can be critical, particularly switchboards. Always ensure that the hole is a close fit to the bolt or screw used on the connection, to ensure good electrical contact and use spring washers. One practical method used to prevent nut or screw creep is to dab on a spot of paint.

In-Line Cable (Butt) Splices. Where cables require connection and a junction box is impracticable, use insulated in-line butt splices. This is more reliable than soldered connections, where a bad joint can cause high resistance and subsequent heating and voltage drop. Use heat shrink insulation over the joint to ensure waterproof integrity is maintained. When heated, some connectors form a watertight seal by the fusing and melting of the insulation sleeve. These can be ideal for bilge pump connections.

Pin Terminals. Pin terminals can make a neat cable termination into connector blocks, however from experience I have found these to be unreliable simply because vibration and movement work them loose and in most cases they do not precisely match the connector block terminal and make an inadequate electrical contact.

Snap Plug (Bullet) Terminals. These are useful where used in cabin lighting fittings. I often use these on all cable ends, female on the supply and male on the light fitting tails. This makes it easy to disconnect and remove fittings.

Wire Terminations. Cable ends should have the insulation removed from the end, without nicking the cable strands. The bare cable strands should be simply twisted, and inserted in the terminal block or connector of a similar size. Ensure there are no loose strands. If you are terminating into an oversize terminal block, twist and double over the cable end to ensure that the screw has something to bite on.

Conductor terminations should not be soldered.

Do not solder the ends of wires prior to connection. In most cases, this is done to make a good low resistance connection and prevent cable corrosion. In my experience, soldered connections cause many problems, with the solder traveling up the conductor causing stiffness. This causes greater vibrational effects at the terminal with resultant fatigue and failure. In most cases, the soldering is poorly done with a high resistance joint being made. A soldered cable end also prevents the connector screw from spreading the strands and making a good electrical contact, causing high resistance and heating. You should use connectors of the correct size for the cable.

Conductor terminations should be marked with a number. The negative and positive cable should be marked with the same number. Identification should be consistent with the wiring diagram.

Always mark cable ends to aid in reconnection and troubleshooting. The numbers should match those on the wiring diagram. A simple, slide-on number system can be used. The stick on adhesive types should be avoided as they generally unravel and fall off as the adhesive fails. If wires are color-coded, still use numbers, as they are easier and much quicker to identify. Commercial shipping uses numbers.

The circuit positive should sequentially match the supply source such as the circuit breaker. The circuit negative should match the positive, and be placed in the same sequential order on the negative link. The numbering convention if unmarked is left to right.

Where connections are made within any area subject to water or moisture, such as bilges, the terminations should be made as far as practicable near the top of the bilge. Connections should be suitably protected against water ingress.

Connections should be made above the maximum bilge water level. Joints should be finished with self-amalgamating tapes, or heat shrink tubing. I have frequently seen connections permanently immersed and fail. In automatic bilge circuits, the live connection also contributes to corrosion problems in some boats.

Plugs and sockets where used for the connection of cables, or equipment should incorporate screw retaining rings, and protective caps to prevent the ingress of water when not in use. They should be rated to a minimum of IP54.

Deck plugs and sockets are often used instead of deck glands and junction boxes at a mast base or as outlets for hand spotlights. Many in use are of inferior quality and prematurely fail. Don't use the cheap and nasty chrome plugs and sockets, they aren't waterproof. The best units on the market are either the Bulgin type units from Index or those from Dri-plug. When using deck plugs ensure that the seal between deck and connector body is watertight. Leakage is very common on wet decks up forward where they are usually located. Ensure that the cable seal into the plug is watertight. It is of little use having a good seal around the deck, and plug to socket if the water seeps in through the cable entry and shorts out terminals internally as is often the case. Most connectors have O-rings to ensure a watertight seal. Check that the rings are in good condition, are not deformed or compressed, and sit properly in the recess. A very light smear of silicon grease assists in the sealing process. Ensure that the pins are dry before plugging in and that pins are not bent or show signs of corrosion or pitting. Do not fill around the pins with silicon grease, as this often creates a poor contact. Keep plugs and sockets clean and dry.

Where connections are made, they should be protected within a suitable junction box, and installed in a protected area.

Junction boxes are the most practical way to terminate a number of cables, especially where access is required to disconnect circuits. To reduce the number of cables radiating back to the switchboard and minimize voltage drops, I use a junction box forward and aft to power up lighting circuits. Terminal blocks are commonly used, and in many cases, the box is too small for the quantity or size of cables. In these cases, the box lid is forced on, applying pressure to the cables. This should be avoided as unnecessary stress is applied to terminations.

Cables terminated within a junction box should enter from the bottom, and be looped to prevent water entering the box and connections.

All cables should enter from the bottom. Junction box upper surfaces should have no openings that permit the entry of water. Cables looped in at the bottom will allow water to drip off, and prevent surface travel to the connections.

Cables within junction boxes should be marked with numbers that correspond to circuit numbers used at the main switch panel.

Cable ends should be numbered to aid in reconnection and troubleshooting. The numbers should match those on the wiring diagram.

The main negative cable should be secured on the engine using a spring washer, to prevent the connection becoming loose from vibration.

The main engine negative cable is prone to vibration from the engine. It frequently comes loose causing starting problems, intermittent equipment operation, interference, and in some instances alternator failures. In most cases, it is simply fastened to a convenient bolt. The mating surface must be cleaned to ensure a good electrical contact, and a spring washer used to maintain tension.

6.18 **Circuit Testing.** Before you energize any circuit you should verify that the circuit is ready, so observe the following:

The insulation resistance between conductors, or conductors and ground of all circuits or the complete installation should be greater than 100,000 ohms. All fuses, circuit breakers and switches are to be closed. Tests should be made with all equipment, lamps and electronic equipment disconnected.

All circuits should be tested to verify that the levels of insulation are satisfactory on the whole system, and on each circuit. Supply circuit breakers should be switched on so that switchboard busbars are included within the test. A multimeter set on the resistance range should be used between the positive and negative conductors. If readings are low, check that a load is not connected.

6.19 **Mast Cabling.** Mast cabling is a common source of failure. Many problems can be avoided if the cables are installed properly. Since masts are generally wired by mast manufacturers and riggers, vessel owners rarely take the opportunity to supervise or specify requirements. There are three major areas of concern in any mast installation:

a. **Mast Base Junction Boxes.** The most common area of failure is the junction box. If mounted inside the vessel, a good water-resistant box should be installed. If mounted externally, and this should only be a last resort, a waterproof box is required. Always leave a loop when inserting cables into the box. If water does travel down the loom, this will drip off the bottom of the loop and will not enter and corrode the junction box terminals or connections.

b. **Deck Cable Transits.** Cable glands are designed to prevent cable damage and ensure a waterproof transit through a bulkhead or deck. A significant number of problems are experienced when water gets in through deck fittings, and I have seen some amazing systems utilizing pipes, hose, and the like. If figure 8 type cable is used, or small, single insulated cables are installed, it is virtually impossible to adequately seal them in cable glands. To overcome this problem, use circular, multicore cables if possible, or use the consolidation procedure described below to make a cable loom that can be put through a deck gland. The Index (Thrudex) cable glands are in my opinion the best on the market. You need to take deck material into account before selecting a gland. Steel decks require a different gland type than fibreglass and foam-sandwich decks.

c. **Cabling.** The following factors should be noted when installing electric cable:

(1) **Cable Types.** The major problem is the use of single insulated untinned cables, generally of an under-rated conductor size. Small conductor sizes cause many voltage drop problems with unacceptable low light outputs as a result. Use 15-amp-rated cable for each circuit.

(2) **Negative Conductors.** Masthead tricolors are normally connected to a dual anchor light fitting. These use a three-wire, common negative arrangement. The same arrangement is used for combination masthead and foredeck spotlights. Never use the mast as a negative return, as I have found on some vessels. Always install a negative wire to each light fitting.

(3) **UV Protection.** All exposed cables should be covered in black, UV-resistant spiral wrapping to prevent rapid degradation of insulation. Small cracks in the insulation allow water to penetrate, which subsequently degrades the copper.

d. **Mast Cable Support.** Cabling must be properly secured within the mast. The weight of a cable hanging down inside a mast causes fatigue through stretching. If the cables are not fully enclosed in conduit which is still a common practice, the internal halyards can whip against them. This will cause severe damage to the conductors in multicore instrument cables or damage the insulation. There are a number of methods for securing mast cables; a combination of all three is best.

(1) **Cable Glands.** Where a cable enters the mast base and exits at the masthead, it should pass through a cable gland. The ideal glands for this are the Thrudex DR1 rectangular units. Once cables have been placed through the neoprene, the gland is tightened and compression around the cables takes the strain. The cables are protected from chafe against the mast entrance hole.

(2) **Messenger Line.** A small messenger line can be installed with the cables and supported at the masthead. The messenger should be tied or taped to the cable loom and then fastened to take the load off the cable ends. The messenger serves as a pull-through for adding or replacing cable. However, once the line is taped to the loom over its entire length, it is impossible to remove and replace single cables.

(3) **Cable Ties.** Where possible, use cable ties to fasten and support cables. The ideal place to do so is where cables come out of the mast to connect lights, radar, etc., which usually gives 3 or 4 fastening points. There is generally sufficient space to insert a cable tie around the cables. A second hole large enough for a tie is required next to the main cable entry to enable tie to be supported. Always use black, UV-resistant cable ties.

e. **Mast Cable Consolidation.** In most cases, the mast is wired with single insulated cables. To put these cables through deck cable glands, you need to consolidate them into a single loom. One method is as follows:

(1) Neatly make a cable loom and hold it in place with cable ties. Keep the loom as circular as possible.

(2) Apply silicone sealant to the loom, and work it through all cables. This will ensure that a solid core is made. If done properly, it will prevent water from traveling down the cable loom.

(3) Apply a layer of black, UV-resistant spiral wrap to the loom. Again, spaces between the wrap should have silicone compound applied to fill any voids. The spiral wrap gives the cable loom a circular shape.

(4) Slide on a length of heat shrink tubing and shrink it in place. This forms the outer sheath.

(5) Use a suitable deck gland, pass the cable through the deck, and connect into a suitable junction box.

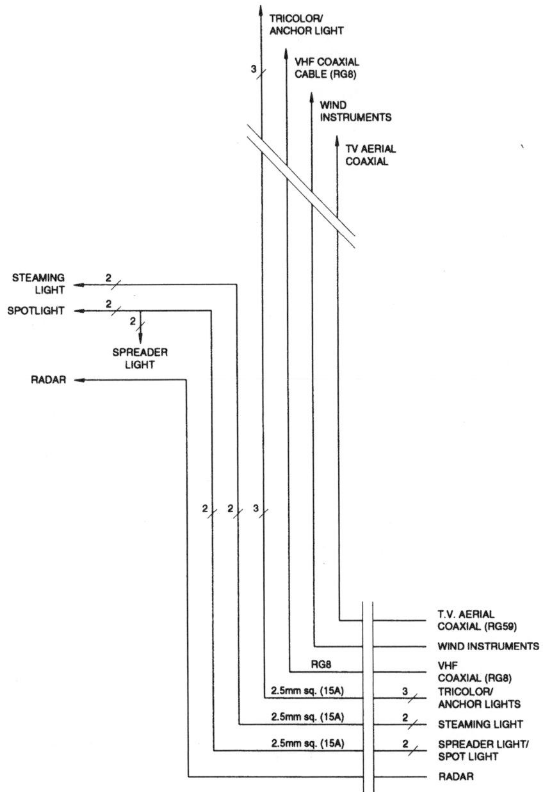

Figure 6-8 Mast Cabling Diagram

f. **Deck Plugs.** Instead of deck glands and junction boxes at a mast base, deck plugs are sometimes used. They can also provide outlets for hand spotlights, or other equipment commonly used. Many are of inferior quality and fail prematurely, often when they are needed most. Don't use the cheap and nasty chrome plugs and sockets as they are not waterproof. The best units on the market are either the Bulgin-type units from Index or those from Dri-plug. When using deck plugs, observe the following:

(1) **Deck Seal.** Ensure that the seal between deck and connector body is watertight. Leakage is very common on wet decks up forward where the plugs are usually located.

(2) **Plug Cable Entrance.** Make sure the cable seal into the plug is watertight. It is of little use to have a good seal around the deck if the water seeps through the cable entry and shorts out terminals internally.

(3) **Connector Seals.** Most connectors have O-rings to ensure a watertight seal. Check that the rings are in good connection, are not deformed or compressed, and sit properly in the recess. A very light smear of silicone grease assists in the sealing process.

(4) **Connection Pins.** Ensure that the pins are dry before plugging in and that pins are not bent or showing signs of corrosion or pitting. Do not fill around the pins with silicone grease, as this often creates a poor contact. Keep plugs and sockets clean and dry.

g. **Mast System Maintenance.** Basic maintenance tasks will reduce mast wiring problems:

(1) **Mast Base Cable Exits.** Regularly examine cables where they exit the mast for signs of chafe. If the cable loom has not been protected with a UV-resistant sleeve, carefully examine insulation for cracks.

(2) **Masthead Cables.** Regularly examine masthead cable exits for chafe. Ensure that coaxial, wind instrument, and power cables have a reasonable loom to allow for shortening and repair.

6.20 Mast Cabling Troubleshooting. Mast wiring faults are common because the mast subjects cables to the worst damaging factors, such as vibration, exposure to salt water, stretching, and mechanical damage. Fortunately, mast wiring is easy to troubleshoot.

a. **Tricolor/Anchor Lights.** If a light does not illuminate, lamp failure is the usual cause. If the lamp is replaced and it still does not illuminate, perform the following tests:

(1) **Test Supply.** Open the mast connection box and locate the appropriate terminals. Using a multimeter on the DC-volt range, check that voltage is present at the terminals with the power on. Many failures are due to poor contacts within terminal blocks, or corrosion of the terminal and cable.

(2) **Continuity Test.** Turn the power off, and with a multimeter set on the resistance x1 range, test between the positive and negative terminals. The reading should be approximately 2–5 ohms with a good lamp installed. If the reading is above that range, the light fitting or connection has failed or the cable has been damaged. The mast cable entry and exit points should be examined first. Internal breaks only occur in masts without wiring conduits. Many tricolor/anchor lights have a plug and socket arrangement, which is an occasional source of trouble.

b. **Spreader Lights.** The above tests are also valid for spreader lights. On many vessels, spreader lights are a sealed beam unit within a stainless steel housing. It is very common to have shorts to the mast as cables chafe through on sharp edges. This problem is notorious for causing circuit leakages and increased corrosion rates on steel vessels:

(1) **Mast Short Circuits.** With a multimeter set on the resistance ohms x1k range, check between the mast and both positive and negative wires. The reading should be over-range. If you have any reading, you have either a short or a leakage from cable insulation breakdown.

(2) **Check Supply.** Open the mast connection box and locate the appropriate terminals. Using a multimeter on the DC-volt range, check that voltage is present at the terminals with the power on.

Lightning Protection

7.0 Lightning Protection. Virtually all classification societies and national marine authorities, ABYC, NFPA and class societies such as Lloyd's lay down recommendations for lightning protection. More than a 1000 people are killed worldwide annually by lightning strikes, and in the US lightning voltage transient damage exceeds 1 billion dollars. The majority of lightning strikes take place between noon and 1800 hours. In Florida some 40% of deaths and injuries are related to water based recreational activities.

The National Weather Service (NWS) in the US gives out continuously up-dated weather information on VHF channels WX1, WX2, WX3.

Thunder is the sound generated during a lightning strike, when the air is superheated to around 54,000 degrees F and then expands faster than the speed of sound creating a shock wave. This travels at 1 mile every 5 seconds and the maximum range for hearing thunder typically is around 5–6 miles. If you see lightning count it out in seconds until you hear the thunder and divide by 5 and you will know the range.

(1) In an electrical storm, stay below decks at all times.

(2) Take a position and plot it prior to shutting down, in case of all electronic navigation equipment being blown.

(3) Turn off all electronic gear and isolate circuit breakers if at all practicable.

(4) Disconnect radio and other aerials if practicable.

(5) Do not operate radios until after the storm passes unless in an extreme emergency.

(6) Compasses should be rechecked and deviation corrections made after a strike. In some cases, complete demagnetization may occur.

7.1 Lightning Physics. Strong updrafts and downdrafts within cumulus and those anvil topped cumulo-nimbus thunderstorm cloud formations generate high electrical charges. The top of the formation develops a positive potential and the lower a negative potential. Lightning occurs when the difference between the positive and negative charges, the electrical potential, becomes great enough to overcome the resistance of the insulating air and to force a conductive path between the positive and negative charges.

a. **Negative Cloud to Ground.** These strikes occur when the ground is at positive polarity and the cloud's negative region attempts to equalize with ground.

b. **Positive Cloud to Ground.** The positively charged cloud top equalizes with the negative ground.

c. **Positive Ground to Cloud.** The positively charged ground equalizes with the negatively charge cloud.

d. **Negative Ground to Cloud.** The negatively charged ground equalizes with the positively charged cloud top.

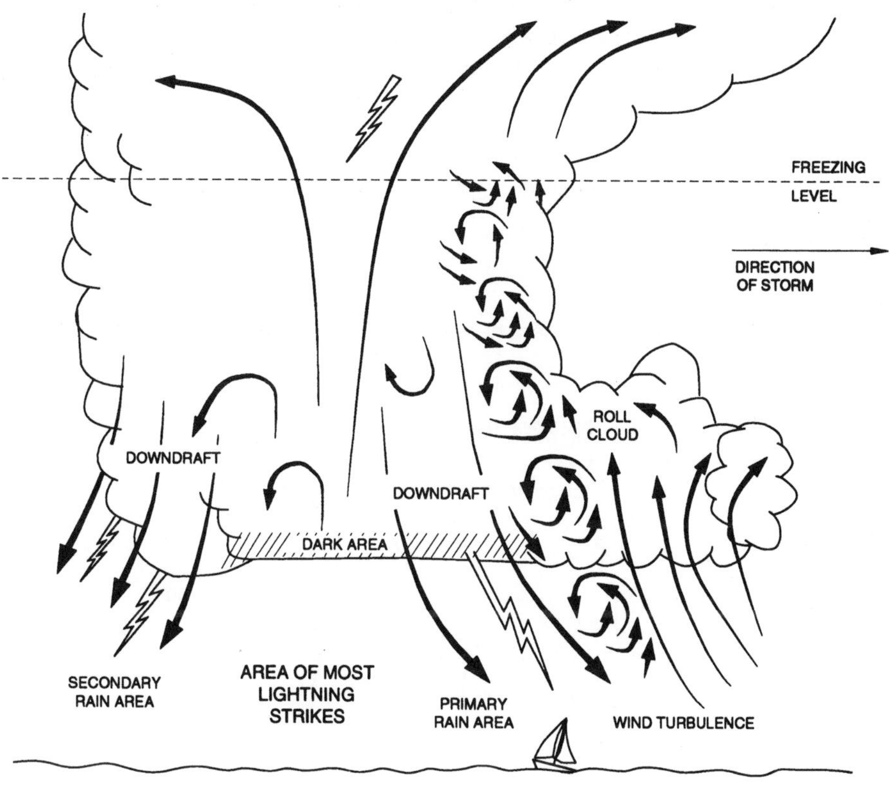

Figure 7-1 Cumulo-Nimbus Storm System

7.2 Lightning Components. Lightning consists of a number of components which form a multidirectional flow of charges exceeding 100 million volts and 200,000 amperes at over 30,000°C for a matter of milliseconds. The positive ions rise to the cloud top, and the negative ions migrate to the cloud base. Regions of positive ions also form at the cloud base. Eventually, the cloud charge levels have sufficient potential difference between ground and another cloud to discharge.

a. **Leader.** The leader consists of a negative stream of electrons comprising many small forks or fingers that follow and break down the air paths offering the least resistance. The charge follows the fork, finding the easiest path as each successive layer is broken down and charged to the same polarity as the cloud.

b. **Upward Positive Leader.** A positive charge rises some 50 meters above the ground.

c. **Channel.** When leader and upward leader meet, a channel is formed.

d. **Return Stroke.** This path is generally much brighter and more powerful than the leader, and travels upward to the cloud, partially equalizing the potential difference between ground and cloud.

e. **Dart Leader.** In a matter of milliseconds after the return stroke, another downward charge takes place along the same path as the stepped leader and return stroke. Sometimes it is followed by multiple return strokes. The movements happen so fast that it appears to be a single event. This sequence can continue until the differential between cloud and ground is equalized.

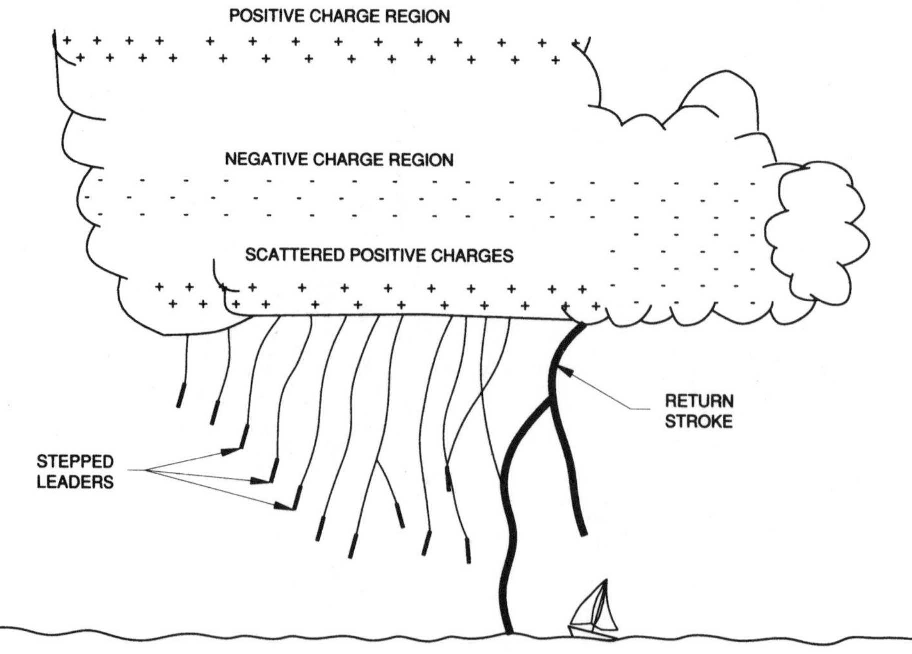

Figure 7-2 Lightning Process

7.3 Lightning Protection Zone. The most reliable protection system is one that grounds any strike directly. You can't protect your boat against a strike occurring but you need to protect the boat systems in the event of a strike. The principles are as follows:

a. **Grounding.** The primary purpose of a grounding system is to divert the lightning strike directly to ground through a low resistance, low impedance circuit suitably rated to carry the momentary current values. This reduces the strike period to a minimum, and reduces or eliminates the problem of side strikes as the charge attempts to go to ground.

b. **Cone of Protection.** The tip of the mast, or more properly a turned spike clear of all masthead equipment, gives a cone of protection below it. The cone base is twice the diameter as the mast height. This protective cone prevents strikes to adjacent areas and metalwork, including stays, rails or other items lower than the masthead.

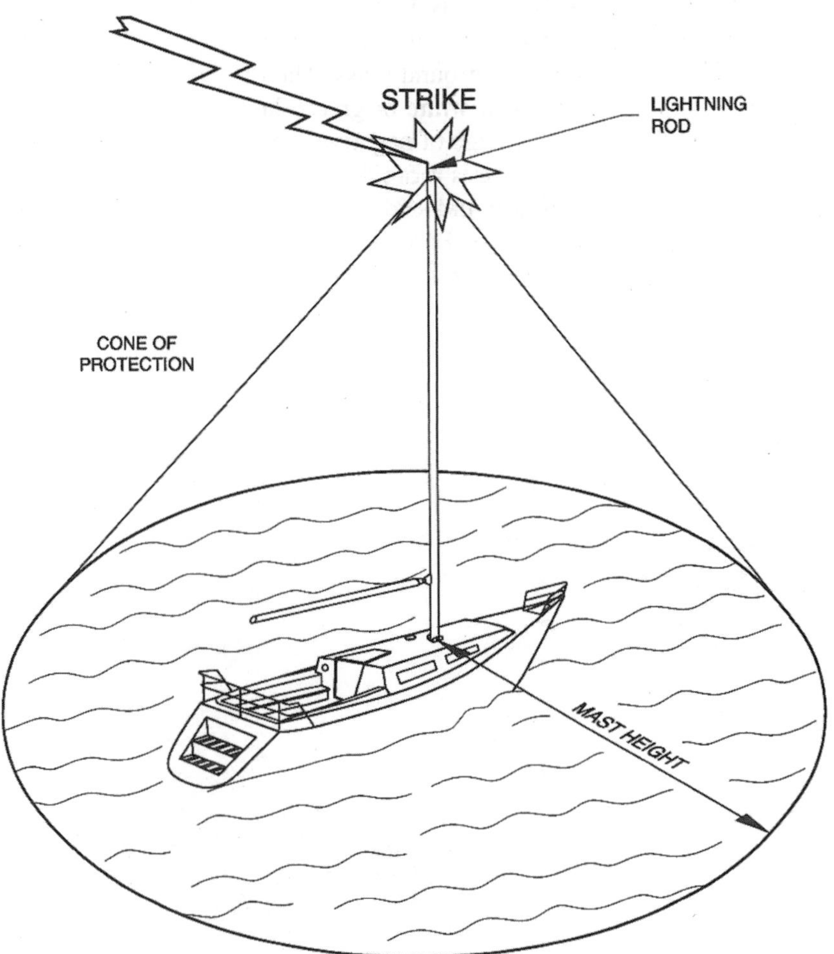

Figure 7-3 Cone of Protection

c. **Electromagnetic Pulse (EMP).** Though insurance companies don't like to accept claims on damage unless you can show total damage to masthead systems, a vessel can have its equipment damaged by a strike within a few hundred meters. A strike sends out a very large electromagnetic pulse, which is a strong magnetic field. This field induces into wiring and systems a high voltage spike that can nearly do as much damage as a direct hit. If you suspect damage from an induced electromagnetic pulse, check with vessels adjacent to yours and get statements to support your contention. Generally all the electronics will be out if this is the case because the mast and any wiring act as a large aerial.

d. **Sidestrikes.** It is common among closely moored vessels and in crowded marinas to have a lightning strike literally jump from vessel to vessel as it attempts to find ground. Usually the strike exits from stays, chainplates, and spreaders. In many cases, the strike goes to water from the chainplates, causing serious damage to hull and fittings.

e. **St. Elmo's Fire (Brush Discharge).** When this phenomenon occurs, it usually precedes a strike, although the effect does not occur all the time. The vessel becomes a large ground mass. The discharge is characterized by ionized clouds and balls of white or green flashing light that polarize at vessel extremities. The discharge of negative ions reduces the potential intensity of a strike. St. Elmo's Fire is more common on steel vessels. Damage to electrical systems is usually induced into mast wiring, as the steel hull itself acts as a large Faraday cage. For amusement, tell the insurance company that your damage was caused by St. Elmo's Fire!

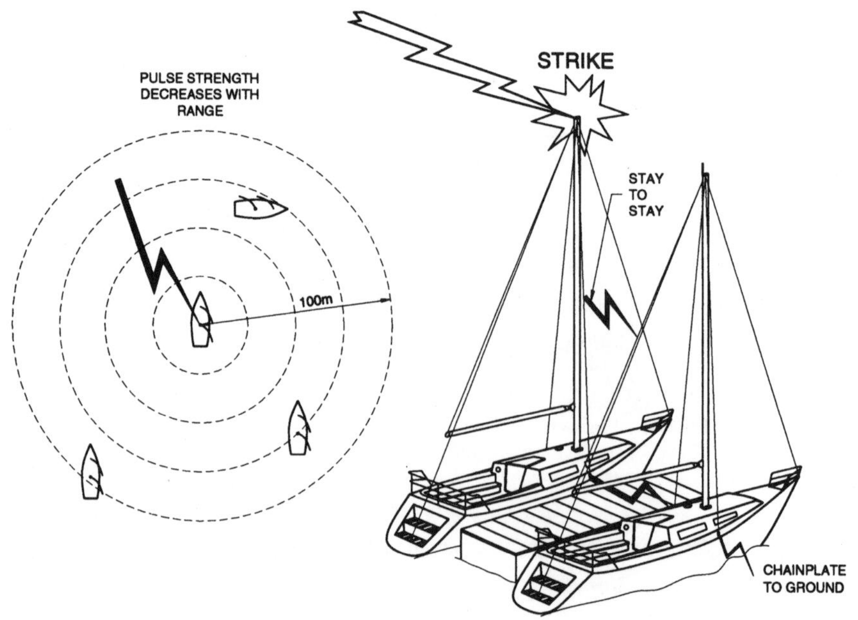

Figure 7-4 Electromagnetic Pulse Effect and Sidestrike

7.4 Lightning Protection Systems. Most classification societies, the ABYC, and other advisory bodies generally recommend lightning protection in the form of a directly grounded mast and spike. There are a few basic elements in any protection system; however they must be done correctly. A range of dissipation devices have come onto the market. These devices are typically brush or "bottle brush" type arrangements. The principle is that all the spikes "bleed" off or dissipate electrons or ions, reducing the differential that may cause a lightning strike. They do not protect the boat in the event of a strike, and safely carry strike energy to ground. Lightning will generally strike the highest point and take the path offering the least resistance to ground. The mast is usually the strike point. Note that a stainless steel VHF whip does not constitute any protection.

(1) The Lightning Conductor. A lightning conductor should be installed at the masthead. This should consist of a turned copper spike of at least 12mm (1/2") in diameter, or a hemispherical dome tipped spike, and project at least 150mm (6") above the highest point.

The first protection element on a sailing yacht is the air terminal, and ideally this should be a copper rod with a turned head, dome shaped, although pointed spikes are also common. To avoid galvanic metal interaction, stainless rods are commonly used, but should be of a thicker section than the more conductive and lower resistance copper. The spike should be at least six inches higher than any other equipment, including VHF aerials; this means a terminal 12–24" in height. Many commercial units (Dynarod and Seaground) have an offset in the rod which, although not being the required straight section, would be satisfactory. The purpose of the point being sharp is that it facilitates what is called point discharge. Ions dissipate from the ground and effectively cause a reduction in potential between the cloud and the sea. In many cases, the strike may be of lower intensity or not occur at all. Note that a stainless steel VHF whip aerial does not constitute any protection. The air terminal is mounted clear of all other equipment and gives a cone of protection below it, to capture the strike. This protective cone prevents sidestrikes to adjacent areas and metalwork, which includes rails or other items lower than the air terminal. Some variations include devices such as the Italian Lightning Protection Device (LPD), essentially a high-performance varistor. This is designed to interact with the electrical charges of the initial stepped leader, when current values are relatively low, and avoid the return strokes. Charges accumulate on the atmospheric electrode and varistor poles. The varistor conducts and the charge condition on the electrode alters and these charges leave when some streamers form to meet the leader.

(2) The Down Conductor. Lightning down conductors should be of at least 100mm^2 (4 AWG) cross sectional area.

The purpose of the down conductor is to safely conduct the strike current through a low impedance, low resistance circuit suitably rated to carry the strike current to the ground point. Also it is to eliminate side flash dangers, to minimize induction into other conductors, and to maintain the strike period to the minimum possible. Much of the damage in a strike can result from heat, as the very large current flow into even a low resistance down conductor cable can act as a large heating element. The chapters on voltage drop are relevant here. It is essential that the cable cross sectional area is sufficient, typically a minimum of 4 AWG to 8 AWG. Conductors should be flexible, compact, concentric stranded cables. The overall resistance of the cable must not exceed 0.02 ohms maximum as electricity follows the path of least resistance, and this reduces side flash dangers if energy looks for alternative paths. This means that if a ground circuit is 2 ohms overall and a communications ground 1 ohm, the energy will tend to divert through the communications ground. In shore installations special purpose tri-axial cables are used and the multiple screens reduce the large radiated fields that are generated; however this is expensive and an option only on large super yachts. The bonding cable to the ground plate should be as straight as practicable without sharp corners as side discharges called corona discharge occur. The minimum recommended radii of cable bends is 8 inches. It is also useful to enclose the conductor internally with PVC flexible conduit normally used in shore electrical systems to increase the insulation levels, as DC battery cable will break down under high voltage conditions.

The total resistance of the grounding circuit from the lightning conductor to the ground plate or hull grounding point should not exceed 0.02 ohms.

A low resistance and low impedance grounding circuit is critical to the performance of the protection system. Any resistance will cause significantly greater heating effects and strike energy will seek shorter and lower resistance ground paths. High resistance circuits contribute to sidestrike activities.

In vessels with alloy masts, the base of the mast should be bonded to the deck and mast step, or the compression post. This should then be bonded to the ground plate or keel.

It is often easier to bond the base of an alloy mast to the mast step, and then bond this to the compression post. The bottom of the compression post is subsequently bonded to the ground plate or keel. Keel stepped masts can be directly bonded to the ground plate or keel with a short and heavy gauge conductor. Wooden masts ideally should have a conductor fastened externally to the mast. Some use a flat copper strip rather than a thick conductor, also bonding the external sail track. There is a view that calls for secondary bonding of the stays and chainplates to the ground point but I do not support this practice. It creates a high resistance and high impedance secondary path down the stays and chainplates which can result in crystallization of the stainless steel and possible loss of the rig under any tension following a strike. One major spar manufacturer actually voids all warranty on masts if they are struck by lightning as the heating effects can alter the metallurgical properties. It is imperative that a single, low resistance grounding system be installed, so that these potentially dangerous alternative methods are not required.

(3) Terminations. The lightning conductor should be terminated at the hull, keel or an immersed ground plate with a minimum area of 0.2 m² (2 ft²).

In any boat the ground plane is considered to be seawater. It is important that strike energy is dissipated to ground with a minimal rise in ground potential through a low impedance grounding system. Steel and alloy boats use the mass of the hull as the ground. In many fiberglass and wooden sailing boats the lightning conductors are grounded to the keel bolts and therefore the keel acts as the ground plate. On most boats, the choice are the sintered bronze radio ground shoes such as from Dynaplate, Wonderbar or Seaground, and if using these, preferably select the largest ones in the 50–100 sq.ft range. It must be noted that NewMar clearly state that their radio ground shoes are not intended for lightning protection. As an alternative up to three smaller ground shoes can be configured in what is called a radial or crow's foot principle. This radial system lowers the overall impedance to allow energy to diverge as each conductor and ground shoe takes a share of current. In a strike, the water permeating the sintered bronze ground shoe can literally boil increasing the local resistance, so any increase in surface areas will help to reduce this effect. The voltage gradients around the shoe will also be lower. Some quality ground shoes use gold-based grease under the shoe fastening bolt heads to ensure a good low resistance connection. Do not use the radio RF ground plate as the lightning ground. The ABYC recommendations call for a minimum one square foot grounding plate of copper, copper alloy, stainless or aluminum on the outside of the hull. In reality few boat owners choose to do this citing drag and corrosion issues which are valid concerns.

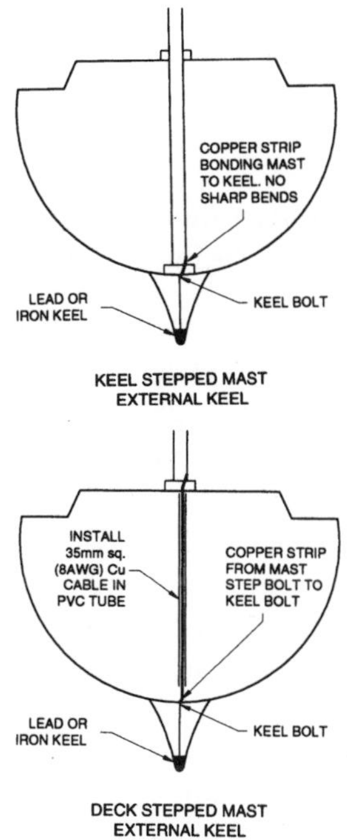

Figure 7-5 Mast Grounding Arrangements

All terminations and connections should be crimped and soldered joints should not be used.

Never use soldered joints alone, as they will overheat and melt during a strike causing further havoc. It is very difficult on large cables to ensure a good low resistance solder joint. After crimping soldering can be run in to enhance the joint; however this is not really necessary. Always crimp the connections and ensure that all bonded connections are clean and tight. All connections must be bolted to the ground point.

A bridge or link should be installed between ground plate bolts or at least two keel bolts to distribute strike current.

It is recommended that you bridge out the two bolts with a stainless steel link to spread the contact area and therefore the current carrying capacity. Links can also be drilled and used to bolt the ground cable connector, as many ground shoes have relatively small bolts designed for RF grounds only.

(4) System Bonding

All metallic items within 2 meters (6 ft) of the mast base or ground point should be bonded to the ground plate.

Some recommendations call for the bonding of rails, stanchions and all large metallic equipment such as stainless water tanks to the lightning ground. It is only necessary to bond internal metallic equipment within six feet of the down conductor and grounding point. The bonding should be made at the point closest to the main conductor. Ground plane potential equalization bonding between systems is designed to eliminate earth loops, differentials and reduce the level of potentially destructive transient currents, and sidestrike activity that can occur when potential differences exist between unbonded grounding systems.

The lightning protection system should not be bonded to the DC negative, radio ground, cathodic protection bonding system, seacocks or through-hull fittings.

There have been many documented incidents where bonding of the cathodic protection system, power supply negatives, grounds and RF grounds have resulted in the vessel sinking as seacocks and through hulls have been blown out, and all communications, electrical and electronics systems destroyed. In a lightning strike there will be a large induced electromagnetic pulse (EMP) into the electrical systems. If other systems are directly connected there will be a higher voltage and actual current flow and this is probably more destructive. On steel and alloy vessels, the hull is the ground plane and the hull acts as a Faraday Cage to minimize EMP effects. On fiberglass and wooden boats ABYC have recommendations for an internal equalization bus to bond systems and minimize sidestrike activity. If a boat has a corrosion protection bonding system; the bonding of equipment close to the down conductor is performed as described; and there is polarized electrical system, then effective systems grounding already exists to dissipate the rise in potential without the requirement for the creation of another one.

About temporary lightning protection systems

An innovative and portable device that incorporates all of the correct lightning protection elements is the strikeshield (www.strikeshield.com). A clamp is connected to the mast to a 1/0 or 2/0 AWG shielded and tinned copper cable. The cable is terminated with a specially designed dissipation electrode that is then dropped into the water. This is a good option in a marina or an anchorage. Some owners have installed gold-plated ground plates. By looking at the metal nobility table in the corrosion chapter you can see that a potentially corrosive situation can arise. If you require an emergency round, use a heavy-gauge copper cable, and clamp it to a stay to cover about half a meter. The other end should be clamped to a ground plate, and then dropped over the side. Do not use chains and anchors as grounds as they are very ineffective.

7.5 Surge and Transient Protection. It is difficult to adequately suppress or control the surges that arise during a lightning strike. Surge protection methods are as follows:

a. **Radio Antennas.** Aerials can draw a strike or have an induced current flow through the coaxial conductor to the radio. All antennas should have arresters fitted, although this is rarely done on boats. Antenna cables can be fitted with a two-way switch, one side to the radio, one to ground. You can buy remote and manual coax switches from NewMar. During a storm or if the vessel is left unattended, place the switch to ground position. Ideally an arrester (Hy-Gain or Dynapulse), or a spark gap device can also be used. Coaxial cable surge protectors (Dynadiverta or Polyphase Corp) can also be used. Coaxial cable surge protectors via RF feeders are used even in shielded cables and tri-ax cables, which will confine most current. Some induction can still occur due to magnetic and capacitive coupling.

b. **DC Power Supplies.** Power supplies should have double pole isolation on both positive and negative supplies. Additionally surge suppression units can be installed which will clamp any over-voltage condition to a safe value, typically around 40 volts. All equipment can have what is called a transient protection device installed across the input power supply connections. These are generally metal oxide varistors (MOV's), and are available from electronics suppliers.

c. **AC Power Supplies.** Efficient clamping and filtering at the power supply point requires surge diverters. The purpose is to limit residual voltages to a level within the immunity level range of the equipment. In 230VAC RMS systems damage can occur with just 700-volt peaks. Typical tolerances of battery chargers are under 800 volts. Some shunt devices can clamp the voltage at less than these voltages but they do not limit the fast wave front of the strike energy (dI/dt) before clamping action starts. In a lightning strike, the rate of current rise can exceed 10kA/μsec, and this can be greater in multiple strikes and re-strikes. Low pass filter technology primary shunt diverters will reduce the peak residual voltage and reduce rate of current and voltage rise reaching equipment. Surge reduction filters (SRF) will provide multi-stage surge attenuation by clamping and then filtering the transients on power input circuits, and these include MOV's. Look at www.yachtguard.com.

Corrosion

8.0 Corrosion. Corrosion can be defined as the chemical deterioration or reduction of a metal or metal alloy due to interaction with the environment. In the corrosion process metal atoms leave the metal to form compounds in the presence of water or gases. This is commonly called rusting. Corrosion is often improperly called electrolysis. Corrosion takes many forms; for boats and basic electrical systems it falls into two main categories: galvanic corrosion and electrolytic or stray current corrosion. For galvanic corrosion to happen, there has to be a potential difference between each of the metals. This principle was discovered in the eighteenth century by Luigi Galvani (which is where the name galvanic comes from). It involved experiments with the nerves and muscles of a frog; they contracted when hooked up to a bimetallic conductor. This developed into the first practical battery cell by Alessandro Volta (which is where the name volt comes from). Stray current corrosion is damage resulting from current flow outside the intended circuit. Both corrosion processes are a result of electric current flow between the two metals in an electrolyte. The result is corroded hulls, propellers, shafts, rudders, stocks and skin fittings.

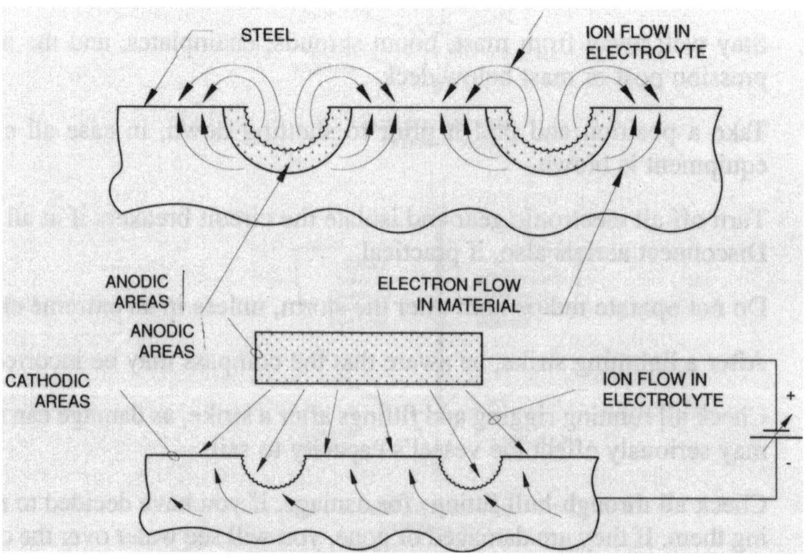

Figure 8-1 Galvanic Corrosion Process

8.1 Galvanic Protection. The British Admiralty introduced anodic protection back in 1824 on copper clad wooden men-of-war. Galvanic corrosion or electro-chemical corrosion is the process that occurs when galvanic cells or couples form between two pieces of metal with different electrochemical potential when they come into contact with each other. If the two metals have the same electrical charge or potential they will not create a cell and so no current will flow and they are called compatible. Ideally a vessel should be constructed so that most metallic items are compatible. If they are different, they must be either isolated or protected. Anodes are the normal protection method for achieving this. Anodes are also called sacrificial anodes because they are sacrificed instead of the hull or fittings. Because they are high on the nobility scale, they tend to corrode faster than other items such as mild steel, alloy, etc. The zinc anode generates an electric current and because the hull effectively has a higher potential, the anode allows current flow through it and bonded items, through the seawater, and back to the hull. The process corrodes the anode proportional to the level of current flow present, while preserving the hull and fittings. The rate of corrosion is affected by several factors. These include the anodic corrosion current level, the water temperature and the water salinity. A basic parameter is derived from Faradays Law, which is that a known current acting for a known time will cause a predictable weight loss of metal. For example, 1 amp applied for 1 year will cause a loss of 10kg (22lbs) of steel. The size of the exposed area of the cathodic metal relative to the anodic metal also will affect the corrosion rate. The corrosion rate varies between metals and the corrosion current within systems is typically rated within thousandths of an ampere, or milliamperes (mA). If galvanic corrosion is to occur there are four basic requirements that must be fulfilled: There has to be a positive or anodic area, called the anode. It will possess the lowest potential and will be the metal that will corrode. There has to be a negative or cathodic area, called the cathode. It will possess the highest potential.

(1) There has to be a path for the current to flow. This is called the electrolyte and this is the water.

(2) There has to be a circuit path for the current to flow and this is any interconnecting connection.

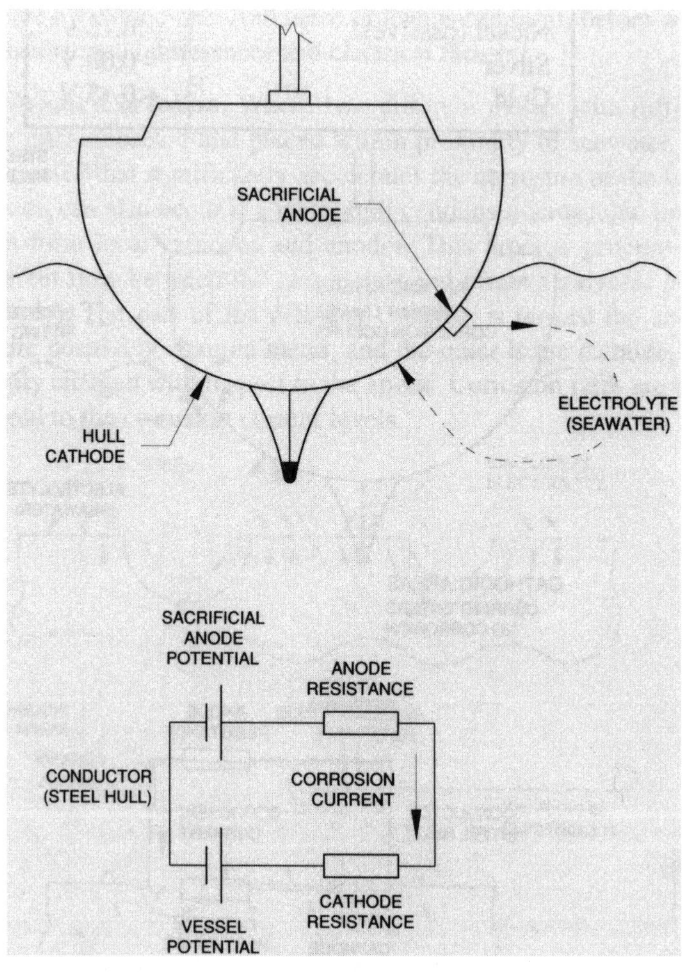

Figure 8-2 Galvanic Corrosion Protection

8.2 Metal Nobility. All metals can corrode, both ferrous and non-ferrous. Base metals such as steel and aluminum corrode more easily than the noble metals such as stainless steel and bronze. All metals can be classified according to molecular structure and these characteristics are listed in a metallic nobility table. The base metals at the top of scale conduct easily, while the noble metals at the bottom do not. The materials with the greatest negative value will tend to corrode faster than those of a lesser potential. The voltage difference between metals will drive current flow to accelerate corrosion of the anodic metal.

Table 8-1 Metal Nobility Table

Metal	Voltage
Magnesium and alloys	- 1.65 V
Zinc plating on steel	- 1.30 V
Zinc	- 1.10 V
Galvanized iron	- 1.05 V
Aluminum alloy castings	- 0.75 V
Mild steel	- 0.70 V
Cast iron	- 0.70 V
Lead	- 0.55 V
Manganese bronze	- 0.27 V
Copper, brass, and bronze	- 0.25 V
Monel	- 0.20 V
Stainless steel (passive)	- 0.20 V
Nickel (passive)	- 0.15 V
Silver	- 0.00 V
Gold	+ 0.15 V

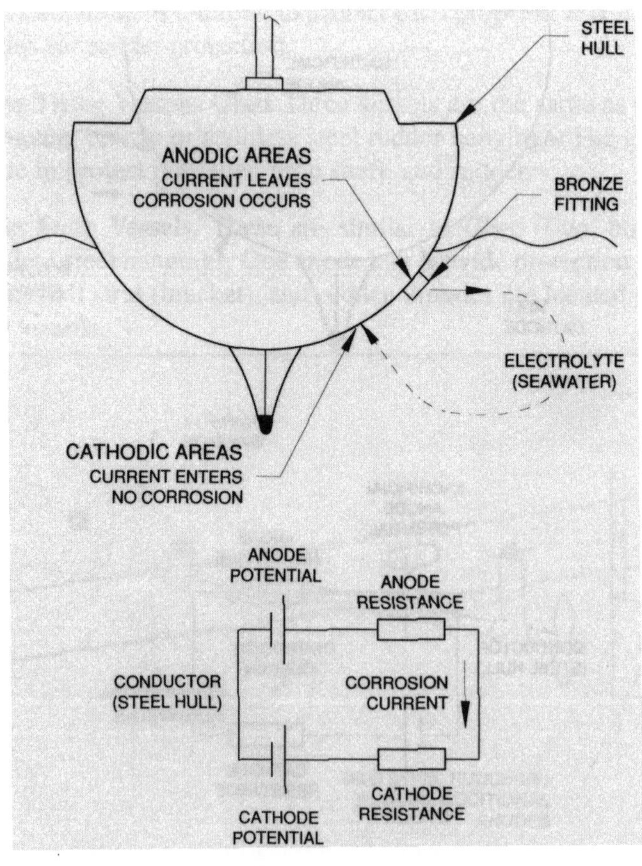

Figure 8-3 Vessel Galvanic Corrosion

8.3 Fiberglass and Wooden Boats. Vessels are generally categorized into some specific groups. The following are typical arrangements based on recommendations by leading corrosion specialists for fiberglass and wooden vessels.

a. **Type A Vessels.** Single screw boats with a short propeller shaft length in contact with seawater. They generally have wooden or fiberglass rudders. Normally only one anode is required for propeller and shaft protection. The main anode should be located on the main hull below the turn of the bilge at an equal distance between the gearbox and the inboard end of the stern tube.

b. **Type B Vessels.** Single or twin screw boats with long exposed propeller shafts supported by a shaft bracket and in contact with seawater. One anode is required for each propeller and shaft assembly. Separate anodes are required for mild steel rudders. Bronze or stainless steel rudders with bronze or stainless steel rudder stocks must be bonded to the same anode.

c. **Type C Vessels.** Single screw boats with long exposed propeller shafts supported by a shaft bracket and in contact with seawater. They have fiberglass rudders and bronze or stainless steel rudder stocks. Normally only one anode is required for propeller and shaft protection. If there are mild steel bilge keels they should have separate anodes affixed.

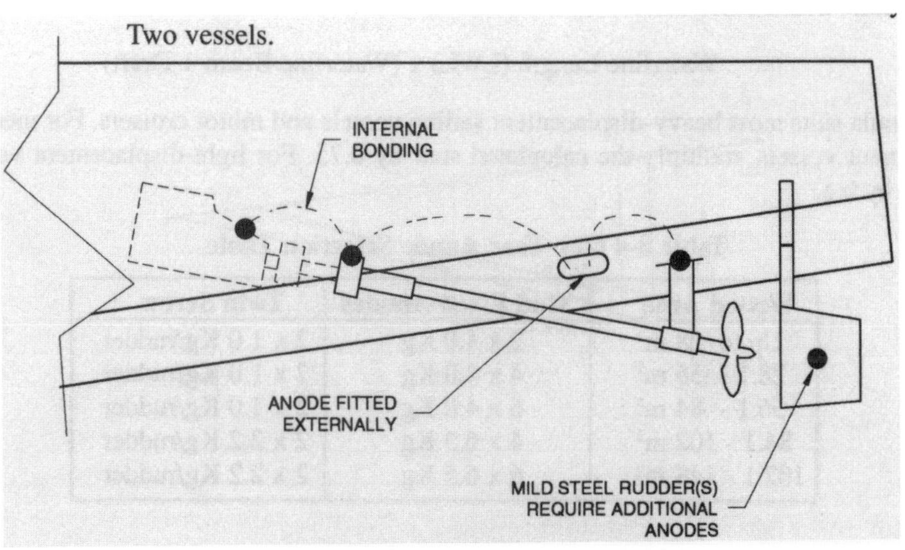

Figure 8-4 Class B Anode Arrangements

8.4 Anode Purity Standards. All cathodic protection systems should use anodes with approved purity standards. Zinc alloy anodes should conform to US Mil Spec MIL-18001J or Australia AS2239-1979. The average sailing vessel uses zinc sacrificial anodes for corrosion protection. They are called sacrificial because they are sacrificed instead of the item they are attached to, such as the hull, shaft or rudder. As they are high on the nobility scale, and have a higher electrical potential, they will corrode faster than other items such as mild steel. Pure zinc anodes are good conductors. If they do not have a standard quoted then exercise caution using them. The zinc anode generates an electric current and as the hull effectively has a higher potential the anode allows current flow through it and bonded items to the seawater and back to the hull. The process corrodes the anode proportional to the level of current flow present, while preserving the base metal such as the hull. Vessels on freshwater rivers and lakes follow the same criteria as boats in salt water; the difference is that anodes are made from magnesium or aluminum. Freshwater has a much greater insulation value than salt, so anodes such as magnesium and aluminum with a higher driving voltage than zinc anodes are required. When a boat moves into seawater or water of a higher salinity, anodes will become more active and should be inspected after only just 14 days.

8.5 Anode Systems. It is essential for the anodes to be of the correct size, in the correct location, and of the correct number for the area being protected. It is quite possible to overprotect the hull and fittings, excessive numbers or larger anodes do not equal improved protection. If your vessel is in warm highly saline waters, you must make more frequent inspections of zinc anodes. The following illustrates typical arrangements recommended by corrosion specialists M.G. Duff Marine for steel, aluminum, fiberglass, and wooden vessels. Note that metal and fiberglass hulls are treated separately. Anode position is not critical, but they must be able to see the parts to be protected. Anode fixing must be above the bilge line internally, and there must be a minimal internal bonding cable run length. These anode sizes are based on propeller sizes and are approximate only.

Table 8-2 Anode Mass Table (Salt and Fresh Water)

Prop Size	Type A	Type B	Type C
10" SW Zinc	1.1 kg	2 x 1 kg	1.1 kg
FW Magnesium	0.3 kg	2 x 0.3 kg	0.3 kg
14" SW Zinc	1.1 kg	2 x 1 kg	1.1 kg
FW Magnesium	0.3 kg	2 x 0.3 kg	0.3 kg
19" SW Zinc	2.2 kg	2 x 1 kg	2.2 kg
FW Magnesium	0.4 kg	2 x 0.4 kg	0.4 kg
21" SW Zinc	2.2 kg	2 x 2.2 kg	2.2 kg
FW Magnesium	0.7 kg	2 x 0.7 kg	0.7 kg
26" SW Zinc	2.2 kg	2 x 2.2 kg	2.2 kg
FW Magnesium	0.7 kg	2 x 1.0 kg	0.7 kg
30" SW Zinc	2.2 kg	2 x 2.2 kg	2.2 kg
FW Magnesium	0.7 kg	1 kg	1 kg
36" SW Zinc	4.5 kg	2 x 4.5 kg	4.5 kg
FW Magnesium	1 kg	1 kg	1 kg
40" SW Zinc	4.5 kg	2 x 4.5 kg	4.5 kg
FW Magnesium	1 kg	2 x 1 kg	1 kg
48" SW Zinc	4.5 kg	2 x 4.5 kg	4.5 kg

8.6 Anode Number Calculations. Calculations are normally based on the wetted surface area of the hull. The main vessel dimensions used are the waterline length, waterline beam, and the mean loaded draft. The area is calculated using the formula:

Waterline Length (LWL) x (Waterline Beam + Draft)

This formula suits most heavy-displacement sailing vessels and motor cruisers. For medium-displacement vessels, multiply the calculated sum by 0.75. For light-displacement vessels, multiply by 0.5.

Table 8-3 One-Year Anode Selection

Wetted Area	Hull Anodes	Rudders
Up to 28 m^2 (300ft^2) SW	2 x 4.0 kg Zinc	2 x 1.0 kg Zinc
Up to 28 m^2 (300ft^2) FW	4 x 1.5 kg Mag	2 x 0.3 kg Mag
28.1 - 56 m^2 (>600ft^2) SW	4 x 3.5 Kg Zinc	2 x 1.0 kg Zinc
28.1 - 56 m^2 (>600ft^2) FW	4 x 3.5 kg Mag	2 x 0.3 kg Mag
56.1 - 84 m^2 (>900ft^2) SW	4 x 4.0 Kg Zinc	2 x 1.0 kg Zinc
56.1 - 84 m^2 (>900ft^2) FW	4 x 3.5 kg Mag	2 x 0.3 kg Mag
84.1 - 102 m^2 (>1100ft^2) SW	4 x 6.5 Kg Zinc	2 x 2.2 kg Zinc
84.1 - 102 m^2 (>1100ft^2) FW	6 x 4.5 kg Mag	2 x 0.7 kg Mag
102.1 - 148 m^2 (>1600ft^2) SW	6 x 6.5 Kg Zinc	2 x 2.2 kg Zinc

Rudder, Skeg, and Bilge Keel Anodes. Anodes for mild steel rudders, skegs, and bilge keels are normally installed directly to steelwork. In most cases, the best solution is to bolt the anodes back-to-back, or they may be welded on if required.

Table 8-4 Rudder, Skeg and Keel Anodes Table

Protection Area	Anode Size
Up to 10 Sq ft Steelwork	2 x 1.35 Kg Strip Anodes
Up to 30 Sq ft Steelwork	2 x 0.90 Kg Anodes
Up to 70 Sq ft Steelwork	2 x 2.22 Kg Anodes

8.7 Anode Bonding. There are a number of factors to consider when fitting and connecting anodes.

Anodes should be fixed in view of the parts and areas they protect, and be bonded to the parts under protection.

Anode positioning is not critical but they must be able to see the parts to be protected. When a zinc anode is 50–75% wasted it must be replaced. White or green halos around zincs or metals indicate stray current is affecting them. Bright zincs indicate excess current flow. A small amount of current also causes paint reactions. Rapid zinc wastage and degree of paint reaction indicate problems that are more serious. Where a boat has moved into freshwater and back to salt the anode will become encrusted with a white crust. This will stop it from functioning and it must be cleaned off. Do not bond ferrous and non-ferrous metals to the same anode; otherwise you effectively create a cell or battery. There are few circumstances where this might occur.

No grounding connections should be made to any skin fitting or plumbing system.

I do not support the practice of connecting every metal item, including through-hull (skin) fittings, and also the method of using stainless wire and hose clamps. It is only advisable to bond the main raw sea water inlet fitting. Any other fittings that are isolated and are connected with rubber or PVC hoses need not be bonded. The current flow in a bonding circuit is very small; any resistance introduced into the circuit from bad connections and cable resistances creates a difference in potential that will cancel any protective measures and may actually create problems. Do not bond the lightning ground system or the down conductor to the anode bonding system. This is at variance to ABYC; there have been several well-documented cases of skin fittings being blown out in a lightning strike and the vessel subsequently sinking.

Equipotential bonding conductors should be green in color or clearly indicate the function so as to avoid confusion with any AC ground.

Many recommendations nominate green to identify bonding cables. Caution should be used, as this is an AC safety ground color in the US. Identifying each termination with the sleeved letters EB (Equipotential Bond) will reduce chances for accidental disconnection. This possibility is real as such bonds also go to the common ground point. Disconnection of an AC ground can cause shocks so this is for clear safety reasons. While some standards allow bare conductors, this can lead to early conductor deterioration and is therefore not recommended.

All cathodic bonding system cables should be installed clear of bilges or other wet areas.

Any bonding cables should be installed well above the bilge line, or any other area that may be subject to water. They should only interconnect the items to be protected, and not indiscriminately bond all items. All interconnections must be of at least a 4.0 mm^2 tinned copper conductor and be bolted to the main bonding connection. Anode fixing studs should be connected to bonded parts by the shortest practical route to minimize resistance. It is critical that bonding be resistance free, therefore a heavy-gauge conductor is necessary. Use 11 AWG (4 mm^2) cable as a minimum.

The total resistance of any cathodic bonding circuit should not exceed 0.02 ohms.

The purpose of bonding is to equalize the electric potential of the underwater metals being connected. It is not to dissipate stray currents on the 12-volt system and spread the surface areas. It is critical that bonding cables be resistance free and therefore the use of a heavy gauge conductor is necessary. When the vessel is hauled out, use a multimeter set on the 1 ohm range and check the resistance between anode and propeller, the maximum reading must be 0.02 ohm. The current flow in a bonding circuit is very small and any resistance introduced into the circuit from bad connections and cable resistances creates a difference in potential that will cancel any protective measures and may actually create problems. Many recommendations call for a bonding loop. In fact, it is better to connect bonded items in a radial arrangement back to the anode bonding bolt to minimize resistance. If one bonding wire is accidentally broken, the majority of the bonding network will not be lost. It is also important to regularly check that the main bonding points are secure and clean.

8.8 Shaft Collar Anodes. Most sailboats have shaft anodes installed. When installing collar anodes to propeller shafts, make sure that the shaft is clean and not covered with antifouling. I have frequently seen this done around boat yards. The collars must be mounted as close as possible to the shaft strut (bracket), typically a clearance of 4–10 mm. Do not put bottom paint on the anode!

8.9 Anodes for Fiberglass and Wooden Vessels. There are a few facts to remember with fiberglass or wooden vessels. Anodes on fiberglass and wooden hulls must have the internal bonding system connected to them. This is a common omission. An anode is working only when some corrosion is visible on it. I have frequently seen anodes mounted on a hull and not connected, and heard those same vessel owners proudly proclaim that there was no corrosion problem.

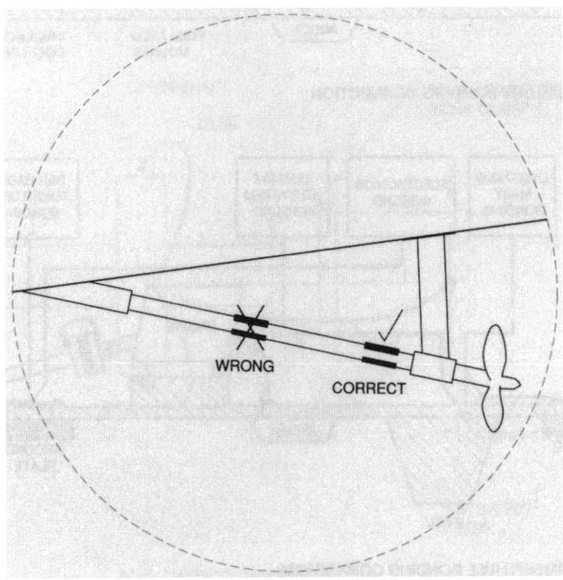

Figure 8-5 Shaft Anode Installation

a. **Propeller Shaft Bonding.** The usual method for bonding propeller shafts, both commercially and in small vessels, is to install a brush system. M.G. Duff's system is called the Electro Eliminator. Essentially, it is a brush system connected to the cathodic bonding system. If such a system is used, the shaft must be kept clean and free of oil, grease, and water. It is often better to bridge the coupling to the engine block and use a collar anode or separate anode bonded directly to the engine block.

b. **Seal Anode Bolt Holes.** Always seal the wood around the anode bolt holes as this can prevent wood electrolysis if an overprotection situation exists.

c. **Shaft Coupling Bonding.** Many engine installations incorporate flexible couplings to the propeller shaft. The coupling must be electrically bridged to ensure proper continuity of the system where the engine is not maintained electrically isolated above the bonding system.

8.10 Anodes for Steel Vessels. On steel boats the anodes are either welded to the hull or properly on brackets. Many also have studs welded to the hull and the anode fastened onto the studs. It is important to ensure that the studs and connections are good and low resistance. It is common to see a stainless nut on a mild steel stud, and the mild steel threads corrode. Use all 316 grade stainless steel. At each slipping take off the anode and ensure threads and contact surfaces are clean. When the vessel is hauled, use a multimeter set on the 1-ohm range and check the resistance between anode and propeller. The maximum reading should be 0.2 ohm. The reading between the anode and the hull should be zero. On ferro-cement yachts, forget the cement component of ferro-cement vessels and treat them as steel vessels. Make sure bronze through-hull fittings are isolated from reinforcing mesh. Some vessels also use a zinc anode (guppy) on a wire over the side while in a marina or on a mooring. This must be properly grounded to the hull and have a low resistance copper cable connected to the anode. This method is quite ineffective and may cause corrosion problems as it is hard to ground properly:

a. **Keel Anodes.** If a steel keel is fitted, an anode may be installed on each side. Follow the calculated wetted surface guidelines so that overprotection does not occur.

b. **Negative Cables.** Install isolated negative return starter motors, alternators, and monitoring gauge sensor units. These ensure that no electrical items are connected to the block. Use double-pole, engine starter isolators. If you do not have a fully isolated engine, make sure that the engine is isolated above the hull, with insulated coupling and engine mounts.

c. **AC Ground.** The AC ground can be bonded directly to the hull. This normally has no voltage flowing in it, except under fault conditions. It is essential that the main hull be part of the AC ground because in a fault condition the hull could become partially live, up to rated voltage. This will not cause premature corrosion of anodes.

d. **Isolation Transformers.** Install an isolation transformer on the shore power supply. This is probably the single greatest protective measure for marina based boats.

e. **Vessel Interaction.** Avoid mooring adjacent to any copper sheathed or aluminum vessels. Do not use steel cables to tie up to shore.

f. **Paint Systems.** Although a vessel's hull is protected by an extensive paint program, isolating metal from the seawater, it is often incorrectly assumed that corrosion cannot occur and cathodic protection is unnecessary. But air holes, or small areas of paint imperfection, occur along weld seams, and paint coatings are damaged due to abrasions from chains, piers, tenders, etc. Cuprous oxides used in antifouling paints can convert to copper sulfide and create a galvanic cell. I have encountered an increasing number of corrosion problems with steel vessels using copper-based antifouling paint. These antifoulings rapidly degrade the anodes' performance, and in most cases I have found corrosion and pitting where the paintwork has been chipped off the hull. Shaft struts (brackets) and propellers have suffered damage also. Wherever possible, use copper-free antifoulants. When in doubt, you should always consult a corrosion specialist. Use the technical services of the paint manufacturer for assistance.

8.11 Anodes for Aluminum Alloy Vessels. Aluminum hulls have different requirements from steel vessels, and it is essential that they be correctly protected. Unlike steel vessels, an overprotected aluminum hull doesn't simply lose paint, it is eaten away by a caustic attack. The recommendations for steel hulls are also valid for the electrical systems on an aluminum vessel.

a. **Material Compatibility.** Insulate or use compatible through-hull fittings. In fact, insulate any equipment made of metals above aluminum on the metal nobility scale (Table 8-1). Avoid bronze fittings if at all possible. I have come across some who are using Lloyd's approved aluminum and plastic fittings, which solves many problems.

b. **Vessel Interaction.** Avoid mooring next to steel or copper sheathed vessels for extended periods. The interaction can be very severe with aluminum and corrode a hull rapidly.

8.12 Corrosion Leakage Monitoring. Leakage to the hull on steel and alloy vessels can be monitored using suitable systems. Verifying the condition of a corrosion protection system requires the use of testing and the measurement of the hull potential. This involves then immersion of a reference electrode called a half cell. This is made from silver/silver chloride which is a silver wire coated with silver chloride. The reference cell is immersed approximately 6 inches into the water away from the hull. The reference electrode is connected to a digital multimeter positive terminal. The negative terminal is connected to the boat ground which is usually the battery negative ground point. The multimeter is then set to the 2 volt DC scale or simple DC volts on most as they are auto-ranging. The value displayed is the actual hull potential. The readings must take account of water temperature and salinity.

8.13 Galvanic Isolators. These devices are designed to provide galvanic isolation of the AC shore ground from a DC bonding system when they are connected. There is a relatively common situation in crowded marinas where small stray DC currents may flow from the AC ground on one boat to the DC bonding system on another and may increase anode corrosion rates. When a boat is berthed at the marina connected to shore power, it effectively has the same AC ground plane as all other connected boats. In boats without inverters or generators, the main ground is effectively the shore ground, as no on-board grounding is usually installed. In effect, the vessel is simply an appliance on the end of an extension cable, much like a boat trailer. In this mode, any DC currents imposed on the AC system will not affect the boat. When the boat has an AC system that is grounded due to an inverter or generator installed on board, the AC ground will be then connected to the DC ground point. This provides a path for DC stray leakage currents to the DC power system, the immersed parts of the boat such as the propeller, and the cathodic protection system. Depending on circumstances a significant current can flow. In addition it also interconnects all boat DC systems through the AC safety ground, effectively creating a large battery cell. The boats with the least noble metals underwater, such as a powerboat with aluminum stern drives, suffer the worst corrosion. If your isolator is possibly faulty you can perform a simple check. The 2 major failure modes are short circuit diodes or they are open circuited. If your digital multimeter is able to show either positive or negative voltages then use the following method. Place the meter on the DC voltage range and check across the input or shore power side of the isolator to the output or boat side. If the meter reads zero the diodes are probably short circuited. If the reading indicates more than 1.0–1.5 volts the diodes are possibly open circuited

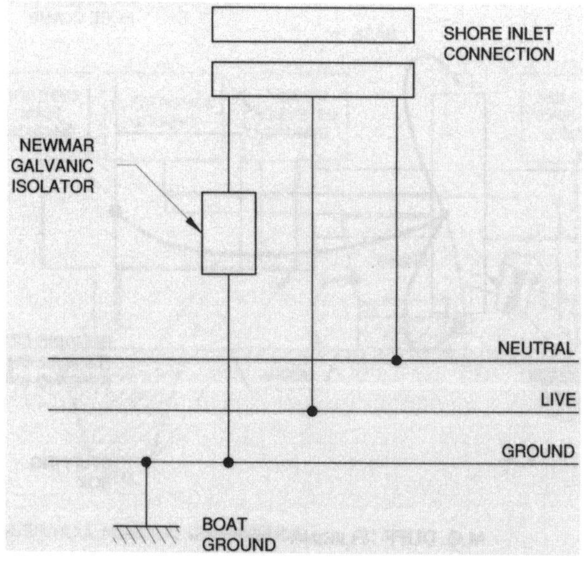

Figure 8-6 Galvanic Isolator System

8.14 Galvanic Isolator Selection and Installation. A galvanic isolator, or zinc saver as it is also referred to, has an electrical circuit that incorporates diodes which are configured to block DC current flow. It also incorporates capacitors which allow the AC to bypass the diodes. The AC flows from the boat ashore in a fault condition and must be able to pass very low AC leakage currents. You must install the correctly rated isolator. A 50-amp supply requires a 50-amp rated isolator, and a 30-amp supply requires a 30-amp rated isolator. ABYC approved units must be able to carry 135% of rated current. In a short circuit condition, the unit will suffer a very high current several times the rated current for a short period until fuse or breaker protection systems operate. Where a boat has two shore power inlets each must have a separate isolator installed. If an under-rated isolator is used, it will possibly burn out under full fault conditions. Isolators must be installed close to the shore power inlet socket although it is most common near the switchboard. They also must be installed in a well ventilated location as they can become very hot when in a fault current conducting mode. The American Boat and Yacht Council (ABYC) recommends that no part of the AC grounding system should bypass the galvanic isolator. This would mean it would not function and allow leakage currents to come aboard. As an isolator is installed within the AC safety ground, monitoring the integrity is of critical importance.

8.15 Galvanic Isolator Testing. Testing is to ensure that it can safely conduct AC fault currents at all times and that the device has not failed. In accordance with ABYC recommendations many isolators have a monitor with an LED indicator that displays the functioning of diodes and capacitors, along with AC safety ground continuity, reverse polarity and more. It must be clearly stated that no shore electrical standards allow any device to be inserted within any grounding or earthing conductor. It is not allowed ashore any place in the world that I know of, so any person having an isolator installed may be in breach of local wiring rules and regulations and so should check this out. If the isolator is faulty or not working the boat may be left without an AC ground if an onboard ground is not used and generally manufacturers have tried to eliminate this possibility:

(1) Disconnect the shore power supply before testing.

(2) Disconnect the incoming shorepower lead, and select the diode test function on your multimeter.

(3) Place a test probe on each terminal of the isolator.

(4) Verify that the reading slowly rises to around 0.9 volt, as the internal capacitor commences current conduction. If the reading rises instantly to 0.9 the capacitor is faulty or the unit does not have one fitted. If the reading is 0.45 volt, one of the diodes is short circuited. If the reading is zero volt then both diodes are short circuited. If the reading is higher than 0.9 volt the diodes are probably open circuited and you should stop test before voltage reaches 2 volts to avoid damage to the capacitor.

(5) Short the two isolator wires together to discharge the capacitor, and repeat the check with test probes reversed.

8.16 Corrosion System Maintenance. Perform yearly maintenance as follows.

a. **In Water Examinations.** Perform the following basic examinations:

(1) **Main Anodes.** Do an underwater check of anodes after six months and check for increased corrosion rates. If vessel has moved into warmer or more saline conditions rates increase. Rapid zinc loss and shiny zincs indicate stray current problem.

(2) **Shaft Anodes.** Check that shaft anode is still on the shaft. Check the anode corrosion rates.

b. **Haul-out Examinations.** Perform the following examinations:

(1) **Anode Replacement.** Replace anodes if more than 75% reduced and check connections.

(2) **Shaft Anodes.** Replace anode if necessary. Check mating surface of shaft anode. Check that it is correctly positioned.

(3) **Bonding Connections.** Inspect the bonding system interconnections to see that they are sound and clean. Remove connections and clean so that the contact resistance is zero.

(4) **Check Bonding System Resistances.** Check bonding resistances between the anodes, the propeller and the hull.

Table 8-5 Corrosion System Troubleshooting

Symptom	Probable Faults
Anode corroded 80%	Replace anode
Rapid anode corrosion	Hull electrical leakage Increased water salinity Increased water temperature Degraded bonding system Moored adjacent to vessel Marina electrical problems No isolation transformer on boat
Paint stripping off keel & hull	Hull over protected (too many anodes) Severe hull leakage (electrical problem)
Paint stripping around studs	Anode stud connection defective
No anode corrosion	Anode hull connections defective Bonding wires broken Impure zinc anode
Propeller & shaft pitted	Inadequate protection Degraded bonding system Shaft anode missing Shaft anode fitted over antifouled shaft Cavitation corrosion

8.17 Electrolytic (Stray Current) Corrosion. Electrolytic corrosion has different principles from galvanic corrosion, and they should not be confused. Protective measures for galvanic corrosion do not protect against electrolytic corrosion. Stray current corrosion will however dramatically increase corrosion rates on under protected hulls and anodes, degrading the galvanic protective system. If the faults are undiagnosed, anodes will rapidly degrade followed by stripping of paint and antifouling. This will often require a complete repainting of the hull from the metal primer upwards. Automotive battery chargers are a common cause of corrosion in boats, particularly small boats without shore systems. Auto chargers often provide no isolation between the AC and DC windings and can energize the negative terminal, which also energizes the boat's grounding system. Portable auto chargers should not be used on boats, and are a frequent cause of stern drive damage.

a. **Electrolytic Corrosion Sources.** Electrolytic corrosion is caused by an external DC current source. This may be from electrical faults on the boat or from shore sources:

(1) **Leakage Currents.** Leakage currents are not nearly as common as many magazine articles suggest. They are mainly caused by leakages across condensation or conductive salt deposits at DC connections or junction boxes, or tracking from main starter motor cables. 24- and 48-volt systems have higher risks then 12-volt systems given the higher potential differences. In some cases, they may also be caused by damaged insulation. In a properly installed electrical system, there are relatively few opportunities for the situation to arise. The most common area is tracking across from engine starter motor and solenoid connections which are often never cleaned of grease, oil and moisture.

(2) **Ground Faults.** Ground faults on AC and DC conductors occur where the cable insulation has been damaged and contact is made with the hull or connected metalwork. In many cases, the fault may not be sufficient to operate protective devices and remain unnoticed for a considerable and damaging period. The most common areas causing faults are where cables enter grounded stainless steel stanchions, alloy masts, engine charging and starter cables. In any area where a cable can contact grounded metal leakage or fault currents can flow. Install a leakage test lamp unit such as that from Mastervolt that handles both AC and DC. This allows the hull to be monitored continuously and any problems found and rectified promptly. It is the standard commercial ship practice, with daily checks and all leakages and faults corrected promptly.

(3) **Shore Power.** Generally AC does not cause corrosion. AC must be rectified to DC to cause corrosion. I have had questions regarding rectification through various fittings on a boat. It is possible although relatively rare that some materials may form such a device similar to a semiconductor diode causing some partial half-wave rectification. Most blame is attributed to faulty AC system wiring but in reality I have encountered very few boats with AC ground faults that can cause this situation. The electrolytic corrosion problem is principally due to interaction between shore and the vessel when the two grounds develop an unequal potential and a DC component flows to the vessel. Variations can exist within potentials of grounding connections at various shore power supply locations within a marina complex.

(4) **Water Current Gradients.** DC potential gradients may exist in the water around a vessel. This will contribute to corrosion, and is caused by variations in salinity and temperature. If paint is chipped off under the bow, a circuit may be created with anodes or hull fittings in another part of the vessel with an area of water of differing potential. This can occur in small marinas with reduced tidal flows to flush out water heated during the day.

b. Corrective Measures. Install a DC leakage test unit so the hull can be monitored continuously, with any problems identified and rectified promptly. This is standard practice on commercial ships.

(1) **DC Leakage Currents.** All connections, and ideally there will be none in wet areas, should be in proper water-resistant junction boxes. They should be placed in dry locations away from any metalwork liable to conduct any leakages to the hull. Install a monitor on steel and alloy vessels.

(2) **DC Ground Faults.** Ensure that all cables are double insulated. Check that all transits through metal bulkheads or stanchions have additional mechanical protection or grommets to prevent grounding.

(3) **AC Shore Power.** Steel and alloy vessels should install an isolation transformer on the shore supply mains. This will provide galvanic separation between the shore and vessel power systems, and the ground systems. The isolation transformer can also be used on fiberglass and wooden boats as well as steel and alloy ones where problems exist. A separate vessel ground is used, connected in a grounded neutral configuration. Another measure is to install a galvanic isolator to prevent leakages of DC stray currents. See the section on galvanic isolators. A polarity indicator will indicate transposed neutrals and grounds. On some switchboards a switch is installed to change over to correct polarity. Do not connect shore power until you have corrected the problem.

8.18 Steel/Alloy Hull Leakage Inspections. It is always difficult to maintain a hull above electrical ground. Moisture and oil residues mixed with salt lower the isolation level. It is important to regularly examine isolation values to ensure that isolation is maintained.

a. **Passive Insulation Test.** This test simply measures the level of resistance between the hull and both positive and negative. A multimeter set on the ohms scale is required. Perform the test as follows:

(1) Turn main power switch off.

(2) Turn on all switches and circuit breakers to ensure that all electrical circuits are at equal potential or are connected in one grid.

(3) Connect the positive meter lead to the positive conductor, and the negative to the hull. Observe and record the reading.

(4) Connect the positive meter lead to the negative conductor, and the negative to the hull. Observe and record the reading.

b. **Passive Test Results.** The test results can be interpreted as follows:

(1) 10k ohms or above indicates that isolation above the hull is acceptable.

(2) A reading in the range of 1k ohm to 10k ohms indicates that there is leakage, and that isolation is degraded. While not directly shorted to hull, leakage can be through moisture or a similar cause. With meter connected, systematically switch off each circuit to localize the fault area and rectify the problem. A common area is the starter motor connections.

(3) A reading less than 1k ohm indicates a serious leakage problem which must be promptly rectified or serious hull damage can result.

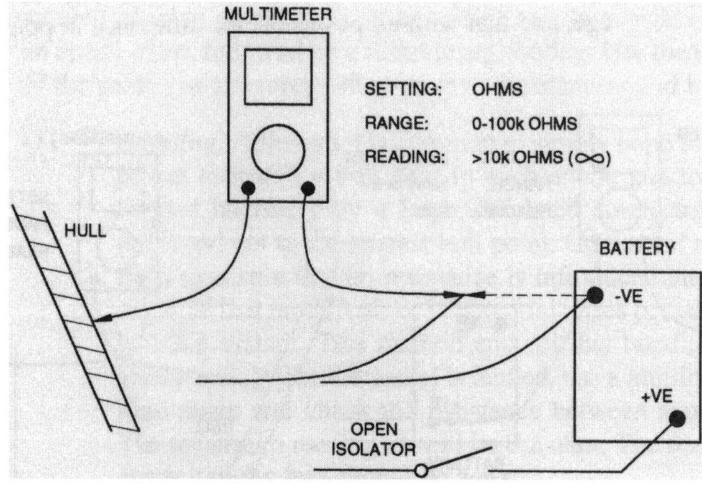

Figure 8-7 Passive Testing

c. **Voltage Insulation Test.** While a passive meter test can show that all is satisfactory, the voltage of a system in use can break down resistances and cause leakage. To properly test the electrical isolation, a voltage test should be performed. With 220/115-volt power systems, this test must be performed using a 500-volt insulation tester. All results must exceed 1 megOhm. This is not recommended for low voltage installations as the insulation values of cables are not rated this high. A low voltage DC tester set at 100-volt DC should be used. Another easier test is as follows:

(1) Turn on all electrical circuits so that all are "alive."

(2) With a digital multimeter set on DC volts, place the positive probe on the supply negative. Place the negative probe on the hull.

(3) There should be no voltage at all. If there is a small voltage, a leakage may exist on the negative.

(4) With a digital multimeter set on the DC volts, place the negative probe on the supply negative. Place the positive probe on the hull.

(5) There should be no voltage at all. If there is a small voltage, a leakage may exist on the positive.

(6) Systematically turn off electrical circuits to verify that there is a leakage, and that with all power off the difference in potential is zero.

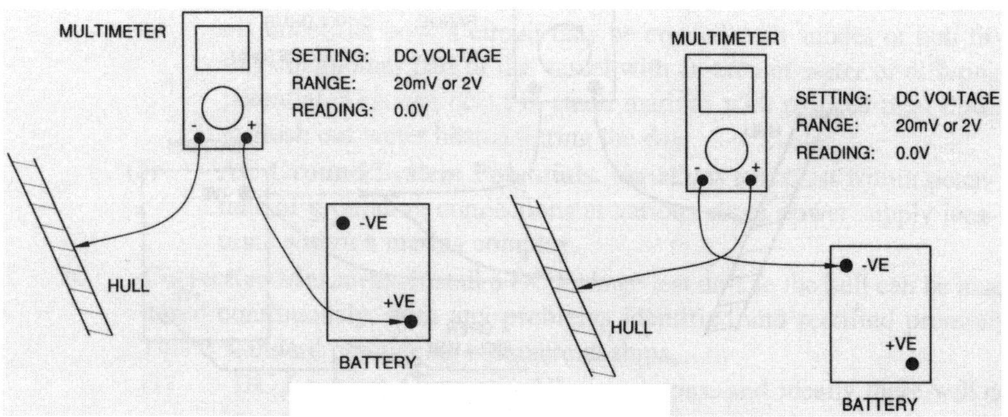

Figure 8-8 Insulation Testing

LIGHTING SYSTEMS

9.0 Lighting Systems. Boat lighting systems are often frustrating, as the selection of suitable lights is always a problem. Aesthetic considerations are of obvious importance, but there are other more important technical factors to consider.

a. **Illumination Area.** The size of the area to be illuminated is one important consideration. Where specific areas are to be illuminated, the factors to consider are as follows:

   (1) **Spot Lighting.** Spot lighting in areas such as bunk reading lights and spreader lights require special considerations. Projected light applications require a reflector or a special lamp. Factors such as beam power and beam angle are important.

   (2) **Large Area Illumination.** When lighting deck areas or the saloon, consider beam angles and light output power.

b. **Illumination Level.** The level of light reaching the working areas on deck or the overall light level below must be sufficient to perform tasks safely. A number of factors must be considered:

   (1) **Background Lighting.** This is generally low power lighting and uses interior surfaces and upholstery to reflect light for unobtrusive and shadowless illumination.

   (2) **Low Level Lighting.** This is localized illumination that does not require levels sufficient to perform work. Typical are night lighting, courtesy lights, and general saloon lighting.

   (3) **High Level Lighting.** This lighting is used in any application where safety or ease of work is important. It includes deck spotlights, hand-held spots, saloon lights, bunk lights, engine space lights and targa/transom lights, to name a few. Ideally, such lights should give shadowless illumination without excessive glare.

c. **Light Colors.** The color-rendering properties of a light source play a significant part in effective lighting. Using lights with the right color-rendering properties can significantly alter the apparent richness of woods, for example.

   (1) **Warm or Soft Colors.** Fluorescent tubes are generally warm soft. The newer, low energy lights tend to have a softer light that strikes a balance between good illumination levels and good color rendering.

   (2) **Cold or Hard Colors.** Halogen lamps and some fluorescent tubes generally have a cold and intense white light.

d. **Power Consumption.** Electrical power consumption is the very first factor to consider. Compare the main light types and make your decision based on the most satisfactory light for a given power consumption. To date, low energy tube lights have the best light-per-amp ratio and I have had great success with them.

9.1 Internal Lights. Lighting systems for cabins may consist of a number of light types. Different lights may be used for different functions, and one of the main criteria is gaining maximum light output for a given power consumption. There are four main types available. Before deciding on lights for below, consider light reflectivity.

a. **Reflectivity.** The level of brightness and the contrast with background must be considered. In a teak-lined cabin, reflected light will be minimal, while a cabin with painted surfaces or light wood will increase overall illumination levels. I have seen some beautiful wood-lined cabins with large numbers of lights fitted in the headliner and additional corner spots. Yet even with many lights on, they are still gloomy with low light levels. Interior schemes that are efficient mean fewer lights with less cable, and much lower power consumption for a given light level. Reflectivity is usually expressed as a percentage. The following range of interiors are typical:

Woods

- Maple and birch: 60%
- Light oak: 40%
- Walnut and teak: 15-20%

Painted Surfaces

- White and light cream: 70-80%
- Pale yellow: 55-65%
- Sky blue/pale grey: 40-45%
- Beige: 25-35%

b. **Fluorescent Lights.** Fluorescent lights are one of the most common lights, but they do have drawbacks that must be considered. DC tubes have a built-in inverter that raises the voltage to a higher AC value. Their elongated shape provides a good lumen/watt ratio with relatively low power consumption, 80% less than incandescent lights for the same light output. Typical output is 65–90 lumens. They also withstand vibration and shock well, and their working life is 5–8 times that of incandescent lights. Components are as follows:

(1) **Inverter.** The inverter in low voltage DC fittings is generally the main cause of failure. In most cheaper light fittings, the quality of the electronics is poor. They also fail in relatively small overvoltage conditions such as when charging voltages rise to 14 volts. Always install fluorescent lights with a voltage input up to 15 volts.

(2) **Tube.** The fluorescent tubes for household use function quite satisfactorily with good quality inverters. If the electronics are of poor quality, the tubes will show blackening in a short period. Tube output varies with temperature. Peak output is normally at 25°C. If hotter or colder, output is reduced.

(3) **Radio Frequency Interference.** Fluorescent lights have a notorious reputation for radio frequency interference. This is due to the quality of the inverter electronics. High quality inverters, such as those from Aquasignal, are suitably suppressed to international standards. Additionally, fluorescent units with high quality inverters are far more reliable.

c. **Incandescent Lights.** Incandescent lights are the oldest and most common light types. The following factors should be considered:

(1) **Power Consumption.** When switched on, power consumption can be 15 times normal (hot) power consumption. The basis of the incandescent lamp is the heating of a filament, therefore much of the energy is dissipated as heat.

(2) **Life Expectancy.** Incandescents are power hungry for the available light output, are subject to damage by vibration and overvoltage, and suffer rapid filament degradation.

(3) **Voltage Limitations.** Overvoltage conditions significantly reduce incandescent lamp life expectancies. Operating at lower voltages extends service life, but seriously reduces light output. For every 5% voltage drop, light reduces by 20%. Many of you are familiar with that yellow glow as the battery voltage decreases. The secret to operating incandescents, especially navigation lamps, is to minimize voltage drop.

d. **Halogen/Xelogen/Xenon Lighting.** Halogen lights were the most common in many vessels due to its higher light output, typically around 20 lumens. They have been supplemented and in some cases replaced by xenon and xelogen lamps. The halogen lamp base is designated as G4 and Xelogen G5.

(1) **Life Expectancy.** Halogen lights belong to the incandescent light category, and are designed for use in commercial installations on a stable 12/24 volts AC power source. When used in DC installations their life expectancy is significantly reduced, and the higher voltages generated during battery charging also reduce life. Xelogen lamps have a service 10 times that of a halogen lamp, and should last up to 20,000 hours. They have a lower operating temperature and while the glass can be handled, it is still better not to. The xelogen lamp is also dimmable.

(2) **Voltage Limitations.** Vibration resistance is relatively poor. Resistance to over-voltage situations is poor. Normally a halogen lamp is operated in commercial applications with a very stable 12-volt AC supply, with maximum life being at around 11.8 volts. Operating on DC, and at charging voltages up to 14.5 volts, life can be seriously reduced.

(3) **Installation.** Halogen lamps also degrade with salt air interaction with the pure silicon glass. Under no circumstances must the glass be handled as salts and impurities off the fingers will degrade the silicon glass and shorten life. Allowances must be made with halogen lights due to the high temperatures generated that can reach 700°C in normal operation. Good ventilation is required to prevent lamp holder or wire reaching a maximum of 250°C. Most halogen fittings also have high temperature wiring.

e. **Low Energy Lighting.** Low energy lights are now commonly installed on many vessels both AC and DC. They operate on a principle similar to fluorescent lights in which an electrical arc occurs between two electrodes, which are located at each end of the tube. The arc is conducted by vaporized mercury, and inert gases such as argon, neon or krypton through a phosphor coated tube. The emitted ultraviolet light makes the phosphor glow and emits visible light. These lights give a very high output for a relatively small power draw. They produce a light only marginally less than a 60-watt household bulb, and have a power consumption of just 16 watts. Most vessels I have installed these on now run just one light for the entire saloon area. Life expectancy is greater than standard fluorescent lights. Similar to halogen or fluorescent lights, they are intolerant to overvoltage conditions, most tolerate voltages up to 17 volts. In AC lighting, the compact plug-in fluoro tubes have an average rated life of around 10,000 hours. These include 2D (square fluoro), short twin-tube, quad-tube, triple-tube. There are screw-in Edison Screw (ES) types also that do not require adapters including twin and quad tube types. Quality light fittings have a PCB with integrated circuit operating at 35khz suiting lamps in the 5–11 watt range. Typical life expectancy is around 8000 hours. Typical outputs and equivalent light outputs for 12-volt units are 5W/520mA for 25W, 7W/680mA for 40W, 9W/850mA for 60W and 11W/1000mA for 75W.

f. **Red Night Lights.** Red lights are very useful in strategic locations. It can take up to 45 minutes for normal night vision to return if the eye is subject to a white light. The typical locations for night lighting are at the chart table and the steering station. Some fluorescent lights have dual tube fittings, and the use of LED cluster lights is effective. Some manufacturers make fittings with red diffusers. I use a small Hella or Aquasignal port navigation light—one at the steering position, and one in the galley—with a minimum-rated lamp of around 5 watts; a large light level is not required. Make sure that the steering position light faces down toward the deck and cannot be seen or may be construed as a port navigation light by anyone outside the vessel. If you have a coach house, a red-and-white unit can be mounted and switched locally.

g. **LED Lights.** Many vessels are now using Light Emitting Diode (LED) electronic lights. The LED is a solid-state semiconductor device that converts the electrical energy directly into light energy. They are low power consumption, vibration resistant, low heat emission and reliable, and can be used virually anywhere on board. LED bulbs are made in clusters of the LED, and they are different to incandescent types as they are polarity sensitive. Correct polarity connection is essential. With current development trends LED lighting will probably make most other boat lighting types almost redundant in 10 years.

9.2 Dimmers and Voltage Stabilizers. Many boats incorporate dimmers on lighting circuits.

a. **DC Dimmers.** Earlier types were variable resistance rheostat types; new technology systems use electronic pulse width modulated (PWM) control. PWM units' features are no heat, high outputs of around 100W at 12 volts, high efficiency with minimal losses of only round 2%. Control modules are also overload and reverse polarity protected. One feature on some models is a soft-start function that limits the initial inrush current, which can improve the life of bulbs. Some modules also have a no-load consumption, typically around 10mA, so light circuits must be isolated or switched off to prevent this when boats are unattended. Modular dimmers such as Cantalupi have short circuit protection, and thermal protection. When installing, the correct output rating must be used, overloading will lead to early failure. Ratings are 2 amps = 12V/24W or 24V/48W; 5 amps = 12V/60W or 24V/120W; 10 amps = 12V/120W or 24V/240; 20 amps = 12V/240W or 24V/480W; 30 amps = 12V/360W or 24V/720W. Dimmers often create electrical noise, and a filter capacitor should be installed as close as possible to the dimmer. Dimmers are reliable and where failures occur, always check the module power input and connections first. Where more than one pushbutton is used, check all first. If all are out the module is the cause; pushbuttons and connection to them are the most common failure point.

b. **AC Dimmers.** Larger yachts having AC lighting systems use different dimmer types. One technique is the use of toroidal transformer units. Unlike normal transformers, there is virtually no noise due to mechanical vibration, or any magnetic hum. The no-load power consumption is around 80% less than standard transformers, and overall efficiency is around 95%. The transformers have primary voltages of 120 volts with secondary of 12 or 24 volts. Capacity ratings are quoted in VoltAmps (VA) and can be in the range of 60 up to 600 VA. Units typically include overload, thermal and short circuit protection. If a unit suddenly stops, the thermal protection is the first possible cause. Some lighting systems also use electronic transformers such as the Aurora types which offer full dimming and a soft start function. The transformers have AC input voltages of 120 volts 50/60 Hz with stable output voltages on 11.5 volts. These supply low voltage AC lighting such as halogens, and an output frequency of typically 30kHz, and power outputs of 20–75 VA. Some units also have a soft start feature, which ramps up the voltage from zero to full output current, as well as the usual thermal and short circuit protection. The units also use toroidal transformers instead of ferromagnetic ones to reduce heat and increase efficiency. Like all AC powered equipment caution must be used when installing and troubleshooting, and they must be switched off when working on equipment.

c. **Lamp Voltage Stabilizers.** Lamps are affected by higher voltages reducing life expectancy. Light manufacturer Cantalupi has developed a voltage stabilizer that can accept a variable input voltage in the range 12.5 to 16 volts at 12 volts or 24.5 to 29 volts at 24 volts. The output is stable at 12 or 24 volts over all voltage ranges up to maximum power rating of 25W to maximize lamp life by least 200–400%. It is a good investment in vessels with large halogen light installations. Devices have short circuit, thermal, overload and reverse polarity protection. A similar device is the IML Bulb Saver which reduces lamp voltage by 1.5 volts and has a 15 amp rating in the range 10–40 VDC. Always mount in an area with good ventilation to assist in heat dissipation. Units should be mounted on a heat sink.

9.3 Deck Lights. There are ranges of lights with different functions that must be installed properly and maintained. Some have safety implications.

a. **Courtesy Lights.** The installation of courtesy lighting in the cockpit and transom areas is very useful but there are points to consider. Many of the lights available are of very poor quality and quickly degrade so always select quality fittings. A newer development is the LED light fittings which consist of high output LED clusters. Units such as the Hella units have 10 LEDs, which is equivalent to a 20-watt incandescent bulb using just 0.16 amp. The cheaper fittings that use festoon bulbs are a constant cause of problems with poor bulb contacts being the main one. I recommend fitting a couple of stern navigation lights facing downwards off the stern pulpit (pushpit) or the stern arch. They provide satisfactory low level illumination, are weatherproof, and are a valuable safety feature when retrieving the dinghy or a crewmember. Again some LED lights are a substitute for this method.

b. **Safety and Working Lights.** The various options are as follows:

(1) **Deck Lights.** Deck lights are essential for everything from fishing, entertainment, security and boarding at night to name the important ones. Halogen and xenon lamps are now an alternative to the incandescent sealed beam types, with increased efficiency and light outputs. Xenon lamps have working lives up to 2500 hours and have internal xenon ballast modules. Anchor area deck spotlights are typically fixed at the spreaders. With stainless pulpits, I have installed a small white navigation light at the pulpit facing down to illuminate the anchor well, and LED lights are a good option. The lights are switched from the steering station. The light is not too bright but it is practical.

(2) **Fiber Optic Lighting.** Fiber optics are used to direct lighting to steps and deck areas. The light sources have long life expectancy at around 3000 hours average. The Fondle Company have Edgepoint for toerails, Deckpoint for winches and cleats, Specpoint for chart tables and engine bays and Navpoint for nav light systems. Light source reliability is quoted as 3000mtb.

9.4 Lamp Bases. Lamp bases are extremely varied and designations are often confusing. The following are many of the more common lamp bases and their designations.

a. **Halogen Lamps.** Lamp socket types include E14, E27, E40, R75, BA95, B15D, G4, G6, and 35.

b. **Incandescent Lamps.** Lamp socket types include E14, E27, E40, B15D, B22D, P28, Candle base E12, Medium base E26.

c. **Fluorescent Lamps.** Lamp socket types include G5, G13, G23, G24, and G32.

9.5 Spotlights. Spotlights are generally confined to spreader and foredeck spot lighting. The very severe environment they are subject to entails careful selection. Usually ratings are given in Candlepower or Lux, with Lux being the amount of light at a nominated target distance, typically this is 100 and 500 yards. A rating may be given then as 330,000 candlepower or 52 Lux at 100 yards and 2.1 Lux at 500 yards. Some units use xenon lamp systems, and have outputs of 1.5 million candelas, and with a 24-volt supply consume 8 amps. The biggest failure area on spotlights is poor wiring and connections. Ensure that the cables are rated for the power of the unit. As they are exposed to weather, keep them covered when not in use, and operate them regularly through complete rotations, as well as pan and tilt to prevent seizing. If fuses blow it is generally due to seizing and overloading. Deck spotlights have more concentrated beams that are around 6–8 degrees compared to general floodlights of around 3–40 degrees. Spotlights should have a clearly defined beam pattern without scattering at the sides.

a. **Hand Spotlights.** Spotlights should have a clearly defined beam pattern without scattering at the sides. The illustration below shows the different beam distributions and light ranges for Optronics Blue Eye spotlights. The ranges shown are for clear conditions and a reflective target. If you want increased power for less-than-ideal conditions choose a higher candlepower rating. Always select a light with a switch for signaling.

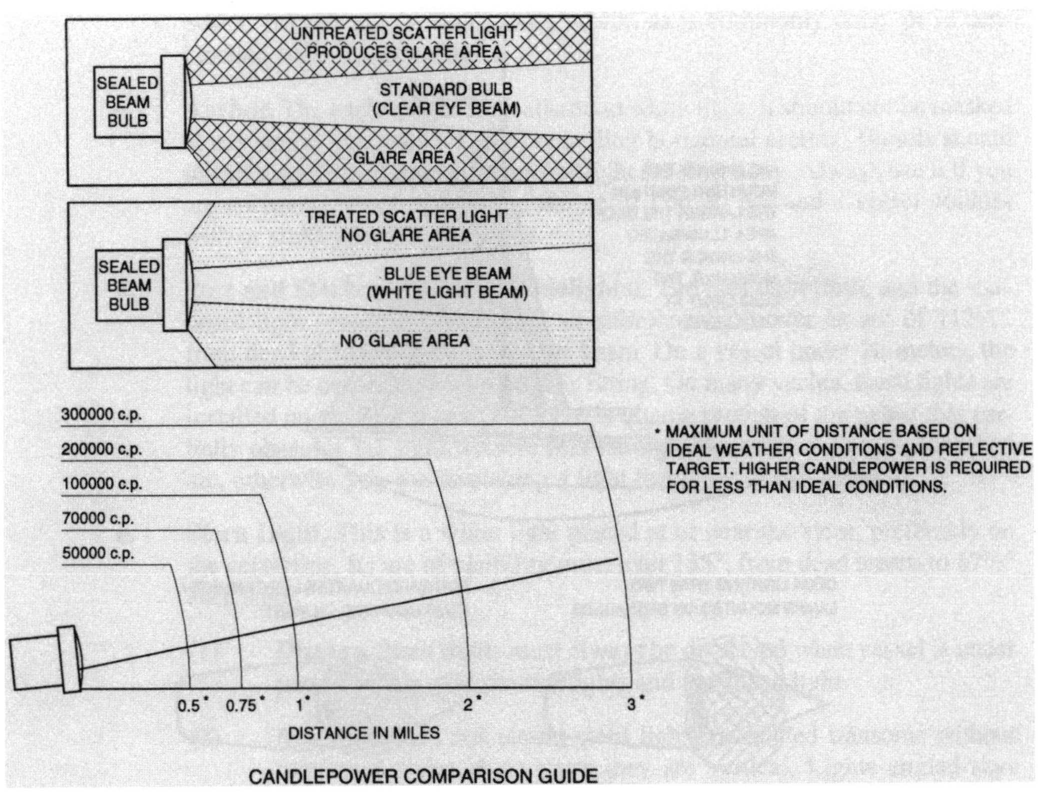

Figure 9-1 Spotlight Characteristics

b. **Spreader Lights.** Spreader lights are invaluable safety equipment. When severe problems are encountered on deck, good lighting allows for safer, faster work, and this reduces deck exposure time. The following must be considered when installing spreader lights:

(1) **Illumination Levels.** Lights are designed to facilitate on-deck safety, without excessive shadows. Do not use navigation lights under the spreaders; their deck level illumination is generally very poor. Use high quality, sealed beam halogen lights rated around 35–50 watts.

(2) **Construction.** Use fittings made of plastic to avoid corrosion. Stainless units are prone to shake apart at the spot welds.

(3) **Installation.** Spreader lights are exposed to weather and subject to severe vibration. It is essential that lights are mounted securely in a location where these factors are minimized. Cables should be of sufficient length and connections should be wrapped with self-amalgamating tape. Make sure that you can change burned-out lamps.

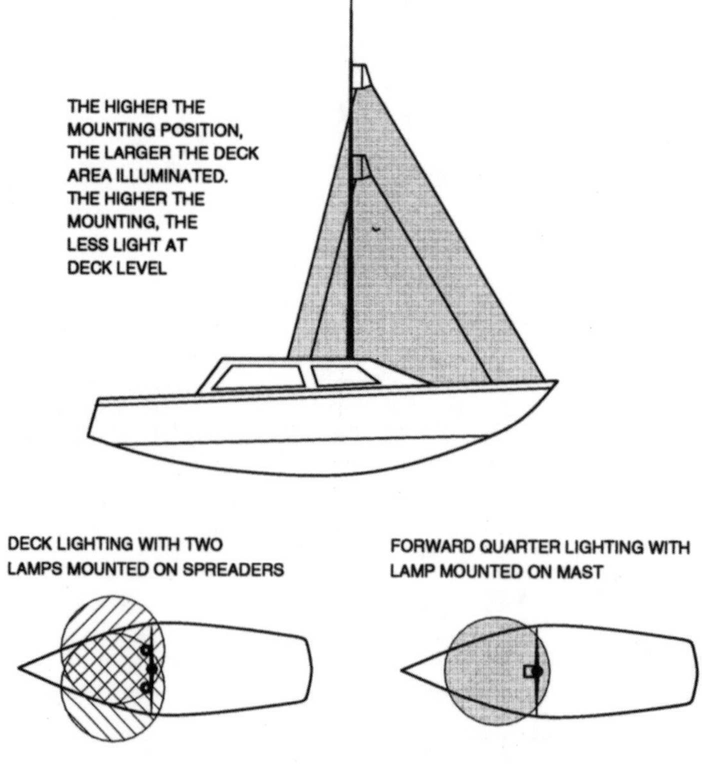
THE HIGHER THE
MOUNTING POSITION,
THE LARGER THE DECK
AREA ILLUMINATED.
THE HIGHER THE
MOUNTING, THE
LESS LIGHT AT
DECK LEVEL
DECK LIGHTING WITH TWO
LAMPS MOUNTED ON SPREADERS
FORWARD QUARTER LIGHTING WITH
LAMP MOUNTED ON MAST

Figure 9-2 Mast Lighting Arrangements

9.6 Navigation Lights. Navigation lights are of the utmost importance, both for safety and for legal, rules-of-the-road reasons. It is amazing how few sailing vessels display the correct lights. My personal survey shows that only about 40% of vessels have the correct lights displayed. It is not sufficient to simply say you have lights installed and turned on, they must be mounted at the correct locations. It is all very well to blame merchant vessels for running over pleasure craft, but if your correct lights are missing, nobody will be able to identify your vessel and status. Special incandescent lamp types are almost standard; LED based lights are now common. The LED has up to a 50,000-hour life expectancy, light output increase of 20% and consume 90% less power than traditional incandescent lights.

a. Legal Requirements. All vessels are required by the International Regulations for Preventing Collisions at Sea to display the correct lights. Failure to comply may void insurance policies in the event of a collision:

(1) Navigation Lights. Lights should be displayed in accordance with the provisions in Part C, Lights and Shapes.

(2) Lights. Lights should be of an approved type and conform with the provisions of Annex I with respect to positioning and technical details of lights and shapes.

b. Tricolor (Under Sail Only). For yachts under 20 meters, the combination port, starboard, and stern light mounted at the masthead is the best solution. Because only one lamp is burning, it does not consume too much battery power. It must not be used under power, as is commonly done, or in conjunction with any other light.

c. Anchor. The anchor light is an all-round white light. It should not be masked at any point. See Annex I, 9(b) regarding horizontal sectors. Vessels should install a combination tricolor anchor light for simplicity. Always use it if you are anchored where traffic is possible. If you do not and a vessel collides with or sinks you, it's your fault.

d. Port and Starboard Lights (Sidelights). The port light (red), and the starboard light (green) must display an unbroken light over an arc of 112½°, from dead ahead to 22½° abaft the beam. On a vessel under 20 meters, the light can be combined into a bicolor fitting. On many yachts, these lights are installed on the bow pulpit, but there is often a section of the pulpit that partially obscures the light. Ensure that the light is visible over the prescribed arc, otherwise you are displaying a light that is technically illegal.

e. Stern Light. This is a white light placed at or near the stern, preferably on the centerline. Its arc of visibility must total 135°, from dead astern to 67½° each side:

(1) Display. Stern lights must always be displayed when vessel is under power, along with the sidelights and masthead light.

(2) Mounting. Do not mount stern lights on angled transoms without mounting plates that ensure they are vertical. Lights angled skywards are very difficult to see and aren't legal.

f. **Masthead Light (Steaming Light).** This is a white light that must be visible over an unbroken arc of 225°, from dead ahead to 22½° abaft the beam on each side. The light must be fixed on the centerline of the vessel, typically on the top of mast. There are also vertical mounting requirements for the masthead light. See Annex I.2, Vertical Positioning and Spacing of Lights:

(1) **Vessels 12 meters LOA or less.** A minimum of 1 meter above the sidelights.

(2) **Vessels 12 to 20 meters LOA.** A minimum of 2.5 meters above the gunwale.

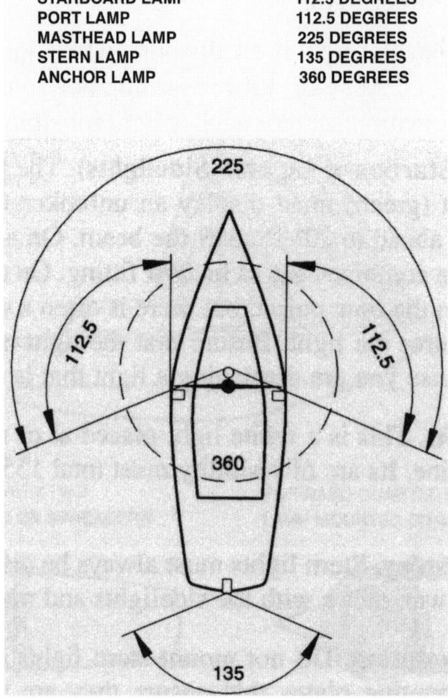

Figure 9-3 Navigation Lighting Plan

9.7 Navigation Light Technical Requirement. Regarding (chromacity) color are covered in Annex I.7: Color specification of lights and Annex I.8: Light intensities

a. **Color.** Color or chromacity are defined by international collision regulations. By purchasing approved light fittings, you know that they meet the requirements.

b. **Lamp Ratings.** Lamp ratings are generally given by the manufacturers and are designed to give the required range and luminosity for which the light is granted approval. Do not increase the lamp rating to increase the brightness or decrease it to save power. If you do alter the lamps and have an accident, your insurance may be invalidated and you could be sued for damages because technically you were not displaying approved navigation lights:

   (1) **Sockets.** Special sockets (typically BA15D, 1156, 1157) are used to ensure that filaments are correctly aligned to the lens and horizontal shades systems.

   (2) **Light Outputs.** Light output and wattage are designed for a high lumen-per-watt ratio.

   (3) **Light Consistency.** The lights are designed to emit an even output through a 360° azimuth.

c. **Visibility.** The required minimum range of visibilities are as follows:

(1)	**Stern Light.**	<12m = 2 NM,	12–20m = 2 NM.
(2)	**Sidelights.**	<12m = 1 NM,	12–20m = 2 NM.
(3)	**Masthead Light.**	<12m = 2 NM,	12–20m = 3 NM.
(4)	**Tricolor Light.**	<12m = 2 NM,	12–20m = 2 NM.
(5)	**Anchor Light.**	<12m = 2 NM,	12–20m = 2 NM.

d. **Approvals.** Navigation lights should all be approved. Most manufacturers issue a certificate with each fitting:

   (1) **National Approvals.** It is important to note that some fittings are only approved by a national ports or marine authority (ABYC or USCG) and may technically be illegal in another country.

   (2) **Approval Certificates.** Always keep the numbered approval certificate with your vessel's files in case of litigation.

e. **Maintenance.** Check lights regularly. Any defects may cause failure or be illegal. Check the following:

   (1) **Moisture.** Check the light's interior for moisture that can degrade lamp contacts or cause a short circuit.

   (2) **Diffuser.** Check the light diffusers for cracks or crazing that will alter the light's characteristics.

DC Electrical Equipment

10.0 DC Electrical Equipment and Systems. There are many important DC electrical powered systems and equipment on board that must be properly maintained. These include the following

a. **Refrigeration Systems.** Basic refrigeration, electrical and mechanical eutectic systems.

b. **HVAC.** Air-conditioning, ventilation, and diesel heaters.

c. **Anchor Windlass.** Windlass selection, installation and troubleshooting.

d. **Electric Furlers and Deck Winches.** Installation and troubleshooting.

e. **Bow Thusters.** Operation, installation, maintenance and troubleshooting.

f. **DC Motors.** How to maintain and troubleshoot DC electrical motors.

10.1 Refrigeration Systems. A well-found galley is important on any yacht, and refrigeration is central to this. I served on some of the most automated and advanced refrigerated cargo ships as Electrical Officer/Engineer and the multi-compressor refrigeration and computerized control systems required a great deal of maintenance. The fundamental principle is that when a high-pressure liquid or gas expands, temperature reduces. In a refrigeration system, a compressor is used to pump the refrigerant fluid, normally Freon, around the system:

a. **Compression.** The compressor pumps the refrigerant vapor. This increases the refrigerant gas pressure, which becomes hot. The high-pressure hot gas then passes through to the condenser.

b. **Condensation.** The hot gas passes through a condenser, which acts as a heat exchanger and releases heat. The condenser is either air cooled by natural convection, a fan, or water passing through coils. Coils are typically made from cupro-nickel/copper. The hot gas condenses into a hot liquid, and passes through to the expansion valve.

c. **Dryers.** A small amount of water vapor will remain in a system regardless of purging and evacuation. Water causes ice formation at the expansion valve resulting in either total blockage or bad operation. The dryer is installed in the liquid line between the receiver and expansion valve serving as both a filter and removing water. The dryer desiccant materials are of silica gel or activated alumina. Flared units enable easy change-out when saturated rather than soldered ones, and removable cartridge type are even better. Internal corrosion begins at above 15ppm, also causing oil breakdown making it acidic, and contributing to motor burnouts in hermetic systems. In Glacier Bay systems a module concept is used incorporating a receiver, accumulator and dryer. The suction accumulator is used to prevent liquid slugging to the compressor.

d. **Sight Glass.** The sight glass allows visual inspection of the liquid. Bubbles indicate low refrigerant levels. All indicators incorporate a moisture indicator. Indicators are chemical and the color changes to pink in R12 systems, blue if safe; in R22 green is dry and pink is wet. The site glass will show bubbles if refrigerant is low, often this is seen at start-up and stop. It will also show restrictions or blocked filter dryers ahead of the sight glass.

e. **Refrigerant Control.** The thermostatic expansion valve (TX valve) regulates the rate of refrigerant liquid flow from the liquid receiver high side into the evaporator low side. It maintains the pressure difference and therefore expansion into the evaporator is in exact proportion to the rate of liquid leaving the evaporator. Flow is regulated in response to both pressure and temperature within the evaporator. The thermal sensing element is placed on the outlet end of the evaporator and where installed also a motor thermal sensing element, which consists of a bulb and capillary. Valves can normally be adjusted for optimum temperature. This pressure reduction causes a fall in temperature of the liquid. The cold liquid then passes to the evaporator. The Thermostatic TX valve is controlled by two conditions, the temperature of the control element and the evaporator pressure. The Automatic TX valves allow refrigerant flow only if the evaporator pressure falls when the compressor operates. The single greatest cause of TX valve failure is dirt, acids, moisture and sludge in the system. All of these will freeze up or jam the valve.

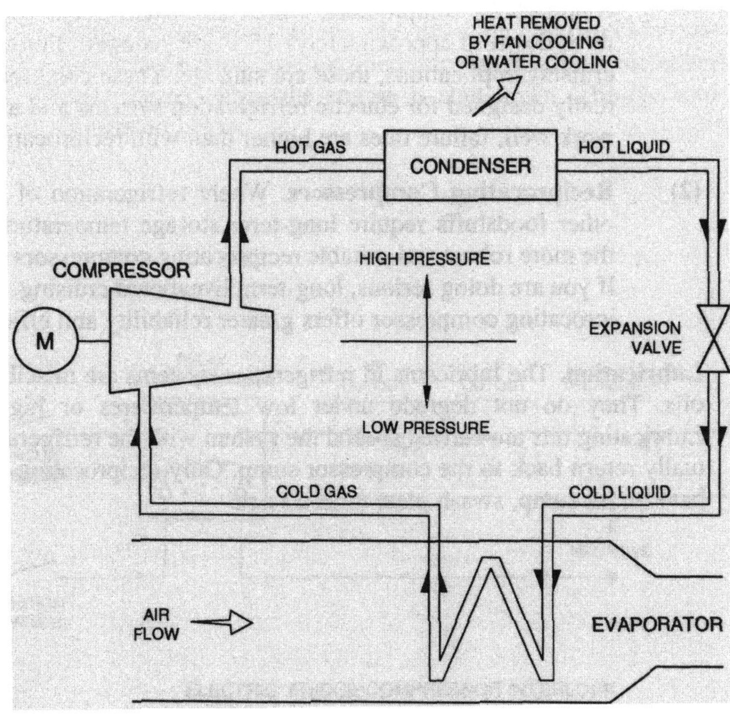

Figure 10-1 Basic Refrigeration Cycle

f. **Evaporator.** The cold liquid passes through to the evaporator cooling surfaces (or eutectic tanks). Heat within the refrigerator space is absorbed by the cold refrigerant causing the air to cool. The absorption of the heat causes the refrigerant liquid to evaporate into a gas. Heat is removed by either conduction on evaporator, radiation, and convection. Isotherm has developed an innovative new system, called the self-pumping (SP) cooling system. A special integrated condenser and through-hull fitting have been developed that replace the galley sink fitting. The movement of the vessel causes water to pump in and out and remove waste heat.

g. **Recycle.** The cold gas is suctioned back into the compressor to repeat the cycle.

h. **Refrigerants.** The most commonly used refrigerant was Freon 12 (R12), and as this is harmful to the ozone layer, like all CFC gases, it is being replaced by Freon R22. HFC-134a is the new standard refrigeration system gas. This gas type is an obvious choice as auto air conditioning systems will all use the same gas, having already made the conversion. You cannot use HFC-134a in an existing Freon 12 system as virtually all the system components are incompatible and will require renewal.

10.2 Eutectic Refrigeration Systems. This is the most common and efficient method of vessel refrigeration. The evaporator is replaced by a eutectic plate or tank. Operation is as follows:

a. **Eutectic Principles.** A eutectic system uses brine or a fluid that freezes at what is called eutectic temperature. Originally brine solutions were used but systems now have an ethylene glycol/water mixture or similar. The mixture has a much lower freezing point than water. Once the mixture is frozen completely (eutectic point) and refrigeration is removed, the tank will cool the refrigeration space, gradually thawing out as it absorbs heat.

b. **Holdover Period.** The period of time that the space will remain within required temperature ranges before refrigeration is required is called the holdover period. When specifying a system, the holdover time and the temperature required are critical to the size of the plates or tanks and the type of eutectic solution required.

10.3 Compressors and Motors. There are five compressor types in common use, the reciprocating, rotary, scroll, screw and centrifugal. Many systems use an engine driven reciprocating compressor, or either DC or AC motor powered compressors with a belt drive. Engine powered units have a belt drive off the engine and the drive pulley has an electromagnetic clutch for operation of the compressor. The most common compressors in use are the reciprocating and swash plate types.

a. **DC Drive Motors.** Glacier Bay DC motor systems have quoted figures for power consumption over 24 hours at 20.5 amp-hours and 39.5 amp-hours for refrigerator and freezer respectively. DC drive motors are typically rated at ½ horsepower, and the most common is that from Leeson. Glacier Bay has developed a low speed motor that operates at lower speed 675 rpm rather than the 1800 rpm Leeson. With 300% greater torque it is nearly 80% heavier. Unlike the Leeson unit with 2 pole magnets and 2 brushes it has 6 poles and 4 brushes and is designed to have an increased service life of 500%. Other features are the ventilated commutator and significantly reduced running temperatures.

b. **Swash Plate Compressors.** These are typified by automotive air conditioning compressors and are satisfactory where temperatures are required down to approximately minus 15°C. For most average applications these are suitable. These compressors are not really designed for eutectic refrigeration systems and although they work well, failure rates are higher than reciprocating units.

c. **Reciprocating Compressors.** The reciprocating compressor consists of cylinders, piston intake and exhaust valves, and connecting rods to the crankshaft similar to an engine. The compressors are driven by belts from the drive motors.

(1) **Daily.** Check operating pressure gauges, temperatures, compressor oil levels, and abnormal noise or vibration.

(2) **Weekly.** Check evaporator and defrost if necessary. Check all valve covers are on and tight.

(3) **6 Monthly.** Check operation of high and low pressure cutout switches. Perform oil sample test. Inspect and clean condenser. Check V-belts and adjust.

d. **Lubrication.** Lubricants in refrigeration systems are miscible, wax free oils. These do not degrade under low temperatures or high pressures. Lubricating oils are carried around the system with the refrigerant and eventually return back to the compressor sump. It should be noted that only reciprocating compressors have an oil sump, while swash plate units do not. On reciprocating compressors check the oil. It should be shiny and clear, have no visible particles, and feel smooth and greasy when rubbed between the fingers. Samples will help determine internal component condition and wear. Wear particles should not exceed in ppm the following: lead (10); copper (10); silicon (25); iron (100); chrome (5); nickel (5); aluminum (10) and tin (10).

e. **Reciprocating Compressor Servicing.** Perform the following:

(1) **Condenser Pressure and Temperature.** High pressures indicate reduced cooling or air in the condenser. Low pressures indicate that refrigerant may be restricted to the evaporator.

(2) **Filters.** Liquid line, oil return, suction line and TX valve require cleaning, as clogged filters will cause restrictions in evaporator supply.

(3) **Moisture Indicators.** If these alter from green to yellow, moisture is in the system and filter dryer requires replacement.

(4) **Leak Detection.** Regular checks should be made, on new installation every month until joints and flanges settle, and are retightened. Refrigerant should be recharged.

(5) **Pressure Switches.** These should be checked and adjusted.

(6) **Condensers.** Open and clean tubes. Check and replace anodes.

(7) **Belts.** Rubber V-belts should be checked and re-tensioned.

10.4 Auxiliary Refrigeration Controls. A few different control devices are essential for safe and efficient operation.

a. **High Pressure Cutout.** The high pressure cutout is to protect against high pressures caused by loss of cooling water, plugged condenser, or in the worst case serious contamination of the refrigeration system with water and air. The cutout is usually wired in series with the compressor contactor or clutch. Typically this is above 75psi in the condenser. To test operation, close the cooling water off and wait until head pressure builds up and activates it.

b. **Low Pressure Cutout.** The low-pressure switch monitors suction line pressure. The cutout operates when gas discharge from the evaporator is too low. Operation of the cutout is indicative of a low refrigerant charge, typically below 30psi. To test the switch, slowly close in the suction valve to activate the switch. Low suction pressures increase compression ratios and can cause compressor damage.

c. **Thermostatic Control.** To test, vary settings and observe cut in and out. Many are now microprocessor controlled and the temperature sensor is located at the holding plate, although some use the general box cooling space.

d. **Clutch Engine Interlocks.** Many electromagnetic clutches are operated from a dedicated circuit breaker on the main switch panel, giving protection on the clutch coil and cabling. It is common to see the switch inadvertently left on and subsequently flatten the batteries, as typical current draw is around 3–4 amps. On some occasions the operating coil can burn out. To prevent this, an interlock should be installed into the ignition system so that the clutch is de-energized when the engine is shutdown.

e. **Reduced Holdover Times.** A common complaint is that holdover times have reduced for the following reasons:

(1) **Warm Foodstuffs.** A refrigerator or freezer system is often pulled down to the required temperature and then a full load of unfrozen food or warm drinks such as a case of beer are dumped in it with the expectation that it will rapidly cool them.

(2) **Climate Change.** More often than not the system worked well in a temperate climate, but the first extended cruise in tropical waters results in a dramatic reduction in apparent efficiency. The refrigerator should be open sparingly, and people new to the lifestyle are probably opening it far more than is necessary and far more often than they did on a normal weekend cruise. Keep access down to the minimum.

(3) **Mechanical Causes.** Engine drive belts are not re-tensioned and as such belt slip under load causes decreased refrigeration.

(4) **Seawater Temperatures.** In many cases, a voyage to warmer waters also has caused changes in condenser cooling efficiency. In many cases the eutectic plate takes longer to pull down, and refrigeration operation times need to be extended.

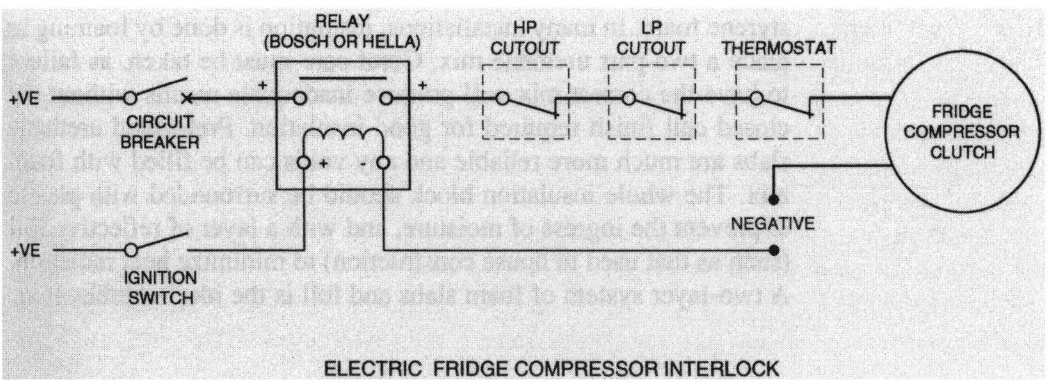

Figure 10-2 Refrigeration Clutch Interlock

10.5 Electric Refrigeration Systems. These systems are typically self-contained. Electric refrigeration requires either AC or DC battery power for operation. The average power consumption is approximately 50 amp-hours for a refrigerator, and approximately100 amp-hours for freezers operating on a 50% duty cycle. To restore battery capacity there is a far greater run time involved than for an equivalent engine driven system, and the installation of either a higher output alternator, and a fast charging regulator should be considered.

a. **Self Contained Electric.** There are a number of self-contained refrigerators in use. They are DC powered and have eutectic holdover plates. Insulation on the units is reasonable but where installed, the surrounding area should be insulated further. Sizes of these units tend to be in the range of 40 to 120 liters. Power consumption of these units average around 35Ah per day, depending on ambient temperatures, and frequency of opening. As many units are built in, good ventilation must be provided to carry away heat from compressor unit. Many units do not function properly as a result of this omission, and a fan and ducting makes a difference.

b. **Energy Utilization.** Some manufacturers have introduced circuitry that enables over-riding the thermostat during engine run periods. If temperatures are down, the alternator can supply loads for an additional pull down period that reduces electrical consumption later. A similar function is used on Isotherm ASU (Automatic Start Up) systems. The controller senses the raised system voltage from the alternator and operates at double the speed to pull down temperatures and maximize the energy available. Another feature of Isotherm systems is the control of compressor speed with respect to refrigeration requirements. The new Danfoss hermetically sealed compressors are used with electronic control on the three phase motor supply. I opted for a freezer model with spillover plate and stainless butterfly vent to adjust cooling in the refrigerator space. Note that the seawater cooling system anode must be removed every 6 months and cleaned, or it will shed enough material to clog the pump suction lines. When shutting down for any period, simply flush the system with fresh water.

10.6 Hermetic Compressors. The majority of electric refrigerator systems use hermetically sealed Danfoss type compressors. The electric motor is sealed within a domed housing along with the compressor. The motor and compressor assembly is supported on a spring suspension system to absorb vibration. These are used as starting torque characteristics are very good. The motor has two windings, one for start and one for running. The motor also has a capacitor, which is the black cylinder on the unit, wired in series with the motor start winding. A starting relay is also used, which may be a current or potential relay switch, located on the outside of the compressor unit. When power is switched on, current passes through the start and run windings. The capacitor alters the phase angle and effectively converts the motor to a two phase motor. When the motor speed increases and start current decreases at approximately 60–75% the switch opens and the start winding and capacitor are disconnected. Motors have a thermal overload.

a. **Compressor.** Hermetic units are typically twin piston reciprocating compressors with a valve plate assembly. Units with fan driven condenser cooling should be cleaned every 3 months. Tube flare nuts and service valve caps should be tight to prevent leaks.

b. **Troubleshooting.** Motor failures are usually due to external causes. Continuity and resistance checks will indicate status of windings. Start winding C to S resistance is approximately 5 ohms. Run winding C to R resistance is approximately 2 ohms. Both windings R to S resistance is approximately 7ohms. Megger R or S to case is at least 1 meg ohm. Testing of both open circuit voltage and on-load voltage also will indicate problems. A difference exceeding 10 volts indicates an overload or motor winding fault. Use a clamp ammeter to check operating current. If the compressor current draw exceeds 7 amps it is close to failure. Check all external control devices first before assuming compressor failure. Discharge the capacitor first, and use an ohmmeter (multimeter set) and test across the capacitor terminals. A short circuited capacitor will indicate zero ohm. A high reading indicates an open circuit. If the capacitor is good the reading will initially go to zero then slowly rise. Regular failures in capacitors are usually caused by slow starts (typical maximum is 3–4 seconds), too many starts, (typical rate is 3–4 starts per hours), low voltages or starting switches are faulty.

10.7. Refrigeration System Installation. There are kits for refrigerator installations, however I would recommend to get a good refrigeration mechanic to install the system:

a. **Insulation.** If a refrigerator system is to be effective and reliable, it must be of sufficient size to meet the expected needs, and be well insulated. Insulation thicknesses should be at least 4 inches or more. Inadequate insulation is a primary cause for inefficient refrigeration systems. Install as much as you can. The ideal insulating material is urethane foam, followed closely by fiberglass wool and then polystyrene foam. In many installations, foaming is done in place using a two-part mix. Failure to have the mix correct will produce inadequate results, without a good closed cell finish that is required for good insulation. Ideally, the use of preformed slabs is much more reliable, and fill any outstanding voids with foam mix. The whole insulation block should be surrounded with plastic to prevent the ingress of moisture, with a layer of reflective foil such as that used in domestic house construction to minimize heat radiation. A two-layer system of foam slabs and foil is the ideal combination. Vacuum insulation panels are a good investment.

b. **Refrigeration Size.** Do not build refrigeration spaces greater than the actual requirements. Far too many build oversized boxes that remain half empty. This is a waste of energy and results in greater installation costs for a system that will not be used efficiently.

c. **Compressor Brackets.** The engine compressor mountings and brackets must be extremely robust to prevent vibration. Make sure that vibration will not fracture any part of it.

d. **Compressor Drive Belts.** Alignment of the compressor and engine drive pulleys is essential to ensure proper transfer of mechanical loads. Belts are usually dual pulley arrangements. Ensure that both belts are tensioned correctly.

e. **Energy Saving Measures.** Energy conservation and efficiency improvement measures can be implemented. Fill any empty spaces in the refrigerator compartment with either blocks of foam, or inflated empty wine cask bladders. This will decrease the refrigerator space and reduce energy requirements. Place all the frozen food at the bottom of the refrigerator and cover it with a mat to retain the cold air.

f. **Battery Voltages.** Ensure that battery voltage levels are maintained. Low battery levels will cause inefficient compressor operation. Do not let the battery level sink to the normal minimum level of 10.5 volts. It takes far more energy and engine run time to charge a nearly flat battery than one half charged.

g. **Ventilation.** Ensure the compressor unit is well ventilated. Install an additional fan and ducting to ensure positive ventilation on non-water cooled units.

10.8 Refrigeration System Troubleshooting. Few boats carry vacuum pumps, bottles of refrigerant, gauge sets and spare parts; in fact to carry out refrigeration work may be a breach of environmental laws if you are not certified. Knowingly releasing Class I (CFC) and Class II (HCFC) substances into the atmosphere can result in severe penalties and imprisonment. Besides working on refrigerated cargo ships I also worked for a while as a refrigeration mechanic repairing shipping container systems. All repairs were done in filtered clean areas. It is highly unlikely that conditions will be suitable for you to properly overhaul and repair compressors. There are few exceptions, and these are largely limited to professional refrigerator mechanics. The best way to avoid problems is to have the system properly installed in the first place. This chapter does not include procedures for the disassembly and checking of compressors, purging and recharging, as you are more likely to do further damage. If after checking the control systems you are unable to rectify problems, call in a licensed refrigeration technician. It is important to determine what is going on in the system. Pressure gauges are used to check system pressures, and thermometers are used to measure evaporator, line and condenser temperatures.

a. **Condensers.** If a condenser is undersized or dirty, internal and external, the head pressure and condensing temperature rise. The higher temperature will make the compressor pump to this higher pressure and temperature. It is important to check and clean condensers regularly.

b. **Refrigerant Loss.** Refrigerant loss is a common fault, and will cause a gradual reduction in cooling efficiency, and eventual tripping of the low-pressure cutout. Low refrigerant levels can be observed in the sight glass and bubbles will be seen. An empty sight glass indicates no refrigerant at all. If all the gas has escaped, and after the leak has been located the system must be purged of air and moisture before gas recharging. You will have to get a qualified refrigeration mechanic to do this. A frost-covered evaporator generally signals that gas levels are okay. If the system is undercharged, refrigerant does not properly liquefy before passing through the TX valve, the effective latent heat is reduced so refrigeration is poor. Some vapor will pass through the TX valve reducing refrigerator control capacity, and the vapor passing at high velocity will increase the wear on the TX valve needle and seat. Air in the system will increase the total head pressure. The refrigerant will then have to condense to a higher temperature and pressure, the cylinder head and exhaust on compressor and top tube of the condenser will all be at higher temperatures. This will also then affect the oil quality. It is important to ensure that caps on service valves are replaced and tight to reduce leaks.

c. **Leak Detection.** Perform leak detection by pressurizing the system and then checking all possible leakage points at connections and fittings. Do not use a torch with HFC-134a refrigerants.

(1) **Halide Torch.** The most common test will require the use of a halide torch. Air is drawn to the flame through a sampling tube. Small gas leakages will give the flame a faint green discoloration, while large leaks will be a bright green color.

(2) **Soapy Water.** A simple method is the use of soapy water, generally dishwashing liquid, and applying to all piping joints with the system running. If a pressurised leak is in the joint, a bubble will form.

Table 10-1 Refrigeration Troubleshooting

Symptom	Probable Fault
Compressor abnormal noises	Low oil pressure Oil foaming Liquid in suction line Coupling misalignment Oil pump faulty Piston rings or cylinder wear Discharge valves faulty Solenoid valve oil return faulty Oil filter clogged Compressor mounting loose Low cooling water flow High water temperature
High condenser pressure	High pressure cutout activated Refrigerant overcharged Cooling water loss High cooling water temperature Condenser clogged Inlet water valve closed
Low condenser pressure	Low refrigerant charge Excess cooling to condenser Piston rings or cylinder wear
Low oil pressure	Oil pressure switch has activated Oil level low Oil pressure too low at regulator Oil foaming in crankcase Liquid in suction line Oil pump defective Bearing worn out Oil filter clogged
Oil level falling	Oil foaming in crankcase Poor oil return Liquid in suction line Piston rings or cylinder wear Solenoid on oil return faulty Oil filter clogged
Reduced or no cooling	Leak in system Clutch connection broken Clutch coil failure High pressure cutout activated Low pressure cutout activated Low refrigerant level Drive belt slipping Thermostat faulty

Table 10-2 Refrigeration Troubleshooting

Symptom	Probable Fault
Slow temperature pull down times	Drive belt slipping Low refrigerant level Compressor fault High cooling water temperature Condenser plugged Low battery voltage Refrigerator space seals damaged High ambient temperature Insulation failure Thermostat faulty
Clutch circuit breaker tripping	Clutch coil failure Clutch cable shorting out Compressor bearing failure
Expansion valve icing up	Dryer requires replacement Low refrigerant charge
High discharge pipe temperature	TX valve fault Discharge valves leaking
Low oil temperature	Oil level low Oil is foaming in crankcase Oil pump defective Bearings worn Oil filter clogged Oil in evaporator
Low suction pressure no cooling water	Low refrigerant charge TX valve frozen up TX valve not operating Liquid line filter clogged Liquid line solenoid valve
Moisture in system	Condenser leaking Compressor gasket failure Compressor bearing failure
Refrigerant gas leakage	Pipe compression fitting Condenser leak Isolation valve leak Damaged piping Valve caps off

10.9 Air Conditioning Systems. Air conditioning is possible on even small boats and is virtually standard on larger vessels. Like refrigeration, air conditioning cools a cabin by transferring heat out. In most marine installations, seawater is used generally for condenser cooling although fan cooled systems are available but are less effective. There are two types of marine air conditioning systeem, the single stage direct expansion and the tempered (chilled) water two-stage type. Manufacturers such as Cruisair, HFL, Climma all offer extensive ranges with options. Under the Montreal Protocol Freon 12 has been replaced by Freon 22. Systems are generally rated in British Thermal Units (BTU) which is the energy required to heat or cool an area. In metric this is Kilo calories (Kcal). The conversion is approximately 4 BTU = 1Kcal. Systems requiring gas charging must be performed by certified technicians using approved (EPA) equipment.

a. **Self Contained and Remote Condensing Systems.** These single stage direct expansion units may be either self-contained or have a remote condensing unit installed within the machinery space. The self-contained reverse cycle system is normally a relatively compact module, such as the Cruisair Stow-Away, that can be installed under a bunk or locker. The modules are rated between 5000–24000 BTU/hr. They are pre-charged with R-22 refrigerant at the factory, are seawater cooled from a remote pump, and have integral reciprocating, rotary or scroll compressor depending on the model. Units also have integral condenser, evaporator, blower, safety switches etc. Many systems have a remotely installed condensing unit within the machinery space, with seawater supply also adjacent. Refrigerant is carried to the air-cooling unit. These units have cooling capabilities in the range 5000 to 60000 BTU/hr.

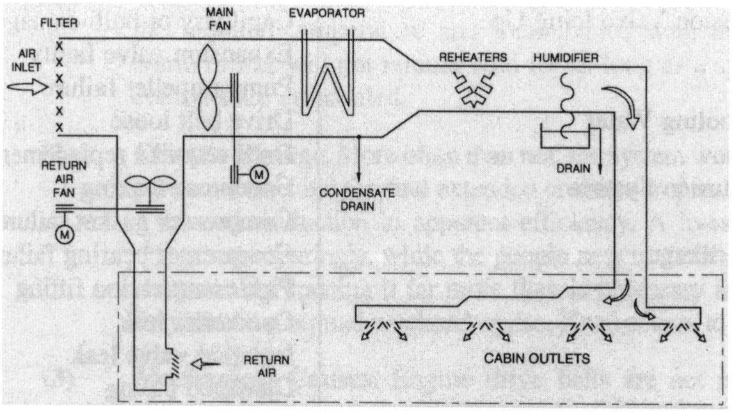

Figure 10-3 Typical Air Conditioning Schematic

b. **Electrical Power Requirements.** Air conditioning system power requirements are as follows:

(1) **AC Systems.** A system normally requires a constant AC power source to operate, so the generator must run continuously. Cruisair quotes as a guide at 117 volts, 1 amp per 1000 BTU/hr, however 1.3 is closer. If an air conditioning system is to be installed on the vessel the generator must take account of the maximum loads. As systems use AC induction motors on the compressor there is a significant start-up current surge that must be allowed for in generator load calculations, typically 3–4 times full-load amps. Hermetically sealed compressors have high starting currents that are reduced by capacitors to around 3–5 times running current.

(2) **Seawater Pumps.** Electrical load calculations should also factor in the seawater pump. Pumps are generally not self-priming and must be positioned at or below the waterline. They are prone to corrosion. Self-priming impeller pumps should be installed where possible. The seawater pump is generally controlled via a relay box, which should be mounted in a dry location. Pump capacities are typically 100gph for a 5000BTU unit up to 250gph for a 12000BTU system.

(3) **DC Systems.** HFL Marine International has a 12- or 24-volt DC system in the "Ocean Cool" series. It has a hermetically sealed compressor and draws 44 amps for a 6000 BTU/h (1500Kcal/h) unit so the engine may have to run to supply the power, and an additional alternator is a good option. The quoted battery capacity is a minimum of 160 Ah.

(4) **Control Systems.** Controls range from simple on and off switches, speed control and thermostat to programmable controllers. These offer timing functions; high and low temperature settings, systems monitoring, fan speed controls and compressor restart time delays, including fault condition automatic shutdown and even automatic dehumidification. Protection and control systems are similar to those in refrigeration systems. There are also variable speed fan controllers that use small variable frequency drives.

c. **Maintenance.** There are a number of maintenance tasks to perform:

(1) **Weekly.** Check the seawater inlet strainer and clean.

(2) **Monthly.** Filters on air cooling units should be checked and cleaned.

(3) **Bi-annually.** Seawater cooling condensers should be cleaned if possible. Where systems have anodes in the cooler they should be checked and replaced.

d. **Troubleshooting.** Air conditioning systems have many faults similar to refrigeration systems.

(1) Check external control equipment such as thermostats.

(2) Check system HP and LP cutouts where installed.

(3) Check evaporator cooling systems are clean and functioning.

e. **Capacity Calculations.** The capacity of the system must be calculated by determining the volume to be cooled. The following are guidelines used by HFL. For ambient temperatures exceeding 30°C add 20%, and for water temperatures exceeding 25°C add a further 20%. This is to maintain 16–22°C. The estimated seawater cooling requirement is for 3.5 gallons per minute for a self-contained unit.

(1) For below decks cu.ft x 14 = BTU. (m^2 x 504 = BTU).

(2) For above decks cu.ft x 17 = BTU. (m^2 x 612 = BTU).

Table 10-3 Air Conditioning Capacity Table

Capacity (BTU/hr)	Below Deck Sq. Ft	Mid Deck Sq. Ft	Above Deck Sq. Ft
6000	90	60	45
7000	115	75	55
9000	165	110	85
12000	200	150	100
16000	267	178	135
20000	335	250	167
24000	405	300	200

10.10 Ventilation Fans. Good ventilation is essential in many areas of the vessel, especially the galley, the engine space, and the cabins. There are a number of ventilation fan options, and all have uses in particular applications. Fans can be classified either as extraction fans or as blowers.

a. **Extraction Fans.** Extraction fans take air out of a space, either to increase natural ventilation flow rates and air changes or to remove heat or fume concentrations.

(1) **Solar Fans.** Solar fans have a small solar cell powering the fan motor. Newer models have a small, solar-charged battery so that the fan can operate at night, the period when it is most required.

(2) **Engine Extraction Fans.** These are used to extract heat from engine spaces. In warmer climates, it is preferable to leave the fan operating for half an hour after the engine stops to reduce heat build-up and stop the increase in lower deck temperatures from radiated heat.

(3) **Ventilators.** These units have two speeds and are reversible, allowing them to be adapted to the conditions inside. At 25 cfm (cubic feet per minute), air displacement is very good, which suits normal cabin environments. Power consumption is also relatively low at only 1.7 amps on the fast setting. The two-speed Vetus units utilize an electronic, brushless motor with a current draw of only 0.2 amp. Air extraction rates are a reasonable 36 cfm.

b. **Blowers.** Blowers push air into a space, and are used either to displace existing air such as in bilge blower applications, or in most cases to direct air in large volumes over specific areas, such as in alternator cooling applications. In-line fans are commonly used in bilge blower applications. Air flow rates are typically around 100 cfm and have a power consumption of around 4 amps. Blowers used in areas where hazardous vapors are concentrated must be ignition proof. They are often used to ventilate engine spaces. Typical air flows are in the range of 150–250 cfm. Power consumption ranges from 4 to 10 amps, which is quite high. In most cases, though, they are run with the engine operating. It is a good practice to interlock the fan to the engine start with a relay to ensure that it always operates and switches off at engine shutdown.

10.11 Diesel Heater Systems. Power consumption figures, heat outputs, and fuel consumption rates for typical Eberspacher models are illustrated in Table 10-4. Heaters have the following operational cycles:

a. **Starting.** Cold air is drawn in by an electric fan to the exchanger/burner.

b. **Ignition.** Fuel is drawn at the same time by the fuel pump, mixed with the air, and ignited in a combustion chamber by an electric glowplug.

c. **Combustion.** The combustion takes place within a sealed exchanger and gases are exhausted directly to atmosphere.

d. **Heating.** Heat is transferred as the main air flow passes over a heat exchanger to warm the air to the cabin. A thermostat in the cabin shuts the system down and operates the system to maintain the set temperature.

Table 10-4 Diesel Heater Data Table

BTU Output	Fuel (liters/hour)	Power Draw
6,100	0.21	40 watts
11,000	0.38	45 watts
15,000	0.57	70 watts
28,000	1.05	115 watts
41,000	1.40	190 watts

e. **Power Consumption.** Typical power consumption is 40 watts (3.33 amps) during running. At start up, the draw can be up to 20 amps for a period of 20 seconds during the glowplug ignition cycle.

f. **Heater Maintenance.** The following maintenance tasks should be carried out to ensure optimum operation:

(1) Check that all electrical connections are tight and corrosion free.

(2) Check exhaust connections and fittings for leaks. Leakages can cause dangerous gases to vent below deck.

(3) Remove and clean the glowplugs. Take care not to damage glowplug spiral and element. Use a brush and emery cloth and make sure all particles are blown out afterwards.

(4) At 2000 hours, take the unit to a dealer and have the heat exchanger decocked and the fuel filter replaced.

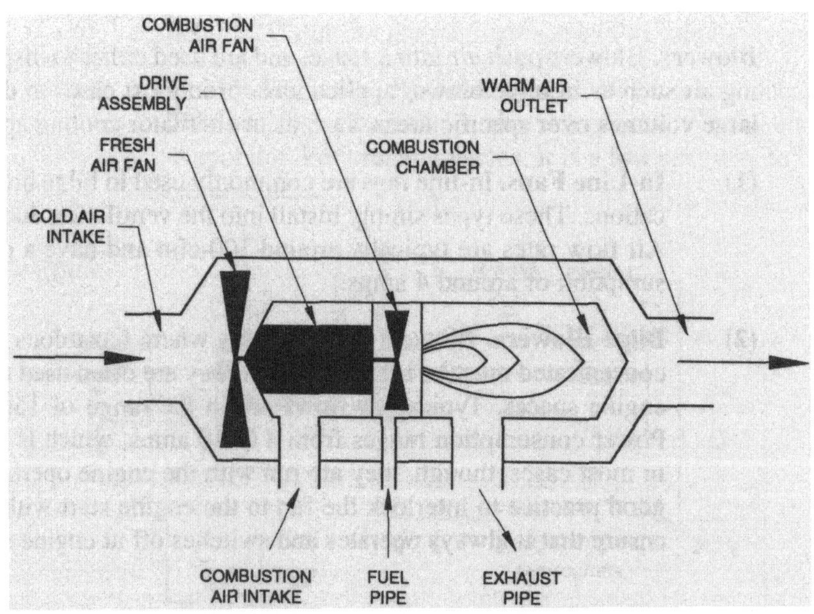

Figure 10-4 Diesel Heater System

Table 10-5 Diesel Heater Troubleshooting

Symptom	Probable Fault
Heater will not switch off	Temperature switch fault
Heater smokes and soots	Combustion pipe clogged Fuel metering pump fault Blower speed too low
Heating level too low	Hot air ducts clogged Fuel metering pump fault Blower speed too low Temperature switch fault
Heat will not start	Supply fuse blown Low battery voltage Blower not operating Fuel-metering pump fault Thermal cutout tripped Fuel filter clogged No fuel supply Glowplug fault Control unit fault
Heater goes off	Fuel-metering pump fault Thermal cutout tripped Fuel filter clogged Fuel supply problem Control unit fault

10.12 Anchor Windlass. It is crucial that anchor windlasses are properly selected and installed. Unfortunately, they are rarely maintained properly and subsequently fail at critical periods. Here are the factors to consider when selecting and installing a windlass.

a. **Windlass Selection.** Choose the windlass based on the weight of anchor chain and the vessel size.

b. **Electrical Installation.** Install correctly rated cables and protective systems.

c. **Electrical Control.** Install a reliable control system.

10.13 Anchor Windlass Selection. If it is a new system, it is prudent to select the correct anchor for the vessel. The CQR, manufactured by Simpson-Lawrence (UK), is one of the most useful and common anchor types, and this will be used as a yardstick. Simpson-Lawrence's selection chart should be used as correct weight selection is critical. Finding the right windlass chain is more difficult than can be imagined. The principal problem is that chain types do not always match the windlass chain lifter. Windlass selection is based on the weight of an anchor and the chain weight. Table 10-6 illustrates a selection of short link chain sizes for a variety of vessel lengths.

Table 10-6 Anchor Chain Weight Selection Table

Vessel Size	10 Meters	12 Meters	14 Meters	16 Meters	18 Meters
Chain Size	8 mm	10 mm	10 mm	13 mm	13 mm
All Chain	40 m	50 m	70 m	80 m	90 m
Rope/Chain	12 m	14 m	16 m	18 m	20 m
Chain Wt	1.42kg/m	2.22 kg/m	2.22 kg/m	3.75 kg/m	3.75 kg/m

a. **Winch Loading Calculation.** Minimum windlass capacity is derived from the following formula, after working out the chain weight for your vessel size:

Windlass Capacity = (Anchor Weight + Chain Weight) x 2

eg., 12-meter vessel has CQR of 35 kg.

Chain Weight 111 kg + 35 kg = 146 kg x 2 = 292 kg

b. **Rated Output.** The windlass in this case must have a rated pull of at least 292 kg. Manufacturers have selection charts to assist in selection. Always add at least a 25% margin to the calculated figure.

c. **Recovery Speeds.** Speeds are typically designed around a figure of 10 meters/ minute at a 100-kg load. The higher the load, the slower the anchor retrieval rate.

d. **Anchor Loading.** A windlass is not designed to take the entire load when riding to anchor, especially in large swells or heavy conditions. As a safety precaution, always transfer the load to a bollard using a rope snubber.

e. **Operational Notes.** When operating the windlass, observe the following:

(1) **Engine Running.** Always operate the windlass with the engine running. The alternator supplies part of the motor load and keeps the motor from impressing a large voltage surge on the electrical system. More importantly, running the engine keeps the voltage from dropping too low.

(2) **Run Times.** In cases where the windlass is used without the engine running, the voltage drop is such that a severe drop in windlass power occurs after a few minutes. A further problem is that the motor may overheat due to the lower voltage, causing winding damage or burn-out. Always pause for 20–30 seconds every few minutes and allow the voltage to recover. If you are having a problem with anchor retrieval, do not continue to load the anchor windlass until it stalls. Stop every five minutes and allow the motor to cool down.

10.14 Anchor Electrical Installation. Anchor windlass performance is frequently reduced by the installation of incorrectly rated cables. Anchor windlass electrical supplies should run the most direct route to the engine starting battery, via the appropriate isolator and protective devices. At full-rated load, significant voltage drops can develop, with a corresponding decrease in rated lifting capacity. The following system components must be specified and installed correctly. The practice of installing a separate battery, either at the machinery space or forward next to the windlass, is not recommended. Use the engine starting battery; it has a high cranking amp rating and is more able to deliver the currents required by a windlass at maximum loads. A deep-cycle service battery cannot cope with these loads without being damaged.

a. **Cabling.** Cabling must be able to cope with large currents over an extended distance. Voltage drop should not exceed 5%. Table 10-7 gives recommended cable sizes for length of cable run, not for vessel length.

Table 10-7 Windlass Cable Rating Table

Cable Length	Current Rating	AWG	Metric	B & S
up to 6 meters	200 amps	4/0	25 mm2	3
up to 8 meters	245 amps	2/0	35 mm2	2
up to 10 meters	320 amps	1/0	50 mm2	0
up to 12 meters	390 amps	0	70 mm2	00

b. **Circuit Protection.** Current ratings vary depending on the manufacturers. Many windlasses have converted DC starter motors on the entire powered range. Typical current loadings are given as 55 amps at no load, 110 amps at half load, and 180 amps at full rated load. Protective devices are as follows:

(1) **Circuit Breakers.** A circuit breaker should be installed on the supply reasonably close to the battery, and easily accessible. Typically, 100- and 125-amp circuit breakers are used. Use DC-rated circuit breakers, not AC ones as many commonly do.

(2) **Automatic Thermal Cutouts.** I would caution against using automatic thermal circuit breakers. They trip automatically in overload conditions and reset; the problem is that you have to wait until they reset, which is usually when you desperately need the windlass.

(3) **Slow Blow Fuses.** ABYC and USCG require a slow blow fuse be installed on the system, and many manufacturers integrate this within the control box. The fuses are normally rated above the windlass rated working current, typically 200 amps for 12-volt systems. Make sure you carry a spare.

c. **Connections.** Connections are a common cause of failures. The following points should be observed:

(1) **Connector Types.** Always use heavy-duty crimp connectors. Do not solder connections as dry joints are commonplace and solder can melt under maximum load. Soldered joints also stiffen up the cables, causing fatigue.

(2) **Insulation.** Put on a section of heat shrink tubing over the entire crimp connector shank and cable to prevent the ingress of moisture.

(3) **Connections.** The lug terminal hole should always fit neatly to ensure maximum contact. Use a spring washer on the nuts to prevent loosening and subsequent heating and damage under load. Coat terminals with a light layer of petroleum jelly.

d. **Performance Curves.** The following curves for Lewmar windlasses graphically illustrate the effect load has on power consumption and hauling speed.

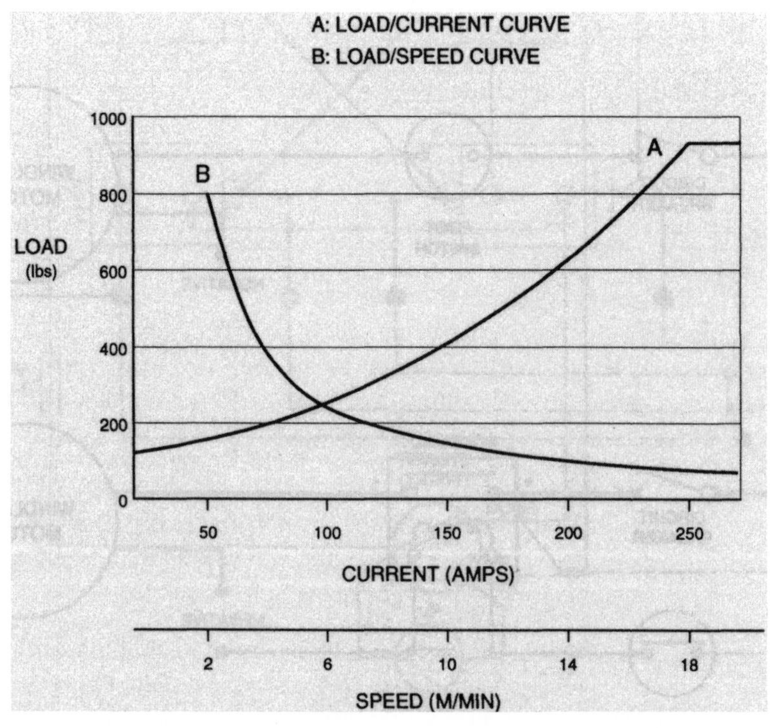

Figure 10-5 Anchor Windlass Performance Curve

(1) The higher the load, the higher the current, until a point is reached where the motor overloads and stalls. The higher the load when the windlass is operated, the shorter the operation time allowed on the motor.

(2) The higher the load, the slower the recovery speed. Hoisting the anchor can take less time and cause less wear and tear on the windlass if you motor up over the anchor and remove chain tension.

10.15 Anchor Windlass Electrical Control. One of the most common failure points in an anchor windlass is the control system. Controls come in the following configurations:

a. **Single Direction Foot Switch.** A foot switch is connected directly in the positive supply to the windlass motor. Foot switches are notorious for filling with water, and usually in this type of control a short develops, or the contacts and spring corrode. Shorting can result in brief, uncontrolled windlass operation and a burned-out switch.

b. **Single Direction Solenoid/Foot Switch.** The foot switch is used to control a heavy-duty solenoid located below decks, which closes the main power supply to the motor.

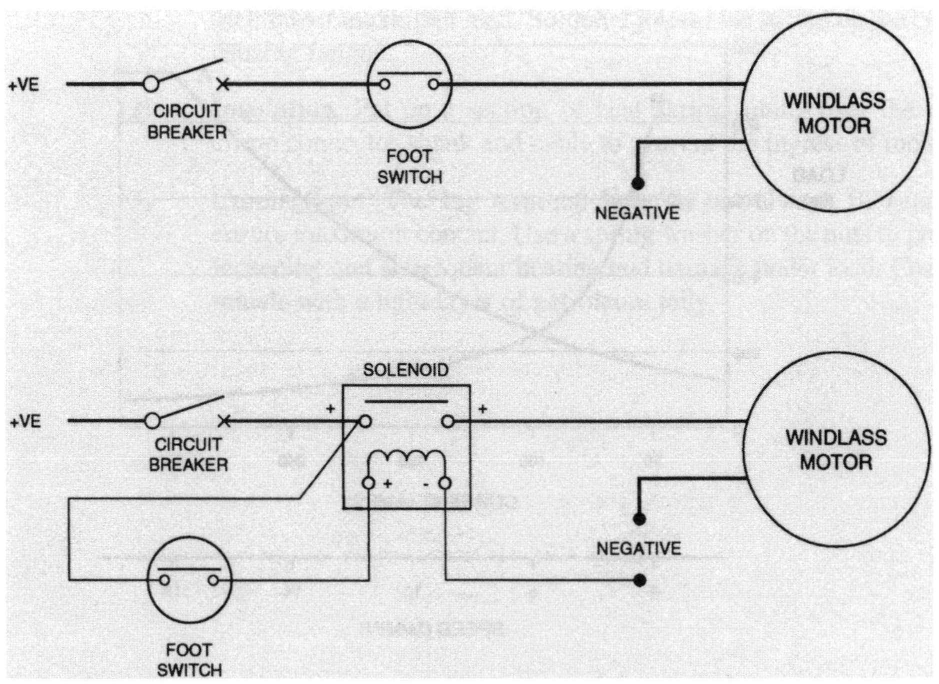

Figure 10-6 Windlass Control Systems

c. **Pneumatic Deck Foot Switch.** These units have a PVC tube connecting the switch to the control solenoid box. Air pressure from the switch operates a microswitch. There have been reports of spontaneous start-ups or shut-offs in extremely hot conditions, which in one case caused serious injuries. The problems were caused by pressure build-up in the air system. Evidently, earlier units are the most prone to trouble and major suppliers such as Lewmar already have a safety air bleed to correct the problem. Carefully follow the proper depressurizing procedures when installing switches.

d. **Dual-Direction Solenoid Control.** A control box consisting of two or four solenoids is used for reversing the motor for both hoisting and lowering. Control is usually by a pair of foot switches and/or a remote panel.

(1) **Power Consumption.** Solenoids typically consume 1 amp each.

(2) **Caution.** Never operate both foot switches together. In fact, many manufacturers specify only the "up" foot control be fitted.

(3) **Protection.** Some control boxes incorporate fuse protection. Fuse failure is rare, but make sure that a spare is in the box for emergencies.

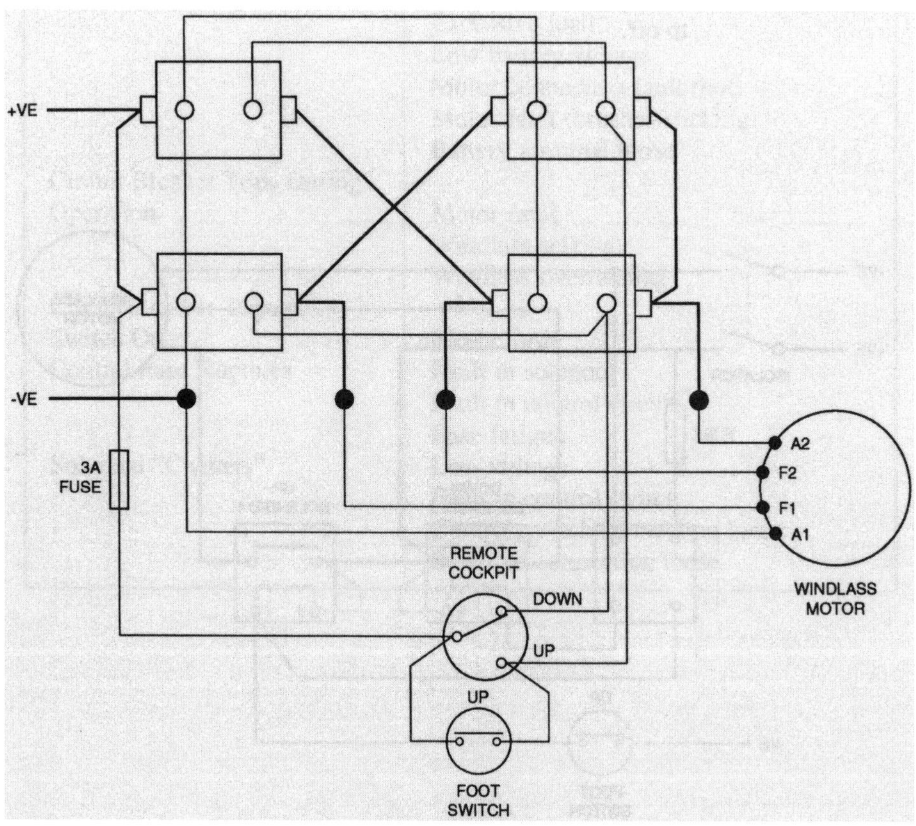

Figure 10-7 Windlass Control Systems

e. **Remote Controls.** Remote control devices take a variety of forms:

(1) **Portable Controls.** These are usually weatherproof control modules that can be plugged into prewired socket-outlet stations. Ensure that the socket remains watertight.

(2) **Radio Controls.** These devices are relatively new and innovative, and work like TV controls. How efficient they are, I am unable to verify.

(3) **Touchpad Panels.** These are touch panels covered with a waterproof membrane. Their reliability is low and I have removed every unit that I installed. Normally, control is achieved through the positive side of relays or solenoids. Many touchpad controls switch the negative so that other foot switch controls on solenoids must also be converted to negative or have relays inserted in the circuit.

(4) **Switch Panels.** The basic, weatherproof pushbutton or toggle-switch remote system has proved to be the most reliable remote-station system. The switches must be waterproof and be spring loaded to off.

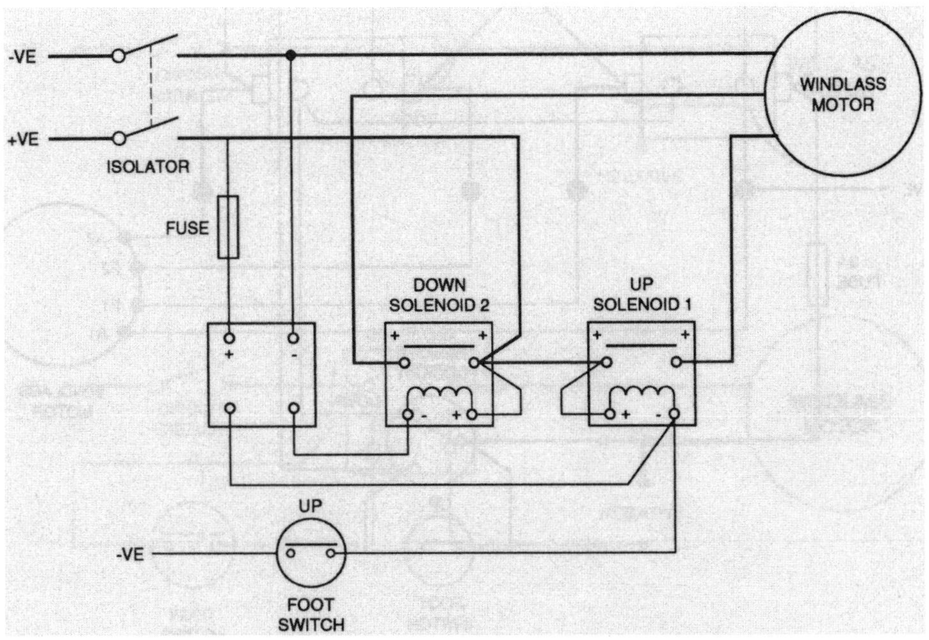

Figure 10-8 Windlass Remote Control Systems

Table 10-8 Anchor Windlass Troubleshooting

Symptom	Probable Fault
Windlass will not operate	Foot switch fault (most common cause) Circuit breaker switched off Isolator switch off Foot switch connection loose Solenoid connection fault Solenoid fault Solenoid fuse blown (if fitted) Motor connection loose Motor fault (sticking brush is common) Motor fault (winding failure) Motor internal thermal cutout tripped Slow blow fuse ruptured
Windlass stalls under load	Excessive load Low battery voltage Motor connection loose Motor fault (brushes sticking)
Windlass operates slowly	Battery terminal loose Excessive load Low battery voltage Motor connection fault (hot) Motor fault (brushes sticking) Battery terminal loose
Circuit breaker trips during operation	Motor fault Windlass seizing Windlass overloading
Circuit breaker trips at switch on	Motor fault
Control fuse ruptures	Fault in solenoid Fault in control circuit Fuse fatigue
Solenoid "chatters"	Low voltage Fault in control switch Control switch connection loose Solenoid connection loose

10.16 Electric Furlers and Winches. Deck winches and furling gear are rapidly being electrically powered and are taking a lot of the muscle out of cruising for the short-handed crew and older husband/wife cruising teams. They are generally treated in the same way as anchor windlass circuits, requiring good circuit protection and correctly sized cables. Winches may be powered from an electric motor or hydraulic power pack. Most furlers operate from a hydraulic power pack. Electrical loads are considerable and for 12-volt systems the following cable sizes are required. Electric winches generally consume far more power than windlasses and careful power supply planning is required. The power source should be a starting battery—the engine start battery can be paralleled with another of equivalent size. Install the battery with the largest possible cold cranking rating.

Table 10-9 Winch Cable Rating Table

Cable Length	Current Rating	AWG	Metric	B & S
up to 10 meters	320 amps	1/0	50 mm2	0
10 to 15 meters	390 amps	0	65 mm2	00
15 to 20 meters	500 amps	00	85 mm2	000

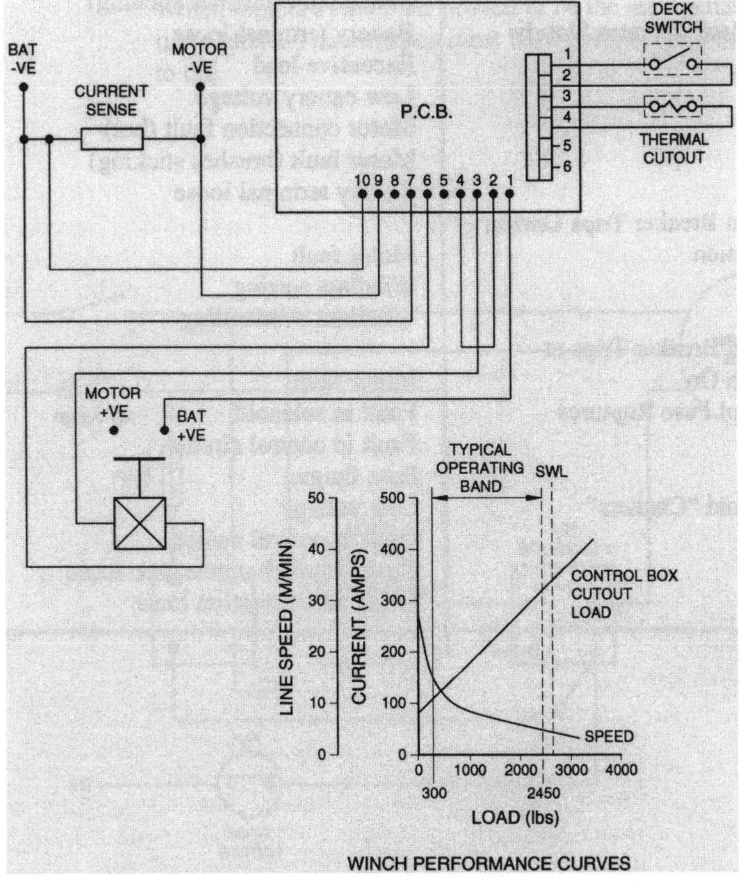

Figure 10-9 Lewmar Winch Control

10.17 Thruster Operations. Bow thrusters are now common on many yachts and offer increased maneuverability in confined areas such as marinas and are invaluable on boats during higher wind conditions. I work a lot with large offshore vessel thruster systems, both fixed and variable pitch, tunnel and azimuth, and while these are somewhat more complex in terms of control systems, the average small boat unit is relatively simple. The major suppliers include MaxPower (www.maxpower.com), Lewmar (www.lewmar.com), Sleipner (www.side-power.com), Vetus (www.vetus.com), Wesmar (www.wesmar.com). There are many factors during both selection and installation that must be considered if thrusters are to be efficient. Basic operational factors to consider when using thrusters are as follows:

(1) Learn and understand the behavior of the vessel in wind conditions.

(2) Avoid using the thruster unless moving very slowly or stopped.

(3) Learn the maximum recommended run times and stay within the limits.

(4) Do not use in short bursts, a single 5-second thruster period is more effective than 5 x 1 second bursts. It causes less strain and overheating on the motor, and uses less battery power.

(5) Always switch off control system when not in use to avoid accidental operation.

10.18 Thruster Ratings. Thrusters are specified in terms of the thrust output capability. It is the thrust and not the output of the electric motor in kW or HP that determines effectiveness. Thrust is a result of the power of the electric motor, the propeller shape and dimensions, the speed in rev/min, and the tunnel efficiency losses. The thrust is typically in the range of 15 to 25 kgf per kW of the electric motor. Thruster output forces are selected so that they equal or exceed the calculated or expected wind thrust forces, or to counter what is also called the sail plane effect. The draft also affects the drift rate, shallow draft vessels tend to have a greater wind effect than deeper draft ones.

a. Wind Pressure. The wind pressure on a vessel has a quadratic increase with wind speed. The equation used by Vetus is pressure P (N/m^2) = $\frac{1}{2}\rho \times V^2$ where ρ is the specific mass of air, and V is the velocity of air in m/s.

Table 10-10 Wind Force Table

Wind Force Beaufort	Description	Speed m/s	Pressure N/m^2 (kgf/m^2)
4	Moderate breeze	5.5 to 7.9	20–40 (2.0–4.1)
5	Fresh breeze	8.0 to 10.7	41–74 (4.2–7.5)
6	Strong breeze	10.8 to 13.8	75–123 (7.6–12.5)
7	Near gale	13.9 to 17.1	124–189 (12.6–19.2)
8	Gale	17.2 to 20.7	190-276 (19.3–28.2)

b. **Wind Draft.** The wind pressure must be multiplied by the wind draft area to calculate the actual wind force. This is determined by the boat surface area, wind speed and wind angle. The lateral surface areas and effects vary with the superstructure size and shape, and the hull freeboard and therefore the hull surface area. The worst case is where the wind is on the beam or at 90º. A factor of 0.75 is used to account for streamlining and a less than flat and rectangular shape.

c. **Torque.** Torque is calculated by multiplying wind force by the distance between the center of effort of the wind and the center of rotation of the boat. The center of effort is dependent on the shape of the superstructure, and while nominally center, it may be either slightly forward or aft depending on the actual location of the accommodation. The center of rotation is dependent on the underwater hull shape. The wind force torque is calculated by multiplying 50% of boat length by the wind force. Torque (T) = Wind Pressure (P) x Wind Draft (D) x 0.75 x Distance between the bow thruster center and the boat pivot point (½ vessel length). Nominally the pivot point is the stern. Thrusters should be installed as far forward as possible to maximize leverage effect around the vessels pivot point. They should be as deep as possible to prevent air being sucked in and maintain maximum water pressure.

10.19 Thruster Types. Thruster types depend on vessel types, sizes, and budgets. Propellers come in single, twin and counter rotation configurations. The latter arrangements are to increase efficiency and reduce cavitation. The range suits a variety of applications. Thrusters can be either DC, AC and hydraulic.

a. **Swing Retractable.** The swing retractable thruster has the advantage of no drag, but the retract system is complicated. The Lewmar type uses a rotating lead screw to activate the arm, and MaxPower uses an innovative folding system. Interlock limit switches are installed in the control circuit for lead screw travel, and retracted position. These are prone to failure and must be maintained.

b. **Vertical Retractable.** These also allow compact installations and reduced drag when retracted. Limit switches are used to limit control of up and down travel and must be maintained.

c. **Transverse Tunnel.** The tunnel thruster is the simplest arrangement and is less complex in terms of mechanical components. Tunnels will cause some drag to the boat although not enough to bother most boat owners.

10.20 Thruster Power Output Table. The table shows figures quoted by Sleipner Sidepower, and are a good general guide to all fixed pitch thrusters. Required thruster force (F) is calculated by dividing calculated torque (T) in Nm by Distance between the bow thruster center and the boat pivot point in meters.

Table 10-11 Thruster Outputs

Thrust (kg/lbs)	Boat Size (ft)	Power (kW/hp)	Voltage (Volts)	Battery (Min CCA)
35/77	22-32'	2.2/3	12	300
55/121	28-40'	3.1/4	12/24	350/175
75/175	35-50'	4.4/6	12/24	50/250
95/209	42-58'	6/8	12/24	700/350
155/341	50-70'	8/10.7	24	600
220/484	60-84'	11.2/15	24	700
285/627	74-100'	15/20	48	2 x 450 (24V)

10.21 Thruster Power Supply. The efficiency of thruster motors is directly related to battery supply voltage levels, the available battery capacity, and the voltage drop in the supply cables during operation. Slow Blow fuses are used as the fluctuating loads must not cause fuse failure in normal service and ratings should cope for up to 5 minutes. The table is a general guide only; all circuits must be measured, and the appropriate cable size installed with the required protection.

a. **Batteries.** A common question is whether to use a separate battery located forward or not. This depends a great deal on boat sizes, and practical space considerations. In smaller boats, it is better to have the supply run off an increased engine start battery bank. This means an additional heavy cable is run forward to the bow thruster, but this makes charging easier. The main engine is always operating when using thrusters. The alternator provides additional power to the battery, with less heavy battery drain, and it is worth considering a high output alternator. In larger vessels, a separate battery bank is preferable, and additional battery charging must be considered. A major cause of reduced thrust is inadequate power availability from the batteries at full load, and maintaining them in optimum condition is essential. If the battery suffers a major voltage droop, the thruster will also suffer a reduction in thrust output. Where separate batteries are to be installed, my preference is for an AGM battery bank, which can deliver the required current, require no maintenance and have a high charge acceptance rate. A separate high output alternator on the engine is a good option and the 24-volt option should be considered.

b. **Supply Cables.** The power supply cables must be rated for the maximum current requirements of the motor and allow for voltage drop, which ideally should not exceed 5% at full load. Voltage drop is a major cause of reduced thrust.

Table 10-12 Bow Thruster Electrical Installations

Thrust (kg/lbs)	Boat Size Ft (m)	Voltage	Cable Size AWG/mm	Slow Blow Fuse Size
35/77	22-32' (10)	12	2/35 mm^2	125 A
55/121	28-40' (11)	12/24	00/70 mm^2	250 A
75/175	35-50' (12-13)	12/24	0/120 mm^2	355 A
95/209	42-58'(12-13)	12/24	0/120 mm^2	355 A
155/341	50-70' (13-19)	24	100 mm^2	355 A
220/484	60-84' (18-30)	24	0/120 mm^2	425 A
285/627	74-100' (25-35)	48	0/120 mm^2	425 A

10.22 Thruster Control and Drive Motors. In general a joystick control lever, or buttons and in some cases a footswitch is used that activates a micro-switch, supplying power to a solenoid, which then closes current on to the motor. Electric motors generate heat, and thruster motors are limited by temperature rise. Generally, most thrusters are rated at between 3 to 5 minutes at full continuous output. Power supplies are critical as high currents cause large voltage drops. A 3kW motor can draw 250 amps or more at initial starting, and less power means loss of thrust. Motors generally have the following safety systems installed:

a. **Overheat Protection.** Once this limit is reached, the motor windings are protected against damage by a winding embedded thermal cutout with heat levels rising to around 100°C. Depending on the amount of heat within the winding, the time to reset the cutout varies from 1 to 5 minutes. To ensure availability the proper use is about 10–30 seconds per maneuver to prevent cutout operation.

b. **Drive Reversal Protection.** Some thrusters incorporate electronic time-lapse protection against sudden drive reversal.

c. **Protection.** Motors and cables are provided with Slow Blow fuses for short circuit, as they must withstand the high instantaneous starting currents. Overload protection is provided by circuit breakers.

d. **Speed Control.** Vetus manufactures an adjustable speed control to ramp up the speed relatively slowly to suit certain 24-volt thruster models.

e. **Series/Parallel Switches.** In many cases, thrusters are only available in 24- or 48-volt versions. In 12-volt boats, two batteries will have to be connected in series to get 24 volts. The switches are usually modular and contain the switching relays.

f. **Motor Testing.** Under no circumstances operate the motor out of water or with load off, coupling disconnected, or propeller off. The series wound DC electric motor will accelerate very fast to a point where it will be seriously damaged.

10.23. Thruster Maintenance and Servicing. Servicing and maintenance requirements are relatively simple. Some thrusters such as the composite leg types from MaxPower have sealed lubricated bearings and have no anodes or oil servicing.

a. **Anodes.** Anodes should be inspected regularly, at least every 6 months. New anodes where installed on the propeller shaft or on the leg should be secured using Loctite. Gear cases are usually bronze; the protection is for casing, propeller blades and shafting.

b. **Oil Tanks.** The gear case oil header tank should be checked and topped up prior to each trip. Side-power specify EP90 oil. Oil consumption usually indicates a leaking seal. If installation is properly done there are usually no problems. However if the tank is installed at the wrong height there will not be enough oil overpressure which leads to water ingress. I have seen this mistake even in commercial installations. A check should also be made on the oil tube to ensure there is no kinks or loops that can cause air locks or affect the oil pressure and flow. Oil should be changed at least every 2 years or in accordance with manufacturers' instructions, usually timed when hauled out of the water. Make sure the oil drain screw is re-tightened securely and that the oil is flowing through before doing so.

c. **Antifouling.** Coat the gear case and propellers, not the seals, anodes or propeller shafts. Use antifouling designed for propellers.

d. **Electric Motor.** Check and tighten motor holding bolts every year. Vacuum out any carbon brush dust, and check the condition of the commutator and brush gear. Check and tighten all electrical connections on the electric motor, and directional solenoids.

10.24 Thruster Troubleshooting. The following are the most common faults.

a. **Thruster Will Not Start.**

(1) **Power Supplies.** Check the obvious causes such as: the isolator is open, the circuit breaker is off, or a Slow Blow fuse is ruptured. Check that voltage levels at the thruster are correct, at no load it should be a minimum of 12.7 or 25.4 volts. If lower check the battery voltage first, and if acceptable then check connections. Check the voltage when trying to run the thruster; if it has dropped lower than 8.5 volts the battery condition is the probable cause and cannot deliver the required power.

(2) **Control Systems.** If the solenoids do not operate a control signal is absent, as solenoids rarely fail. Check the voltage at the solenoid to confirm this. Check the power supply and protection (fuse) to the control panel, and if good then check the control cable connections and control panel outputs. Retractable thrusters also may have interlocks on the retract systems. Check that the thruster is down completely and the interlock limit switches are operating.

(3) **Motor.** If the voltages are correct at the electric motor the thermal cutout switch may be faulty, or the motor brushes are sticking. Turn off the power, open and manually check that the brushes are moving freely within the brush holders.

b. **Reduced Thrust.** The most common cause is reduced voltages caused by battery failure or loose connections causing voltage drops. The brush gear can also cause reduced thrust problems, the brushes and commutator should be checked. If the electric circuit items are fine, the thruster may be fouled with marine growth.

c. **Thrust Failure.** If the thruster stops during operation, check the protection equipment such as fuses, thermal cutouts and circuit breakers first. If the motor is operating with no thrust check that shear pins and flexible couplings have not separated due to the propeller jamming on debris. Large offshore vessels have even known whales to be sucked in!

10.25 DC Motors. Most installed pumps and machinery have DC motors installed, and most are maintenance free. Where larger motors are in use such as thrusters, windlasses, winches, refrigeration compressor drives, starter motors and motor driven generators, the question of proper maintenance becomes critical. My commercial seagoing career started on 220-volt DC systems, and the first lesson was that DC motor performance and reliability is directly related to effective preventive maintenance. Routine inspection of DC motors on thrusters, furlers, starters, windlasses and propulsion systems consists of rigorous and thorough inspection. This should be done every 6 months. Inspect the carbon brushes for chipping, grooving, uneven wear and loose or frayed wire connectors. Compare the length of brushes and replace the set of brushes if worn. Check that brushes move freely within the brush holders. Check that the spring pressure is correct by simply pulling the brush back and snapping it against the commutator. Make sure that the wire brush connectors are clear of any moving parts. Check that brush tail connections are tight. Inspect the commutator. Inspect the field winding connections, the electrical connections and mechanical fasteners inside the motor and connection box. Check the motor interior for condensation, access covers and gaskets.

10.26 Commutators. The condition of a DC motor can be determined by observing the condition of the commutator.

a. **Good Commutator Surfaces.** Good commutator conditions can be found by observing the copper surface patina or surface markings. This skin that develops is made up of oxide and graphite. A light tan film indicates that the machine is performing correctly. A mottled surface is characterized by random film patterns on commutator segments and is normal. In the case of slot bar marking, the film is slightly darker, and occurs in a definite pattern that relates to the number of conductors per slot. Heavy film indicates that this condition is acceptable if uniformly found over the entire commutator.

b. **Commutator Deterioration Signs.** The following signs and causes indicate degrading motor performance and require attention.

(1) **Streaking.** This condition indicates the start of metal transfer from commutator to brush. Light brush pressures, a light electrical load, an abrasive or porous brush in use or dust contamination can cause the condition.

(2) **Threading.** A fine line threading on the commutator surface occurs when an excessive quantity of copper transfers to the brushes. If severe the commutator will require resurfacing and brush wear will be rapid. Light electrical loads, light brush pressures, porous brushes, or dust contamination can cause the condition.

(3) **Grooving.** Grooves in the brush path are caused by abrasive brushes and dust contamination.

(4) **Copper Drag.** A build-up of copper material at the trailing edge of a commutator segment is caused by light brush pressures, vibration, abrasive brushes and contamination.

(5) **Pitch Bar Marking.** Low or burn spots on the commutator surface are caused by poor armature connections, unbalanced shunt fields, vibration or abrasive brushes. The number of marks equates all or half the number of poles.

(6) **Heavy Slot Bar Marking.** Etching of the commutator segment trailing edges is caused by poor electrical adjustment, electrical overloads or contamination. The pattern relates to the number of conductors per slot.

10.27 DC Motor Cleaning. Regular cleaning of DC motors is essential to reliability and long service life. Use a portable vacuum cleaner and soft bristle brush to dislodge dust and other material. Disconnect the power before starting.

a. Clean the brush boxes and brushes, and ensure that the brushes move freely within the brush holders. Clean the accessible field and armature windings.

b. Clean the commutator and commutator risers. Use a small soft brush and clean out any build-up of dust in between the commutator segments. Build-ups have the effect of shorting out the insulation between the commutator segments. *Do not polish or clean commutator with emery paper.* Clean the lower part of the motor.

10.28 Carbon Brushes. The carbon brush must have good commutating and contact characteristics, good mechanical strength and wear properties, a resistance to sparking and a suitable contact voltage drop. Brushes are normally manufactured from hard carbon, and natural or electrical graphite. Metal graphite brushes are used for slipring applications as they have lower contact voltage drops. If the brush wear is abnormal this may be caused by very low humidity, abrasive dust, intermittent loads, a commutator surface without a properly developed skin, incorrect brush grades, jammed brushes and excess sparking. Sparking is caused by poor machine commutation, due to wrong brush types and grades, incorrect brush pressures, badly undercut commutators, or excessive vibration or overloading. Typical brush pressure is in the range 170–210 g/cm^2. If the brush pressure is too low the contact voltage drop increases and brush wear will increase due to burning. If the pressure is too high there will be increased friction and increased mechanical wear. Check the pressure on all brushes using a small spring balance. The wear surface of the brush can also indicate performance. A very shiny surface indicates excessive friction or brush movement. A brush should always be semi-bright and have a surface covered with small pores. If brush replacement is required, it must be done correctly or considerable damage will be done to the commutator.

(1) Ensure that you use the correct brush for the machine.

(2) Use a very fine grade strip of sandpaper slightly wider than the brush. Reverse it so the abrasive surface is under the brush. Move it back and forth around the commutator so that the carbon brush is shaped to that of the commutator.

(3) Use a vacuum cleaner and extract all the dust out of the machine to prevent accumulations of abrasive materials. Never use emery cloth, as this scratches the commutator surface, and the conductive particles lodge in the commutator segments causing shorts and arcing.

Table 10-13 DC Motor Troubleshooting

Symptom	Probable Fault
Windings overheating	Motor overloading Run time excessive Ventilation insufficient High ambient temperature
Excessive commutator sparking	Motor overloading Oil on commutator Brushes sticking Brush pressure too low Brushes worn Commutator dirty Commutator damaged Excessive brush dust buildup
Motor overloading	Excessive mechanical load Bearings binding
Excessive current draw	Excessive mechanical load Bearings binding Valve closed (if a pump load) Electrical connection fault
Excess motor noise and vibration	Bearing failure Motor hold down bolts loose Motor load transmitting vibration Misaligned coupling Coupling damaged and out of balance Brushes bouncing on commutator

Water and Sewage Systems

11.0 Water Systems. The modern sailing yacht has several water systems that must be properly planned, and maintained. A failure in the fresh water or sewage system makes life uncomfortable. A failure in the gray water systems such as the shower drain and sink units is an inconvenience. A failure in bilge pumps systems is dangerous and renders the boat unseaworthy. Fresh water is the one essential, and having the capacity to make it rather than find ports and marinas to refill offers more options, including longer voyages. Watermakers probably pay for themselves when offshore cruising with reduced marina fees and trips to suitable ports. A typical arrangement is illustrated below. Look at Jabsco (www.jabsco.com) who also own FloJet and Rule.

- Pressurized Water Systems
- Desalination Systems
- Bilge Pump Systems
- Sewage and Shower Drain Systems

11.1 Pressurized Water Systems. Water is the one essential, where water is everywhere but none is fit to drink. A system is easy to install, but certain basics must be considered.

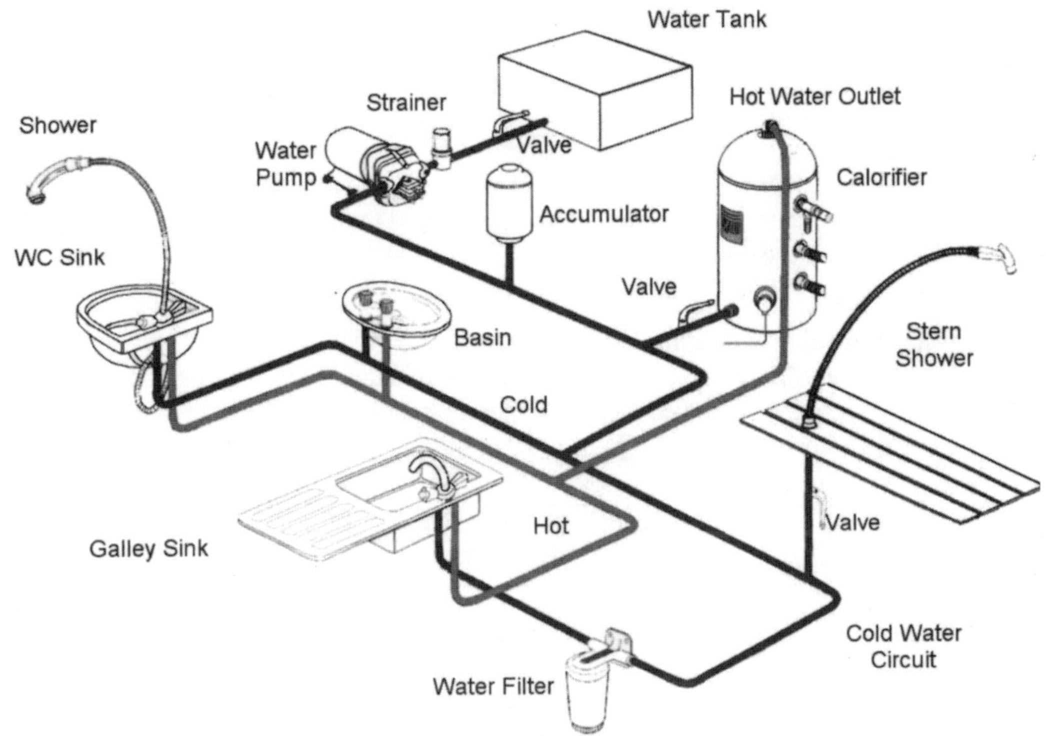

Figure 11-1 Pressurized Water System *(Courtesy Cleghorn Waring)*

11.2 Water Tanks. It is good practice to have two separate tanks for water stowage. Before filling a tank, transfer the remaining water to one tank. The new water can be put in the tank without contaminating water you know to be good. Then, if the water is of poor quality and you have to dump it, you do not lose the whole lot. Toxic by-products from bacteria are characterized by unpleasant smells. Cleaning regimes should be undertaken at least twice a year to ensure the integrity of your water.

a. **Cleaning.** The tank should be scoured by hand with a brush if accessible, but do not use excessive quantities of detergent.

b. **Flushing.** Fill and flush out the tank at least three times.

c. **Disinfection.** New water and the tank must be disinfected to prevent bacterial growth. Water chlorination is easily accomplished by adding a solution of household bleach in the quantities of 5 to 100 of tank contents. Let some amount run though all outlets to disinfect all parts of the system. Then top off the tank and allow to stand for four hours. Re-flush the system another three times. Now add vinegar in the ratio of 1 liter to 50 liters of system capacity and allow to stand for two days. Refill with fresh water and flush three times again. The tank is now ready for use and will maintain potable water quality for several months. An easier and quicker way is to use Puriclean, Aquatabs or a similar brand, which will clean and purify the tank. After filling the tank and adding the cleaning solution, always let it stand a few hours before flushing.

11.3 Water Pressure Pumps. The primary purpose of the pump is to supply and pressurize the water from the tank. A pump is selected based on the number of outlets to be supplied and the flow rate required. If the pump is incorrectly rated for the system the flow will drop off when another outlet is opened. A significant pump development is the Sensor-Max VSD (Variable Speed Drive) water pump from Jabsco. The pump unit has a Hall Effect pressure sensor for the water flow and varies the motor speed to maintain constant pressure. This dispenses with the need for an accumulator and pressure switches. The VSD controls the motor speed using a PWM DC motor controller.

a. **Diaphragm Pump.** These units are the most robust and are designed for multi-outlet systems. They are more tolerant to dry running conditions, self-priming, relatively quiet in operation and have built-in hydraulic pulsation dampening.

b. **Impeller Pump.** These units normally have a pump with bronze casing and nitrile or neoprene impeller. They are less tolerant to running dry and they are self-priming.

Table 11-1 Water Pump Data Table

Model	Current	Flow l/min	Max. Head	Cut-in	Cut-out
Jabsco					
44010	4.0 amps	9.5	1.2 m	10 psi	20 psi
36800	6.0 amps	12.5	1.5 m	10 psi	20 psi
Flojet					
143-12 V	3.9 amps	12.5		20 psi	35 psi
143-12 V	6.0 amps	17.0		20 psi	35 psi
Whale					
EF 2.0612	3.9 amps	7.0		16 psi	32 psi
EF 2.1012	4.2 amps	10.0		16 psi	32 psi

11.4 Pump Wiring. Observe the following when installing water pumps.

a. **Cable Sizes.** Ensure that the cable is rated for maximum current draw and voltage drop as voltage drop problems are very common.

b. **Connections.** Connections are normally made directly to the pump motors; make sure that the crimps are done properly. Also put some Vaseline over the connection to stop moisture getting to the connection lugs. Where wire tails require a butt splice, ensure that they are crimped properly. Some pumps with integral pressure switches require connection directly to one side; make sure this is properly done.

c. **Bonding.** Some metal cased pumps, such as wash down pumps must be bonded. This should be taken to the negative polarity of the supply. In the event of a positive short to the case, this will ensure circuit breaker trips.

11.5 Water System Strainer. The strainer is installed in the water suction line to protect the pump from damaging sediment and particles from the storage tank.

a. **Element Cleaning.** It is essential to clean regularly the stainless steel element. Blockages are most frequent when commissioning a new vessel, or after refilling an empty tank. I have seen a number of vessels where the element has been removed because the owners were tired of cleaning blockages. The result will be early pump failure. It is good practice to clean the system.

b. **Bowl Seals.** After cleaning the element, ensure that a good seal is made with the transparent inspection cover. Imperfect seals can cause air being drawn into the system. Ensure that the seal is in good condition, and apply a smear of Vaseline.

11.6 Water System Accumulators. An accumulator is an essential part of any water system. The basic principle is that air will compress under pressure and the water will not. The accumulator is a tank filled with air that fills to approximately 50% with water when the pump operates. After the pump stops running, the compressed air provides pressurized water stored within the accumulator. It serves two functions, the first being a pressure buffer or cushion which absorbs fluctuations in pressure. The effect is to operate quietly, and the pump pressure switch is able to reach the cut-off pressure, which increases the life of the pump, motor and the pressure switch. The life of the pump is extended, as the accumulator will prevent the pump operating as soon as the water outlet is opened. If a large accumulator is installed, the pump will operate less, and where large demand systems are installed, larger accumulators are required. There are two types:

a. **Non-pressurized.** These units are typically plastic cylinders, which are installed upright within the system. These also have a cock at the top to vent off air within the water system. With tanks that do not have bladders to separate the air and water, the tank must be drained every few months, as the air gradually disappears and the tank no longer functions as an accumulator, there must be air inside.

b. **Pressurized.** These accumulator types have an internal membrane or bladder that can be externally air pressurized with a bicycle pump, or are factory pre-pressurized with nitrogen. At installation the following procedure must be performed: turn the pump off; open outlets and release system pressure; using a car tire pressure gauge, release nitrogen until pressure falls to 5 psi below pump cut-in pressure. If too much pressure is relieved, use a bicycle pump to increase to correct pressure.

11.7 Water Filters. Filters should be fitted to all drinking water outlets. In most cases, this is the galley outlet. A filter will remove small particles, off tastes and smells caused by tank water purification chemicals, as well as some bacteria. These will form in pipes and tanks during extended periods of inactivity. You can use the tank sterilization procedure and flush the system out. Always install a filter with easily replaceable filter elements and replace promptly at the stated service life. The Whale types are a unit replacement. Jabsco have the Aqua Filter. Always cleanse the water system before installing a new filter. Filters are generally made of activated carbon. Improved water quality is possible from filters that use porous ceramic, which removes all particles and detectable bacteria. A good filter should always provide a test report issued by an appropriate authority. It should be rated for the expected flow rate (in gallons/liters) and should be renewed at the due date. A filter is never a substitute for clean tanks.

11.8 Hot Water Calorifier. The calorifier or hot water system is becoming one of those hard-to-do-without luxuries. It is not difficult to install or incorporate into a water system, and it will even function as an additional water reserve. The term *calorifier* is used as most marine hot water systems heat from built-in coils (calorific transfer) supplied from heated engine cooling water or on the old tramp ships I once served on, steam. It makes sense to utilize all the available energy consumed by the engines. The following should be noted:

a. **Heating Coils.** The majority of units are fitted with a single copper heating coil. Beware of the cheaper imported units, as the coils are very small and have only one or two turns. Quality calorifiers will have several turns installed to ensure good heat transfer rates.

b. **Electric Elements.** Calorifiers should also incorporate an auxiliary electric heating element for mains AC heating capability.

   (1) **Ratings.** Element ratings should not exceed 1200–1800 watts due to electrical supply limitations of shore power and small generators, unless you have a reasonably high output generator set.

   (2) **Thermostats.** A thermostat is also essential for controlling temperature and preventing overheating and therefore over pressure conditions.

c. **Pressure Relief Valves.** All calorifiers should have a pressure relief valve. The valve should be regularly operated manually to ensure that it is not seized, and to eject any insects or debris from the overflow pipe.

d. **Valves.** The inlet of a calorifier should always have a non-return valve fitted to prevent the heated and expanding water in the tank from back flowing into the cold water system and pressurizing it.

e. **Insulation.** Ensure that the calorifier has a good insulation layer or cover to avoid wastage of heat. If the engine is run every alternate day, good insulation will keep it warm over the extended period.

f. **Mounting.** The calorifier must be mounted with the coil on the same level as, or below, the engine cooling water source. This is because the engine pump must circulate water through a longer system, which introduces resistance and could overload the pump.

g. **Air Locks.** There must be no air locks in the system as these also go through the engine cooling system and effect cooling. The calorifier must always be installed lower than the engine water filling point.

h. **Hose Connections.** Use heat resistant rubber hoses to connect up heating circuit. Ensure that air locks cannot form in the hoses. Ensure that all hose connections have double hose clamps.

11.9 Diesel Hot Water Heaters. The diesel hot water system is now becoming commonplace on vessels. This unit can also be part of a central heating system. Companies such as Eberspacher and Webasto have very efficient systems. The Webasto is illustrated below. The typical operational cycle is as follows:

a. **Starting.** Cold air is drawn in by an electric fan to the heat exchanger/burner. This is normally from the engine area.

b. **Ignition.** Fuel is drawn in at the same time by the fuel pump from the main tank and mixed with the air. The fuel is ignited by an electric glowplug in a combustion chamber.

c. **Combustion.** Combustion takes place within a sealed exchanger and the exhaust gases are expelled to atmosphere.

d. **Heating.** An integrated water pump circulates the water through the heat exchanger and subsequently to the calorifier and heating radiators. A thermostat in the cabin shuts the system down and operates to maintain set temperature. Eberspacher have developed an automatic quarter heat control to reduce unnecessary cycling, thereby improving fuel economy.

11.10 Water Pipes and Fittings. Water pipes should be of a high quality material that is suited to both hot and cold water. Observe the following when selecting and installing piping:

a. **Pipe Standard.** The piping should be non-toxic, suitable for potable water systems, and must not be able to support microbiological growth. There are two types:

(1) **Semi-Rigid Piping.** Whale offers a color-coded, semi-rigid pipe system. Ensure that the pipe is not kinked. Where tight bends are required, install a bend. They also have a water piping system called WaterWeb.

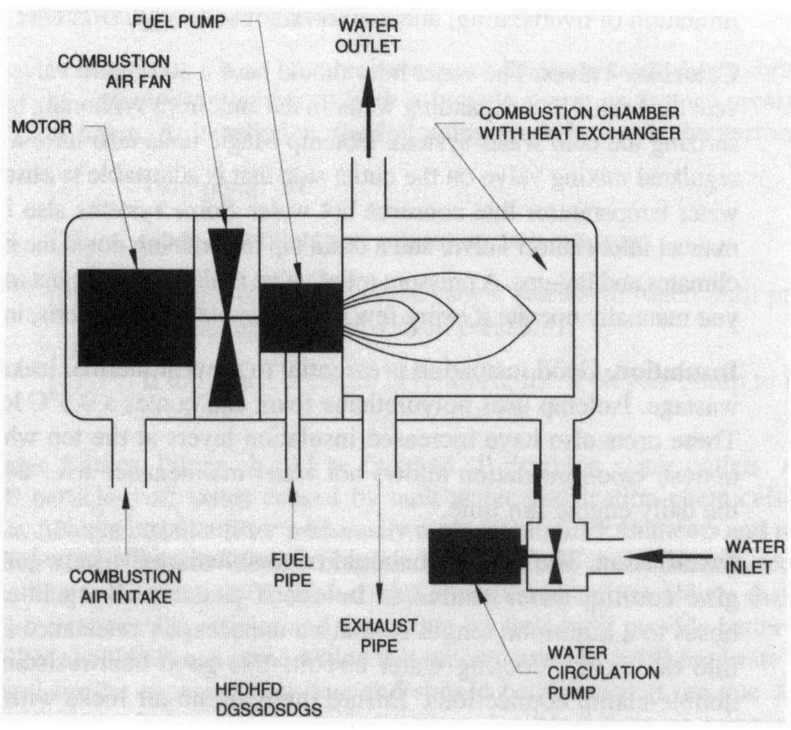

Figure 11-2 Diesel Hotwater System

(2) **Flexible Hose.** Hose is the most common piping. Ensure that it meets required standards. Hose is prone to kinking, so installations should be done with care.

b. **System Pressures.** Piping must be able to withstand the water system pressures. Whale piping is rated at 60 psi and 90°C. When installing piping over longer runs, larger pipe diameters are required to reduce the friction losses. Table 11-2 illustrates pipe diameters.

c. **Fittings.** Fittings must be able to withstand system pressures. Nuisance leakage can be avoided. Where plastic hoses are used, generally PVC T-joints are installed with clips. The Whale or Acorn systems are excellent, and they are easy to install and service.

d. **Outlets.** There are many different taps, valves, and shower heads on the market. Always select good quality items, and choose only those that are compatible with the whole plumbing system. This makes finding spare parts easier. Reputable names include Whale and Jabsco. If you are using a non-flexible, permanent shower head, opt for one of the domestic, low water consumption fixtures.

e. **Connections.** Ensure that all piping or hose connections are double clamped. Acorn and Whale fittings should be firmly tightened.

Table 11-2 Recommended Pipe Diameters

Pump Port Diameter	Hose Diameter
6 mm	13 mm ID
10 mm	13 mm ID
13 mm	16 mm ID
19 mm	25 mm ID
25 mm	25 mm ID
38 mm	38 mm ID
50 mm	50 mm ID

11.11 Water System Winterizing. In colder climates, proper winterization is essential to prevent damage from freezing. Perform the following protective measures:

a. **Remove Pump.** If possible, remove the entire pump and store in a dry place.

b. **Drain System.** The most practical precaution is to totally drain the water system, including the pump and accumulator. Do not use antifreeze solutions in the potable water system.

Table 11-3 Water System Troubleshooting

Symptom	Probable Fault
Will not prime (No discharge)	Restricted inlet Restricted outlet Air leak in suction line Pump diaphragm ruptured Debris under flapper valves Pump housing fractured Strainer clogged Valve closed Kink in water pipe or hose No water in tank Clogged one-way valve Discharge head too high Low battery voltage (pump slow)
Pump will not operate	Circuit breaker tripped off Pump connection loose or broken Pressure switch fault Motor fault Pump seized
Pulsating water flow	Restricted pump delivery
Pump cycling on and off excessively	System pressure leak Water outlet leaking Accumulator problem
Pump will not switch off	Water tank empty Pump diaphragm ruptured Discharge line leaking Pressure switch fault Debris under valves
Low water flow and pressure	Air leak on pump inlet Strainer clogged (common) Pump impeller worn Pump diaphragm ruptured Pump motor fault High discharge head Pump improperly rated

11.12 Desalination Systems. Cruising to foreign places is half the fun, but unfortunately when you get there, the water is often scarce or not fit to drink. As a result, watermakers are becoming more popular on many vessels as they give you a greater degree of freedom. Onboard water resources are limited, and this affects maximum cruising ranges. The most practical system is the reverse-osmosis desalinator, as evaporative systems require long-term engine use for reasonable economy. It must be stressed that water should not be made within 10 miles of a coastline or within inhabited atolls in the Pacific. These are generally polluted to levels well above World Health Organization (WHO) recommendations and this pollution can be carried into the tanks with product water. Principles are as follows:

a. **Reverse-Osmosis Principles.** In natural osmosis, when fresh and salt water are separated by a semi-permeable membrane, fresh water flows through to the salt water side. To reverse this process, salt water is pressurized to force the fresh water out through the membrane. Sea water is pressurized by a priming pump and filtered to remove particles. Then pressure is increased with a high-pressure pump, which forces fresh water through the membranes. The membranes are housed in a high-pressure casing.

b. **System Components.** The osmotic membranes are the heart of any system. Membrane quality is the key to a good unit; cheaper units with poor quality membranes are usually very expensive because of the high maintenance and replacement costs. Pumps can either be engine driven or AC-shore power driven. Power consumption can be up to 2 kilowatts. Seafresh specifies a minimum generator capacity of 3 kilowatts for starting currents and approximately 1.5 kilowatts for running currents. Well-designed systems incorporate prefilters for the salt water. Prefilters typically have a rating of 50 microns, followed by a second filter of 5 microns.

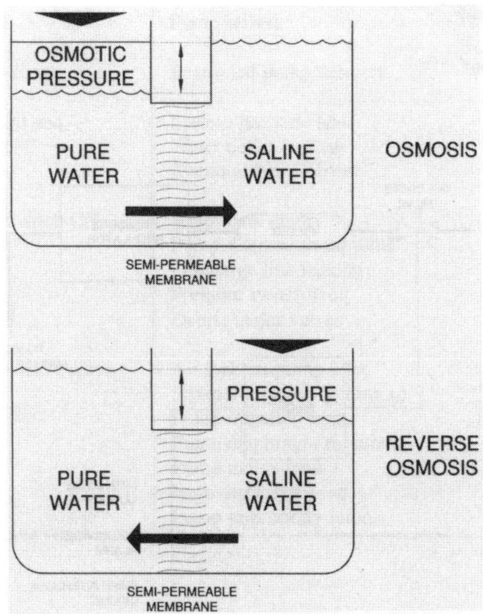

Figure 11-3 Reverse-Osmosis Process

c. **Monitoring and Control.** The Sea Recovery (www.searecovery.com) system is a typical system and the operation is as follows:

(1) Raw seawater is supplied through the seawater inlet valve and sea strainer to the booster pump suction.

(2) The seawater is then pressurized to 20 psi by the booster pump and supplied to the media filter. The 5 micron pre-filter and oil water filter remove sediments, suspended solids, silt and oil. This water is pressure monitored with a gauge and low-pressure switch, which stops the system when low pressure is detected.

(3) The water is pressurized to around 900 psi by the high-pressure pump and regulated by a valve. This is also monitored and controlled by a high pressure switch.

(4) The pressurized water enters the reverse osmosis membranes, which forces out the salt and minerals. A salinity probe monitors the product water quality, which also adjusts for water temperature. The brine flows through a monitor and is then dumped through a discharge valve.

(5) The product water is monitored, passed through a charcoal filter and UV sterilizer and is then sent to the potable water tanks.

(6) The system also has an automatic fresh water flushing system, which flushes the system to reduce membrane fouling.

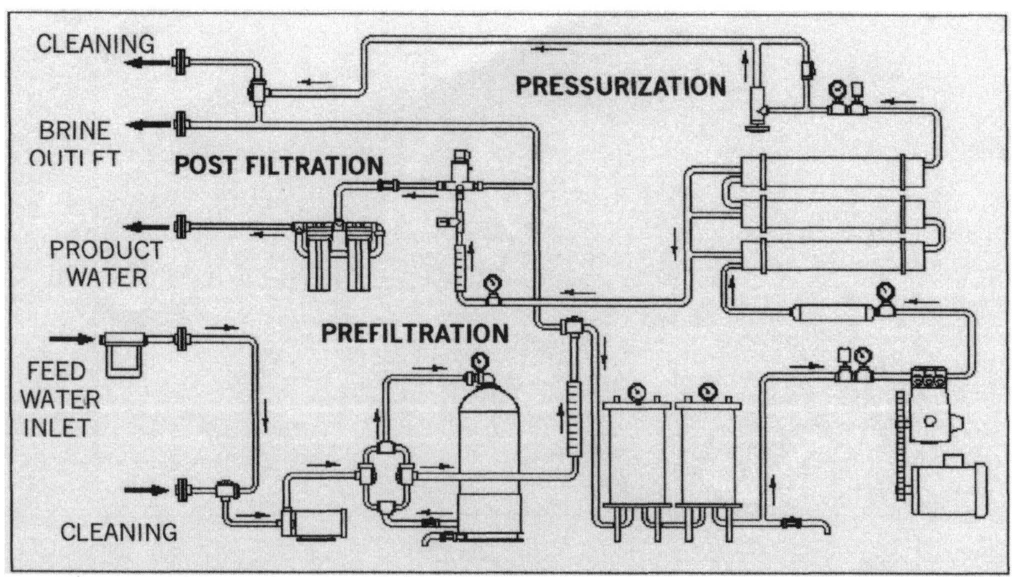

Figure 11-4 Watermaker System *(Courtesy Sea Recovery)*

d. **Installation.** Space considerations are always of critical importance, and many systems such as Sea Recovery, Spectra, SeaFresh and Pur have solved this problem. Through hull fittings are required for raw seawater intake and the overboard brine discharge. It is not good practice to take the input from auxiliary engine or generator water inlets as this may starve the system of water.

e. **Outputs and Membrane Correction Factors.** In a good system, salt rejection rates are typically 99% in the ph ranges 4 to 11 and operating pressure range of 700–900 psi. In these conditions output is unaffected by pressure and temperature. Where temperatures and pressures change correction factors must be applied to ensure improved production rates.

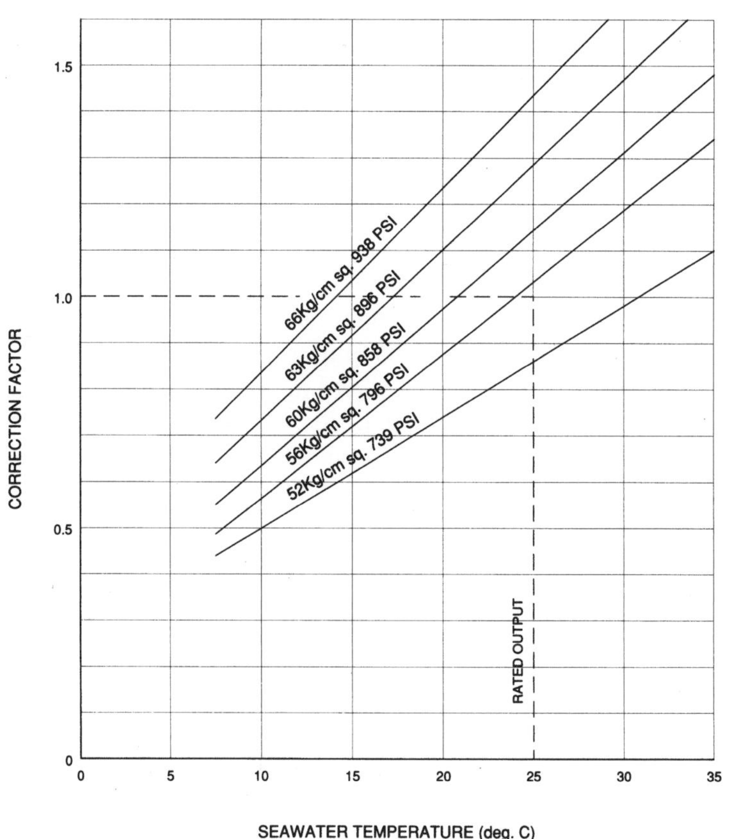

Figure 11-5 Desalinator Temperature Correction Factors.

f. **Water Treatment.** Membrane fouling is a common problem caused by organic molecules, suspended solids, bacteria and algae, as well as minerals etc. Pre-filters take out only partially some of the material. Cleaning is time consuming and requires the use of chemicals. The Zeta Rod system significantly increases the intervals between cleaning and also the permeate rates, lowers feedwater pressure and transmembrane pressure drops. This entails the use of an electrode in the feedwater line. The rod effectively forms a capacitor with the water and its impurities, and the piping and vessel walls. The system charges particles that ultimately form sludge and scale to the same polarity. This causes repulsion and prevents bonding and formation of deposits in piping and membranes. The units use a high voltage to generate the electrostatic field. Current consumption is low. Systems reduce the output to zero if a short circuit develops.

g. **Maintenance.** These are typical maintenance procedures:

(1) Clean inlet strainer at the same time as the engine strainer.

(2) Pre-filters can be washed 5–6 times before replacement. This equates approximately 80 hours operation in clean waters.

(3) Disinfect membranes to prevent biological fouling for any shutdown period exceeding 14 days using recommended biocides. Failure to do this will significantly reduce output and damage membranes. Never allow membranes to dry out.

(4) Check pressure pump oil levels and renew every 500 hours.

(5) Check and re-tension rubber drive belts every 6 months.

(6) Clean membranes when output drops below 15% of rated output or when product salinity increases. This is due to the build-up of grime, biological material and mineral scale. Do not open the pressure vessel to do this. Clean according to the manufacturer's recommendations. This usually entails the use of alkaline and detergent cleaning for removal of organic material, and acidic cleaning for removal of mineral scale.

Table 11-4 Desalinator Troubleshooting

Symptom	Probable Fault	Corrective Action
Low water flow	Blocked strainer Blocked prefilter Membranes fouled Pump belts loose	Clean strainer Clean or replace filter Clean membranes Tension belt correctly
No product-water flow	Pump stopped Circuit breaker tripped	Check circuit breaker Check drive belts Clutch coil fault
Circuit breaker tripping	Pump clutch coil failed Clutch wire grounding out Pump seizing	Replace winding Repair connection/wire Repair pump Overhaul valve Overhaul pump
Low working pressure	Relief valve leaking Pump fault High pressure loss Dump valve jammed open	Examine for leaks Repair valve
Product water salty	Fouled membranes Excess working pressure	Clean membranes Decrease pressure

11.13 Bilge Pump Systems. Bilge pumps play a crucial safety role in any vessel, yet many owners tend to get the cheapest units and install them improperly. Bilge pumps should be of the highest quality available, installed correctly and regularly maintained. The following factors should be considered when selecting and installing electric bilge pumps. There are two basic types, the submersible pump and the centrifugal pump. Submersible pumps are notoriously unreliable and cannot be maintained or repaired.

a. **Head.** Head pressure is related to the height that the water must be lifted to. All pumps have maximum head figures for a particular model.

b. **Flow Rate.** Most bilge pumps are listed with flow rates, which are designated as gallons (or liters) per hour or per minute. Electric pumps with bronze housings are rated up to a maximum of 11 gal/min (50 l/min).

c. **Impellers.** Pump impellers come in a number of different compounds. Choose the correct type for optimum life and efficiency. Centrifugal pumps should never be operated dry for more than 30 seconds; they are designed to be lubricated by the pumped liquid. Operating without liquid generally means a ruined impeller. Impeller types are as follows:

(1) **Neoprene.** These are typically found in bronze pumps (Jabsco) and are suitable for bilge pumping in temperatures ranging from 4°C to 80°C. Use at the outer temperature limits reduces performance and service life. They must not be used to pump oil-based fluids as the neoprene impeller can absorb oil compounds and expand. On the next start-up, the binding impeller is destroyed. Always flush out a line if oily fluids are used.

(2) **Nitrile.** These are designed for pumping fuel, but they are also suited to pumping oil- and fuel-contaminated engine bilges in temperatures from10°C to 90°C. Use at the outer temperature limits reduces performance and service life. Nitrile impellers have a flow rate 30% lower than neoprene impellers, so they should not be used in any high temperature applications.

d. **Submersible Pumps.** These pumps are by far the most common. It is important to always buy and install the very best quality you can. Cheaper pumps are unreliable and tend not to last. Pumps have the following general characteristics:

(1) **Motor Rating.** Motors are rated continuously, but the bilge water normally assists motor cooling while pumping.

(2) **Motor Type.** Motors generally use a permanent magnet motor, which means no brushes.

(3) **Dry Running.** Pump impellers are not damaged by dry running, though motors require water to cool them.

11.14 Bilge Alarms. The main bilge should have a separate visual and audible alarm to indicate levels above the normal operating range of automatic systems. Most boats have automatic bilge pumping systems. If the pump is running and cannot keep up with the water inflow, an additional alarm will indicate the high bilge level. If the float does not operate and the bilge starts to fill a separate alarm will also indicate the condition. I have experienced both conditions and additional protection is prudent.

Where bilge pump control circuits incorporate automatic operation (e.g float switch), caution should be given to the risks of pollution by uncontrolled or unmonitored discharge of oily bilge water.

The majority of bilge pumps have a float switch incorporated to enable automatic unattended operation. Where pumps are running in this mode you should be aware that uncontrolled discharges of oily bilge water into the water might render you liable for stringent penalties and fines. Consideration should be given to installing suitable filtering equipment.

(1) **Pollution.** The wilful or accidental discharge of oily wastes into harbor and coastal waters will carry heavy fines. It is the environmental responsibility of all boat owners not to discharge any waste into the sea. Any bilge that can have oil in it must never be fitted with an automatic pumping system. Filter systems are a good investment. All vessels have obligations under MARPOL.

(2) **Controls.** Automatic switches are notoriously unreliable. If the float switch stays on the bilge pump may burn out and probably ruin a set of batteries by totally flattening them. There are a number of activation devices, which are explained below.

Float Switches. Float or level switching devices may use a number of different operational principles:

(1) **Mechanical Floats.** This is by far the most common device and probably the most reliable, if the float switch is of high quality, if it cannot be fouled by the pump cable, and if the bilge is free of debris. This circuit diagram, provided for steel vessels, isolates the positive supply to the float switch. This minimizes the common and serious risk of corrosion problems if a leakage occurs.

(2) **Solid State Devices.** These include ultrasonics, conductive probes, etc. While some appear to work well, there are a great number of failures. Some cheaper units can cause electrolytic corrosion problems. If the probes are fouled or coated with oil, they often don't work. Some have a delay feature that requires the presence of water for 15 to 20 seconds before they activate. This prevents the pump from start-and-stop cycling in rough water. I have not found any of these devices to be too reliable.

(3) **Optical Devices.** These devices are quite new and resolve many of the problems normally encountered with units using probes. The pump units are controlled by an innovative optical fluid switch that emits a light pulse every 30 seconds. If the lens is immersed in water, the light beam refracts and the beam's change in direction is sensed by a coating inside the lens. This triggers the pump. Time delay circuits can be adjusted for periods of 20–140 seconds so that the pump will continue draining the bilge after water clears the sensor. I have tried some of these devices and I find them very good.

(4) **Vacuum Devices.** These are relatively old but simple devices which work reliably. They depend on the pressure of water in a tube to activate a switch via a diaphragm.

(5) **Ultrasonic Devices.** In practice, these devices have had limited success and I would not recommend them.

11.15 Bilge Pump Installation. Bilge pumps must be installed as follows to operate correctly and reliably:

a. **Location.** Mount the pump or suction line in the lowest part of bilge. It is best to keep this a short distance from the bottom to avoid drawing in bilge sediments.

b. **Strainer.** Always install a strainer on the suction side of centrifugal pumps. Submersible pumps have a strainer as an integral part of the base, but these are rather coarse. It is quite common for bilge debris to jam the impeller.

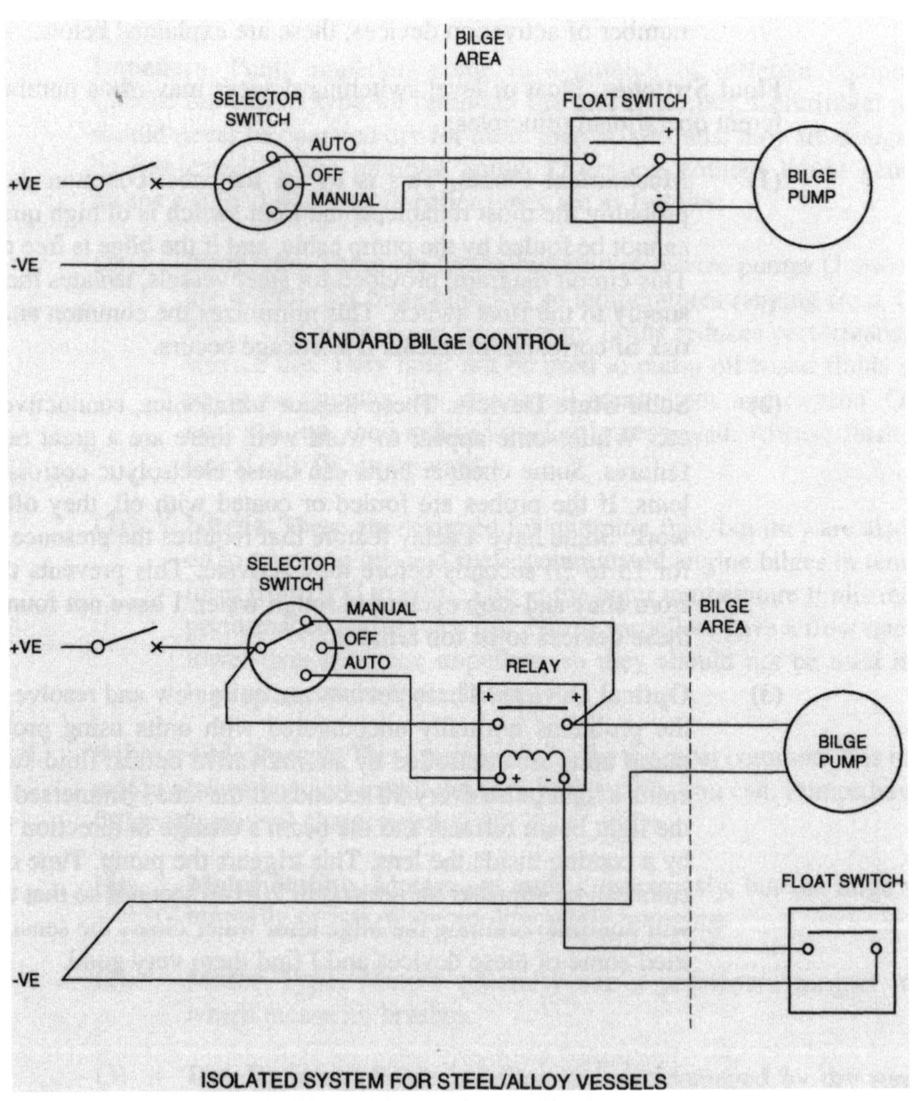

Figure 11-6 Bilge Control Schematics

c. **Discharge Piping.** Select flexible hose that will not kink. Many pumps are rendered ineffective due to kinks or constrictions in the discharge line. Always use two hose clamps on every hose connection as a safety precaution. The discharge should be as far above the waterline as possible so it will be clear even when heeled.

d. **Electrical Connections.** If the cable is long enough, make connections above the maximum bilge water level. I recommend soldering each connection, to cover the joint with heat shrink insulation, and to cover the entire cable with heat shrink insulation or wrap it in self-amalgamating tape. This will generally prevent the joint from interacting with salt water and failing. The circuit must be fused on a circuit breaker rated for the cable size, typically 15 amps. Always run the pump after installation to ensure that pump rotation is correct.

11.16 Bilge Pump Maintenance. Regular maintenance is essential for reliable pump operation. Regularly clean bilges of sediment and debris. Run pumps every month with water in the bilge. Many bilge pumps can seize after months or even years of disuse.

Table 11-5 Bilge Pump Troubleshooting

Symptom	Probable Fault
Low water flow	Strainer blocked with debris Pump impeller fouled Suction hose kinked Suction hose blocked with debris Suction line has air lock
Pump will not operate	Circuit breaker tripped Float switch fouled (usually with debris) Float switch connections corroded off
Pump will not switch off	Float switch jammed Float switch fouled by debris Float switch mounted too low Float switch connection short circuited
Circuit breaker tripping	Pump impeller seized Bilge area connections short circuited Pump winding fault

11.17 Sewage Systems (Marine Sanitation Devices). Many sewage systems are being altered from hand pump toilets to electrics. The stringent requirements for holding tanks and pump out systems have required careful considerations in systems planning. In the US the Clean Vessel Act of 1992 is the primary legislation. In many countries, similar legislation is being introduced. It is important to comply with laws pertaining to illegal discharges. An MSD is any equipment for installation on board the vessel that is designed to receive, retain, treat or discharge sewage, and any process to treat the sewage. Politics and confusion are prevalent in many places relating to the use of devices, and boaters should make sure they understand the requirements. The USCG certifies MSDs, which fall into 3 categories:

a. **Type I MSD.** This device uses chemicals to disinfect sewage. The discharge must be free of visible solids and meet standards for bacterial content. The sewage must be macerated to break up solids. The fecal coliform bacteria output must be at least 1000 colonies per 100 milliliters or less. A system that meets this standard is the Raritan Lectra/San. It operates using both salinity and electric current that consists of electrodes to break down the seawater to form chlorine (hypochlorous acid), a chlorinating agent that kills bacteria and disinfect the sewage. After treatment, the acid recombines to reform as salt water. The Raritan PuraSan MSD uses a solid tablet made of chlorine that produces a halogen solution to treat waste. Both systems draw around 45 amps for 3 minutes on each flush, which has considerable impact on the electrical system power requirements. These units cannot discharge into No-Discharge zones. The SeaLand SanX can be used in all waters but installation requirements are considerably more than the others and may suit larger vessels. The SanX injects a chemical disinfectant agent into the treatment tank to mix with the macerated waste.

b. **Type II MSD.** This is a device similar to Type I with a higher level of treatment and higher quality discharge. The fecal coliform bacteria output must be at least 200 colonies per 100 milliliters, with suspended solids of less than 1000 parts per 100 milliliters. The best-known devices are those from Microfors and Galleymaid. Cost and installation are considerably greater than Type I devices.

c. **Type III MSD.** This is a holding tank. This means no discharges within nominal limits. Emptying is usually via a deck fitting and tanks have a vent line overboard.

11.18 MSD Systems. MSD systems have several process components, and various sensors and electrical elements. Some MSD units are PLC controlled, so correct operation of all input sensors and output devices is critical for proper operation. The first stage treatment tanks contain level sensors. The second stage sedimentation tanks do not have any components. The macerator pump, sludge pump and discharge pump all require routine maintenance. Backwash systems have both a water pump and solenoid valves. The disinfection system has both a flow control system with chemical feed pump, and chlorination units have a power supply. Ensure that no toilet chemicals incompatible with sodium hypochlorite are being used on the boat. Normal troubleshooting principles apply, check power supply, auto and manual selection which may be wrong, PLC may require reset, and then check inputs and outputs to level switches or solenoids. Operational checks include the following:

- **a. Daily.** Check macerator pumps, discharge, backwash and sludge pumps and electric motors for unusual noises and vibration. Check all hose and pipe connections for leaks. Check that the chlorination liquid (bleach) reservoir is full. Verify that the pump pressures are normal where they are installed.
- **b. 6 Monthly.** Check control panel for moisture ingress and corrosion, and ensure cover is closed tightly. Check and tighten all connectors every year. Replace desiccant crystals if installed.
- **c. Yearly.** Check treatment and sediment tank anodes if installed. Check that solenoid valves are operating correctly, and remove to check and clean. Grease bearings only in accordance with operating hour requirements, typically every 10,000 hours.

11.19 Toilet Systems. Toilets come either as a manual system, or as an electric unit with integral macerating function. Manual systems use a piston rod type pump to pump out the combined sewage and seawater. Electric pumps replace the piston rod pump with an impeller type pump and macerator that effectively liquefies waste. These systems use a lot more water to ensure all materials are flushed out properly. There are many power vessels with vacuum flush systems such as the VacuFlush (www.sealandtechnology.com) and Environvac. They use little water, and operate with a vacuum pump. These units have vacuum ejector pumps or generators and discharge pumps that require routine inspection and maintenance, along with automatic valves, vacuum and level sensors, and the control system with alarms. 12-volt vacuum pumps have typical power consumptions of 4–6 amps and large boat systems have pumps up to 1/2 hp. Discharge pumps consume around 6 amps. Look at www.leesan.com.

a. **Electrics.** One of the biggest problems with toilets is the failure to install adequately sized cables to the units or allow for voltage drop. The PAR unit consumes 18 amps and requires a heavy-duty cable rated at around 30 amps. As toilets are always located in a wet shower area, ensure that all electrical connections are taped up with waterproof self-amalgamating tapes. Always allow sufficient cable length to pull the toilet out, as it will be difficult to disconnect the motor. Check the motor connections monthly to ensure no corrosion is occurring. Lightly coat the terminals with silicon or Vaseline. I have started to respray motors with an additional paint layer to seal and prevent water seeping into the motor housing flanges, as corrosion occurs here easily. Before installation, remove each bolt and apply anti-seize grease.

b. **Waste.** It is essential that only normal waste be put through the toilet. To quote that readily available plaque for marine heads, *"Don't put anything in the bowl that you haven't eaten and already digested."* Macerator cutter plates are easily jammed or damaged by putting cigarette and cigar butts, rags and sanitary towels down the bowl. Cleaning macerators is the most unpleasant task on a vessel so it is well worth making the effort.

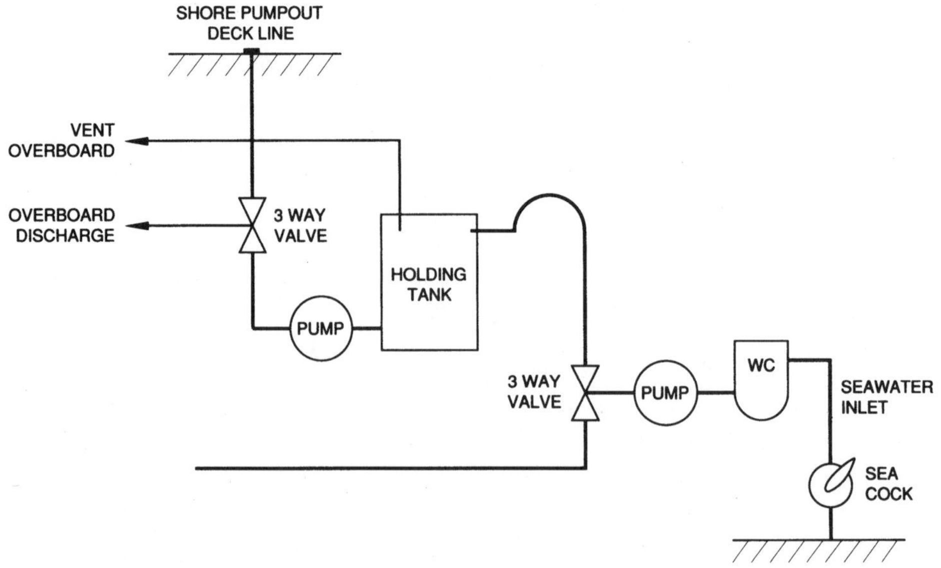

Figure 11-7 Typical Sewage System

c. **Macerators.** Macerator pumps are usually connected to the holding tank discharge and are used to pump out waste to shore facility tanks or overboard. Units grind waste to 3mm size, and are self priming. It should be remembered that pumps are not rated continuously, and run times should not exceed approximately 10 minutes. Heavy-duty models are available for larger systems and greater pump-out capabilities. Jabsco Models are given in Table 11-6. Flow rates are given at maximum and normal heads. After pumping out tanks, flush out macerator pump with clean water to expel any debris that may cause bacterial build-ups.

d. **Troubleshooting.** Noisy or vibrating pumps discharge pressures and volumes have a number of possible causes. Pump or pipes may be clogged; impeller or cutter may be worn. Reduced speeds can be due to wiring and connection faults. High viscosity liquids and altered pump axial clearances affect flow rates and pressures. Noises are often caused by water damaged bearings, pump cavitation and loose impellers.

Table 11-6 Jabsco Macerator Pump Specifications

Pump Type	Port Size	Current	Flow Rate
21950 -1603	Inlet 1½-2"	10 amps	36 l/min @ 1.5m
(12 mins rated)	Outlet 1"		19 l/min @ 6.1m
22140 - 1421	Inlet 1½-2"	25 amps	40 l/min @ 1.5m
(60 mins rated)	Outlet 1"		30 l/min @ 9.8m

11.20 Shower Drain and Sink Systems. Shower drain systems and sinks require specific pumps. The common pumps in use are as follows:

a. **Diaphragm Pumps.** Jabsco have a range of purpose designed shower diaphragm pumps such as Par Max 3 (170 gph) and Par Max 4 units (190 gph) that eliminates the sump pump and float switch. The pumps are self-priming and have four chambers connected directly to the drain outlet and a strainer installed in-line on the suction side. The pumps have power consumption rates of around of 5–8 amps at full load. The Whale Gulper 220 is also very reliable and effective and I have installed one of these myself. They can also run dry, and strainers are recommended to prevent blockages. Pumps are also generally repairable with the appropriate spares kit.

b. **Submersible Pumps.** These plastic units are common. Rule and others manufacture fully integrated sump, suction filter, pump and float switch. The sump pump units usually have a check valve to prevent back siphoning and a clear cover for inspection.

c. **Centrifugal Pump.** Some manufacturers such as Flojet recommend a centrifugal self-priming pump. These have a flow rate of around 12 L/min and use 3.6 amps at full load. This offers the chance to match all the pumps, so that bilge, wash down and shower pump motors are all interchangeable.

d. **Electrical.** Par Max pump motors have an integral thermal overload trip protection. If they trip on internal overload determine the cause of the trip. Typically it is overloading due to bearings seizing or pump jammed with debris and overheating.

e. **Maintenance.** Shower and sink drain pumps are prone to rapid filter clogging due to hair, soap residues and other debris. In automatic float switch units, hair and solidified soap often causes the float to stick. The filters should be checked and cleaned weekly.

f. **Water Conservation.** This water-saving shower idea has worked well on many vessels, including my own. It allows long showers, with low water consumption. The options are that either Solarbag water can be recycled through a water-saving shower head or come directly from the hot water system. The sump suction has a filter to strain suds and hair, and then recycles water back through the system. When finished, water is diverted overboard and you rinse off using the same procedure.

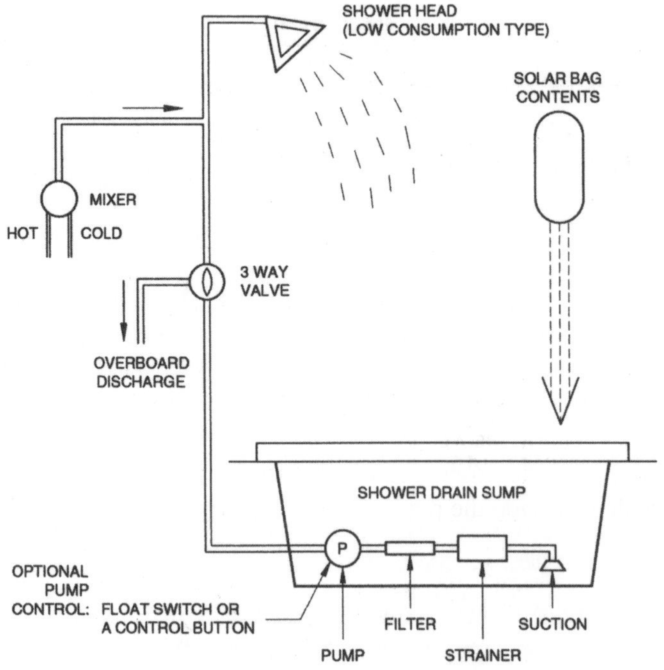

Figure 11-8 Shower System

Engine Electrical Systems

12.0 Engine Starting Systems. A typical engine electrical system includes the battery; the engine control panel; the wiring loom; the preheating system; the starter motor and solenoid; shut-down solenoids; instrument sensors and transducers, and the alternator. There is a basic sequence of electrical functions that take place when starting the engine. When the key switch is turned to <ON>, this closes the circuit to supply voltage to the control circuit, and generally initiates alarms. When no audible or visual alarms occur, no power is on. When the key switch is turned to the <Preheat> position, this manually or automatically energizes the heating glowplugs or heating elements. When the key switch is turned to <START> or the engine <START> button is pressed, voltage is then applied to the starter motor solenoid coil. The solenoid pulls in to supply main starting circuit current through a set of contacts. The contacts when closed supply current to the starter motor positive terminal. This turns over the starter motor to start the engine.

12.1 Electric Engine Starters. Essentially the electric starter consists of a DC motor, a solenoid, and a pinion engaging drive. The DC motor is typically series wound. It provides the high initial torque required to exceed the friction and inertia, (such as oil viscosity), and cylinder compression and accelerate the engine to a point where self-ignition temperatures and combustion starts (typically in the range 60-200 rpm depending on whether glowplugs are used). The pinion and ring gear transmit the starter motor torque to the flywheel. The drive gear pinion has a reduction gear of around 15:1.

a. **Solenoids.** The solenoid is essentially a large high current relay that consists of coil and armature, moving and fixed contacts. The solenoid is mounted directly to the starting motor housing, which reduces cables and interconnections to a minimum. When the solenoid coil is energized by the starting circuit, the solenoid plunger is drawn into the energized core and this closes the main contacts to supply current to the starter motor. On some starters the solenoid also has a mechanical function. The solenoid activates a shift or engaging lever to slide the overrunning clutch along the shaft to mesh the pinion gear with the flywheel. When engaged the starter motor then turns the engine, so meshing occurs before starting.

b. **Starter Motors.** The motor consists of four poles shoes or magnets. Some motors use permanent magnets. The poles are fitted with an excitation winding which creates the magnetic field when current is applied. The rotating part called the armature also incorporates the commutator. The four carbon brushes provide the positive and negative power supply. There are four basic DC motor types in use and they are based on connection of the field windings. The field windings are connected either in series or parallel with the armature windings.

(1) **Shunt (Parallel) Wound Motors.** The motor operates at a constant speed regardless of loads applied to it. It is the most common motor used in industrial applications. It is suited to applications where starting torque conditions are not excessive.

(2) **Permanent Magnet Excited Motors.** The permanent magnet starter offers the advantages of reduced weight, physical size and generates less heat than normal field type starters. Current is supplied via the brushes and commutator directly to the armature. Another feature is that a reduction gear is used, which allows faster speeds and increased torque.

(3) **Series Wound Motors.** On this type of DC motor the speed varies according to the load applied; speed increases with load decrease.

(4) **Compound Motors (Series/Shunt Wound).** This configuration is often used on large starter motors. It combines the advantages of both shunt and series motors, and is used where high starting torques and constant speeds are required.

c. **Pinion Engaging Drives.** The pinion engaging drive is located within the end shield assembly of the starter. It consists of the pinion engaging drive and pinion, the overrunning clutch, the engagement lever or linkage and the spring. When the motor operates the drive gear meshes with the ring gear or flywheel teeth to turn the engine, and then disengages after starting. The overrunning clutch has two important functions. The first is to transmit the power from the motor to the pinion, and the second is to stop the starter motor armature from over-speeding and being damaged when the engine starts. Pre-engaged starters generally use a roller type clutch, while larger multi-plate types are used in sliding gear starters.

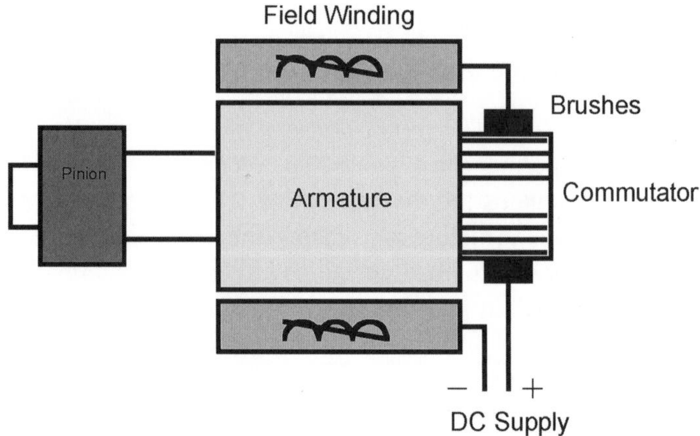

Figure 12-1 Starter Motor

12.2 Starter Types. There are several types of starters in use, the most common being the overrunning clutch starter, and the inertia-engagement Bendix drive is now less common. There are four basic groups of starter motors:

a. **Pre-engaged (Direct) Drive Starters.** The most common type of starter motor is the solenoid-operated direct drive unit and the operating principles are the same for all solenoid-shifted starter motors. When the ignition switch is placed in the Start position, the control circuit energizes the pull-in and hold-in windings of the solenoid. The solenoid plunger moves and pivots the shift lever, which in turn locates the drive pinion gear into mesh with the engine flywheel. When the solenoid plunger is moved all the way, the contact disc closes the circuit from the battery to the starter motor. Current now flows through the field coils and the armature. This develops the magnetic fields that cause the armature to rotate, thus turning the engine.

b. **Gear Reduction Starters.** Some manufacturers use a gear reduction starter to provide increased torque. The gear reduction starter differs from most other designs in that the armature does not drive the pinion directly. In this design, the armature drives a small gear that is in constant mesh with a larger gear. Depending on the application, the ratio between these two gears is between 2:1 and 3.5:1. The additional reduction allows a small motor to turn at higher speeds and greater torque with less current draw. The solenoid operation is similar to that of the solenoid-shifted direct drive starter. The solenoid moves the plunger, which engages the starter drive.

c. **Sliding Gear Drive Starters.** These two-stage starters have either mechanical or electrical pinion rotation. The electrical units have a two-stage electrical pinion-engaging drive. The first stage allows meshing of the starter pinion without cranking the engine over. The second stage starts when the pion fully travels and meshes and then allows full excitation and current flow to the starter motor. The first stage of the mechanical units has a solenoid switch, which pushes forward the pinion engaging drive via a lever. When pinion meshing occurs current is applied to the starter via the solenoid switch.

d. **Bendix Drive Inertia Starter.** The Bendix friction-clutch mechanism drive was developed in the early 20th century. It uses a drive friction clutch, which has a drive pinion mounted on a spiral-threaded sleeve. The sleeve rotates within the pinion, and moves the pinion outwards to mesh with the flywheel ring gear, and the impact of this meshing action is absorbed by the friction clutch. The engine once started turns at a higher speed and drives the Bendix gear at a higher speed than the starter motor. The pinion then rotates in the opposite direction to the spiral shaft and disengages. The drive pinion being thrown out of mesh and then stopping is a common fault. Always wait several seconds before attempting to restart as the drive mechanism may be damaged. Another fault is when the pinion does not engage after the starting motor is energized, and a high-pitched whine is emitted from the starter. Turn off the ignition immediately as the unloaded DC starter motor will overspeed and be seriously damaged. Problems can be minimized by ensuring that the sleeve and pinion threads are clean and lubricated so that the pinion engages and disengages freely. The Bendix gear, shaft, bearings and end plates can be cleaned of dried grease with WD40 and oiled with a fine sewing machine 3 in 1 oil.

12.3 Starter Installation, Maintenance and Troubleshooting. Starter installation is generally limited to two factors. The first is being mechanically secure, and the second is that the attached cables are of the correct rating and that terminal nuts are properly torqued up so that they do not work loose. In addition, the negative cable should also be attached as close as practicable to the starter. Starter motor design is generally robust as it must withstand the shocks of meshing, engine vibration, salt and moisture laden air, water, oil, temperature extremes, high levels of overload etc. Preventive maintenance is essential to ensure reliability:

a. **Shaft Corrosion.** A common problem especially on idle vessels is the buildup of surface corrosion, or accumulated dirt on the shaft and pinion gear assembly. Lack of lubrication causes seizure or failure to engage. It is good practice to remove the starter every 12 months, clean and lightly oil the components according to makers' recommendations.

b. **Starter Motor Maintenance.** Problems often occur with seized brushes, and this is primarily caused by lack of use. Always manually check that the brushes are moving freely in the brush-holders, and that the commutator is clean. Remove all dust and particles using a vacuum cleaner. Wash out with a quality spray electrical cleaner if badly soiled. Follow the DC motor maintenance procedures. Under no circumstances clean and polish the commutator with any abrasive materials.

c. **Starter Troubleshooting.** Many are familiar with the silence and loud click when the start solenoid operates but the starter fails to turn over. The main causes are due to a poor negative connection, a poor positive connection caused by loose or dirty connections, or a solenoid plunger sticking and not closing fully preventing the main contacts from closing.

12.4 Preheating Circuits. Some engines will not start without preheating, requiring extended starting turnover times which may overheat and damage the starter.

a. **Glowplugs.** Direct Injected (DI) engines commonly have glowplug heaters installed within each cylinder. They preheat the air in each cylinder to facilitate starting. In cold weather, this will dramatically decrease the electrical power requirements to start the engine.

(1) **Activation.** Prior to engine starting, the plugs are activated for an operator selected time period, or interlocked to a timer which is typically in the range 15 to 20 seconds.

(2) **Power Consumption.** The glowplugs can draw relatively large current levels for a short time. If your battery is low, allow a few seconds after preheating before starting as this enables the battery voltage to recover from the heater load.

b. **Air Intake Heaters.** These grid-resistor heaters are installed in the main air intake of IDI engines and there is normally only one heating element.

c. **Pre-heater Control.** Many preheating circuits have relays, either timed or un-timed. Timed relays are often a common failure cause. It is advisable to have a straight relay with a separate switch, and simply preheat manually for 15 seconds and then start the engine.

12.5 Pre-heater Maintenance. The following maintenance tasks should be carried out to ensure system reliability:

a. **Electrical Connections.** Pre-heater glowplug connections must be regularly checked if they are to function properly. The connections must be cleaned and tightened every six months.

b. **Cleaning.** The insulation around the glowplug connections must also be cleaned. It is a common fault to have oil and sediment tracking across to the engine block with a serious loss of preheating power.

c. **Glowplug Cleaning.** The plugs should be removed and cleaned yearly. Take care not to damage the heating element.

12.6 Pre-heater Troubleshooting. The following faults are the most common on preheating systems:

Table 12-1 Pre-heater Troubleshooting

Symptom	Probable Fault
No preheating	Loss of power (fuse failure) Connection fault on engine to first plug Relay failure Connection on ignition switch disconnected Terminal short circuiting to engine block
Partial preheating	One or more glowplugs failed Glowplug interconnection failure Dirt around glowplug causing tracking

12.7 Engine Starting Recommendations. These are some useful engine electrical and starting system recommendations.

The power supply to the engine starting system should have an isolator installed as close as practicable to the battery in both the positive and negative conductor. The isolator should also be accessible. Short circuit protection is not required. The isolator should be rated for the maximum current of the starting circuit.

The start cables should have an isolator as close as possible to the batteries, and accessible for isolation purposes. It should be rated for the maximum current starting circuit.

The main starting circuit positive and negative conductors should be rated so as not to exceed 5% voltage drop at full rated current.

The main start cables should have minimal voltage drop at full rated current. Cables should be kept as short as possible and as large as possible to minimize losses and maximize power availability. Voltage drops mean less at starter and slower turning speeds, with harder starting.

12.8 Engine Starting System Planning. The typical starting system comprises the following elements:

a. The DC positive circuit, which includes connections at the battery, the isolator or changeover switch, the solenoid connection, and solenoid contacts, the starter motor (which includes several components such as brushes, brush gear, commutator, bearings, windings).

b. The DC negative circuit, which includes connections at the battery, engine block, the cable back to the battery, the engine block, and the meter shunt if fitted.

c. The engine control system from the panel and which includes key switch, stop and start buttons, wiring harness, connectors, fuses etc.

d. The preheating system, which includes heating elements and interconnections, relays, connectors.

e. The battery.

12.9 Engine Starting System Configurations. There are several engine starting configurations and arrangements:

a. **Remote Battery Isolators.** Many boats have simple mechanical isolation switches to isolate the engine starter motor power supply. In many cases remote isolation circuits have relay type isolators. The control relay may be operated from a separate switch or interlocked to the main key switch, so that when the switch is turned the power is applied.

b. **Two Pole Engine Systems.** In many engines which have dual pole isolated systems, two battery isolation relays are installed, one on positive and one on negative. The relay coil is connected to the alternator D+ terminal and this energizes the coil when the alternator is operating. In remote isolation relay systems one relay can be used to energize both switches.

c. **Parallel Battery Starting Systems.** Some vessels have a 12-volt power system and a 24-volt engine system. The batteries are configured through the relay so that the parallel connected batteries are series connected to 24 volts when the engine start switch is operated.

d. **Parallel Connected Starters.** Larger engines use two starter motors, which keep the size down. The system uses a large capacity double acting relay to supply current to both starter motors simultaneously.

12.10 Starting System Failure Mode Analysis. There is a total of 14 connection points plus the solenoid coil, the starter motor, the battery, and the key switch that can impact on the starting system. Each point represents a single point failure with subsequent total system failure, with no apparent redundancy. If you persist with turning over an engine that will not start, you may also burn the starter motor out. There are other less common scenarios.

a. **Starting Circuit.** Relocate and connect the main negative cable to, or close as possible to, the starter motor. This maintains 2 connections but takes the engine block out of the circuit, and generally reduces voltage drop in the circuit.

b. **Maintenance Factors.** Perform the recommended maintenance on all critical equipment and systems.

c. **Starter Motors.** Starter motors have low failure rates, as actual operating hours are relatively low. Failures frequently occur with seized bearings, or stuck brush-gear. Regular operation reduces failures. In addition this generates heat, which assists in displacing moisture within windings. Starter motors should be cleaned or overhauled on a regular basis, ideally not exceeding two years.

12.11 Engine Starting System Diagrams. The following are simplified wiring diagrams for a variety of engines. Always check the diagrams supplied in the operator's manual for your specific engine model. Make sure that you have the correct circuit diagram for the installed engine and laminate a copy. Wiring varies considerably, even between older and newer engine models. The following table gives equivalent color codes for various manufacturers.

Table 12-2 Engine Wiring Color Codes

Purpose	US Codes	Yanmar	Bukh	Volvo	Perkins	Nanni
Ignition start	yell/red	white	blue	red/yell	white/red	brown
Ignition stop	black/yell	red/blk	black	purple	blk/blue	white
Preheat		blue	yellow	orange	brown/red	orange
Negatives	black	black	black	black	black	black
Alternator light	orange	red/blk	green	brown	brown/yel	green
Tachometer	grey	orange	yellow	green	blk/brown	blue
Oil press. gauge	light blue	yell/blk	green	light blue	green/yell	grey
Oil warning lt.		yellow/wh	brown	blue/wh	black/yell	grey/red
Wtr. temp. gauge	tan	white/blk	brown	light brn	green/blu	yellow
Wtr. temp. lt.		wh/blue	yell/gr	brown/w	blk/lt green	yell/red

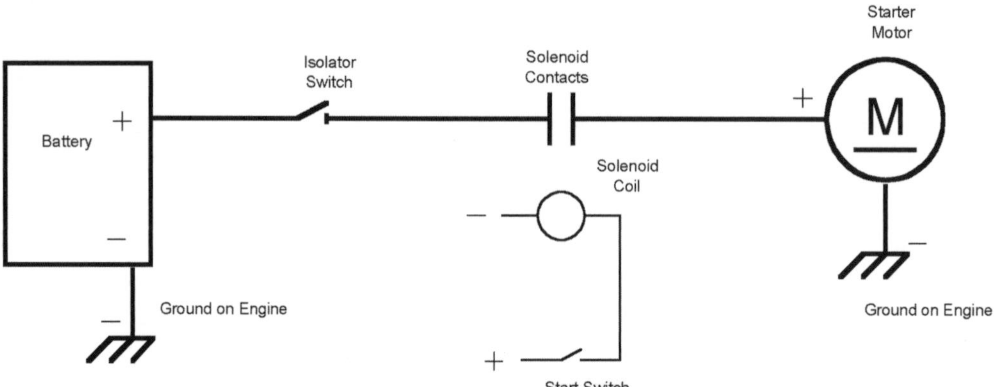

Figure 12-2 Basic Engine Starting Circuit

a. **Yanmar Starting System.** The simplified circuit diagram for the starting system of a typical Yanmar engine is illustrated below.

Table 12-3 Yanmar Troubleshooting

Symptom	Probable Fault
Engine will not start	Start button fault Key switch fault Starter solenoid connection off Stop solenoid seized Stop button jammed in Negative connection fault
Engine will not stop	Stop solenoid seized Stop solenoid connection fault
No preheating	Air heater connection broken Key switch fault Negative connection fault Glowplug connection fault

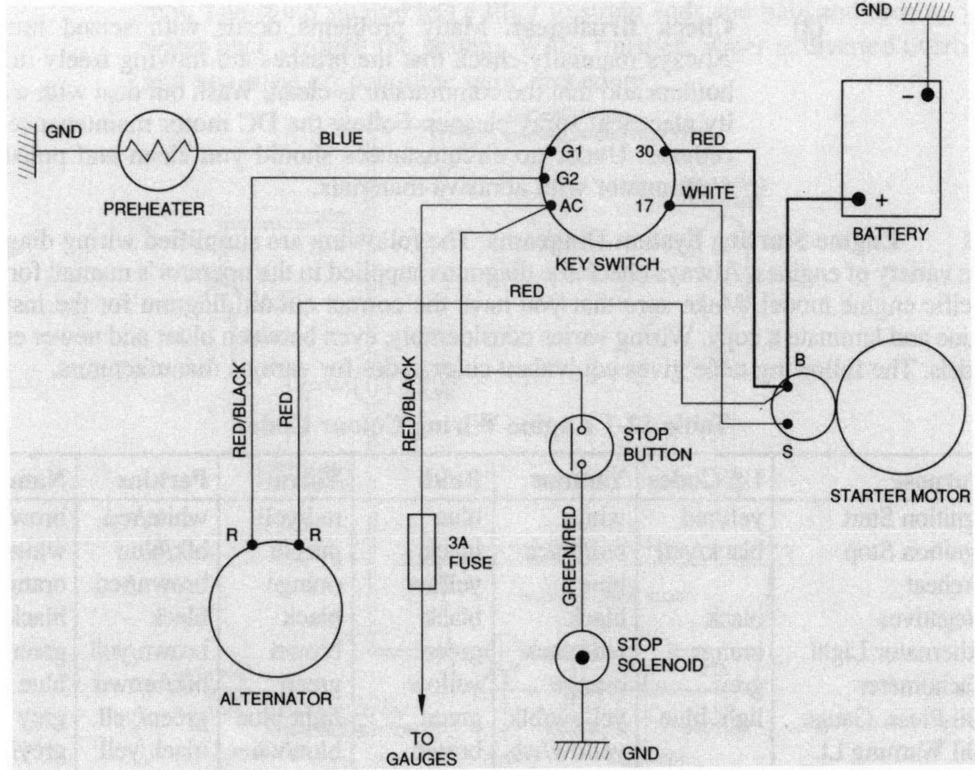

Figure 12-3 Typical Yanmar Engine Starting System

b. **Nanni Starting System.** The simplified circuit diagram for the starting system of a typical Nanni engine is illustrated below.

Table 12-4 Nanni Troubleshooting

Symptom	Probable Fault
Engine will not start	Start button connection off Start button fault Control fuse failure Starter solenoid connection off Stop solenoid seized Stop button jammed in Negative connection fault Loom connector fault Stop solenoid connection fault
Engine will not stop	Stop solenoid seized Loom connector fault Fuse failure
No preheating	Key switch fault Negative connection fault Glowplug connection fault Glowplug timing relay fault Fuse failure

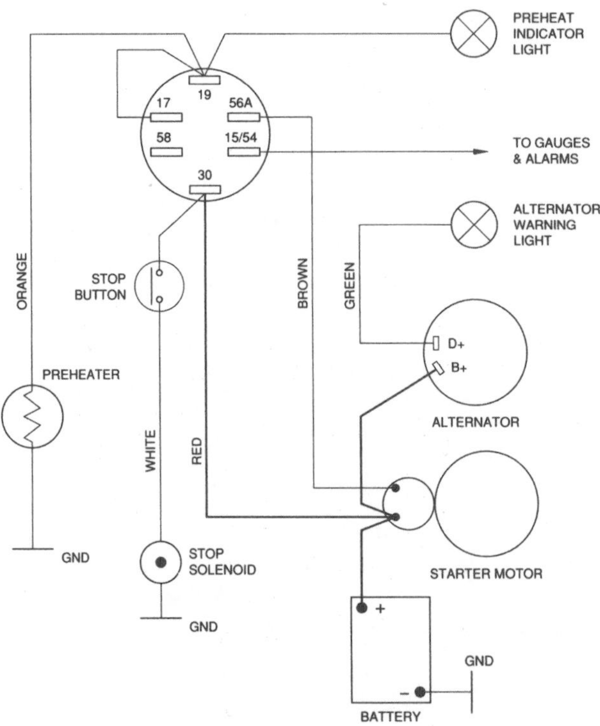

Figure 12-4 Typical Nanni Engine Starting System

c. **Perkins Prima Starting System.** The simplified circuit diagram for the starting system of a typical Perkins Prima engine is illustrated below.

Table 12-5 Perkins Troubleshooting

Symptom	Probable Fault
Engine will not start	Start button connection off Start relay fault Starter solenoid connection off Stop solenoid seized Stop button jammed in Negative connection fault Stop solenoid connection fault Stop solenoid seized Stop solenoid connection off Loom connector fault
Engine will not stop	Diode failure Earthing relay fault Key switch fault Negative connection fault Glowplug connection fault
No preheating	Relay fault Earthing relay fault

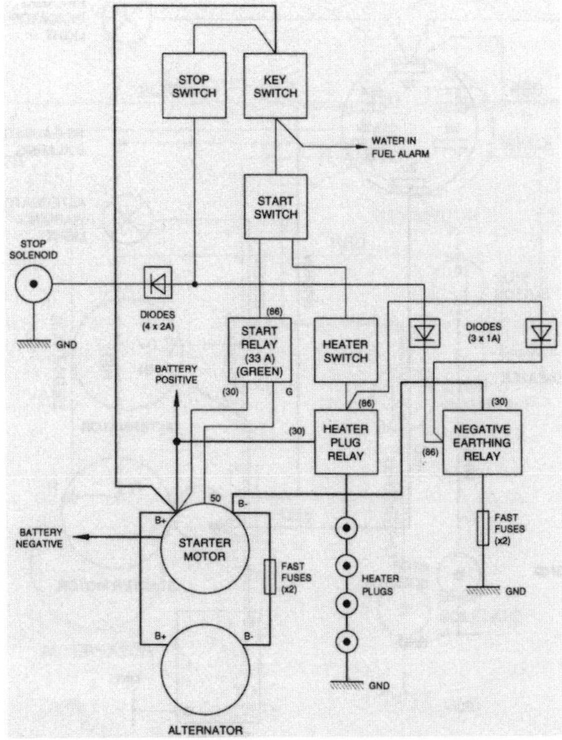

Figure 12-5 Typical Perkins Engine Starting System

d. Bukh Starting System. The simplified circuit diagram for the starting system of a typical Bukh engine is illustrated below.

Table 12-6 Bukh Troubleshooting

Symptom	Probable Fault
Engine will not start	Start button connection off
	Start button fault
	Control fuse failure
	Starter solenoid connection off
	Stop solenoid seized
	Stop button jammed in
	Negative connection fault
	Loom connector fault
Engine will not stop	Stop solenoid connection fault
	Stop solenoid seized
	Loom connector fault
	Fuse failure
No preheating	Key switch fault
	Negative connection fault
	Glowplug connection fault
	Fuse failure

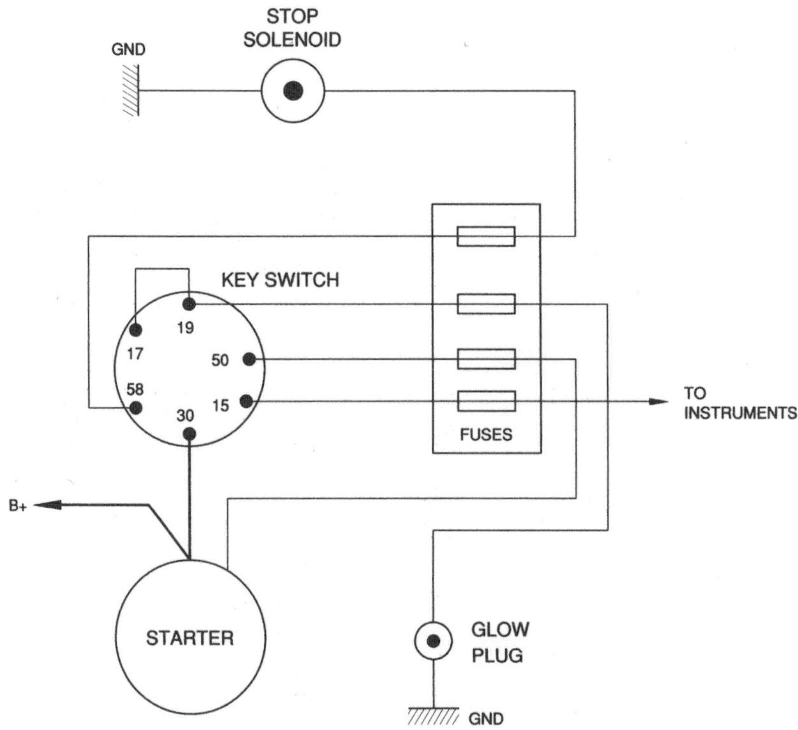

Figure 12-6 Typical Bukh Engine Starting System

e. **Volvo Starting System.** The simplified circuit diagram for the starting system of a typical Volvo engine is illustrated below.

Table 12-7 Volvo Troubleshooting

Symptom	Probable Fault
Engine will not start	Start button connection off Start button fault Control fuse failure Starter solenoid connection off Stop solenoid seized Stop button jammed in Negative connection fault Loom connector fault Stop solenoid connection fault
Engine will not stop	Stop solenoid seized Loom connector fault Fuse failure
No preheating	Key switch fault Negative connection fault Glowplug connection fault Fuse failure

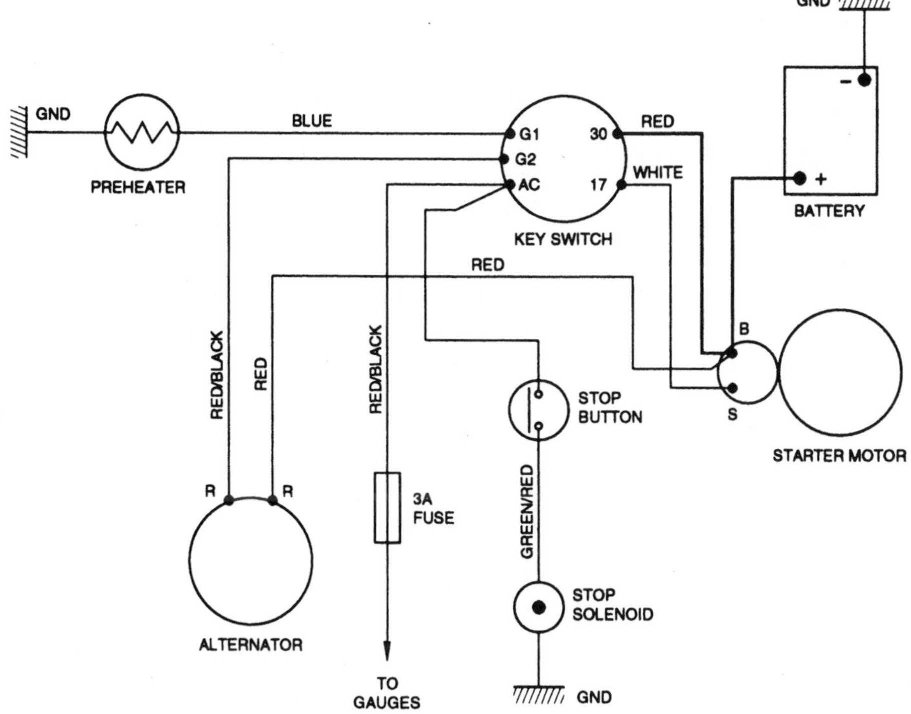

Figure 12-7 Typical Volvo Engine Starting System

Table 12-8 Engine Electrical Troubleshooting

Symptom	Probable Fault
Engine will not start	Flat battery Control power failure Stop solenoid jammed Loose corroded terminals Faulty start circuit relay Fuse blown control system
Starter will not crank (power on solenoid operates)	Starter brush jammed Starter bearings seized Starter mechanical failure Starter windings failed
Low cranking speed	Low battery voltage Battery terminal loose Starter motor fault
Generator will not hold load	Fuel filter clogged Air filter clogged Air in fuel system Governor fault Voltage regulator fault
High temperature (no mechanical fault observed)	Defective temp gauge or sender
Low temperature (no mechanical fault observed)	Defective temp gauge or sender
Undercharging	Excess loads on auxiliaries Engine low speed idles Connections loose Defect battery Alternator defect Drive belt loose
Battery water consumption increase	Overcharging High ambient temps High charge rate
Will not charge	Loose or corroded connections Sulfated batteries V-belt loose

12.12 Engine Alarms and Instrumentation. Instrumentation is crucial to ensuring that engines operate correctly within the designed parameters. Instruments may consist of a bank of discrete analog meters, or an integrated system with digital and visual screen displays; most manufacturers have such systems. The latter is becoming more prevalent, and consists of trend analysis, alarm set-point management, alarm logging and other advanced features. Check all sender unit terminals and connections regularly along with a test of all alarm functions, preferably before you start your trip.

a. **Oil Pressure Alarms.** A pressure alarm either is incorporated into a gauge sender unit or is a separate device. It consists of a pressure sensitive mechanism that activates a contact when the factory set pressure is reached. It is grounded to the engine block on one side, and activating it grounds the circuit setting off the panel alarm. To test the alarm circuit simply lift off the connection and touch it to the engine.

b. **Water Temperature Alarms.** These stand-alone alarm devices consist of a bimetallic element that closes when the factory set temperature is reached. To test simply remove the connection from the sender terminal and touch on the engine block to activate alarm. The sensor has two terminals, "G" is used for the meter, and "W" is used for the alarm contact. In many boats, damage is often done because the alarm did not function or was not noticed. The first reaction is often "What's wrong with the alarm?" instead of "What's wrong with the engine?" It is good practice to add a very loud audible alarm, as some of the engine panel units are difficult to hear at times over ambient engine noise.

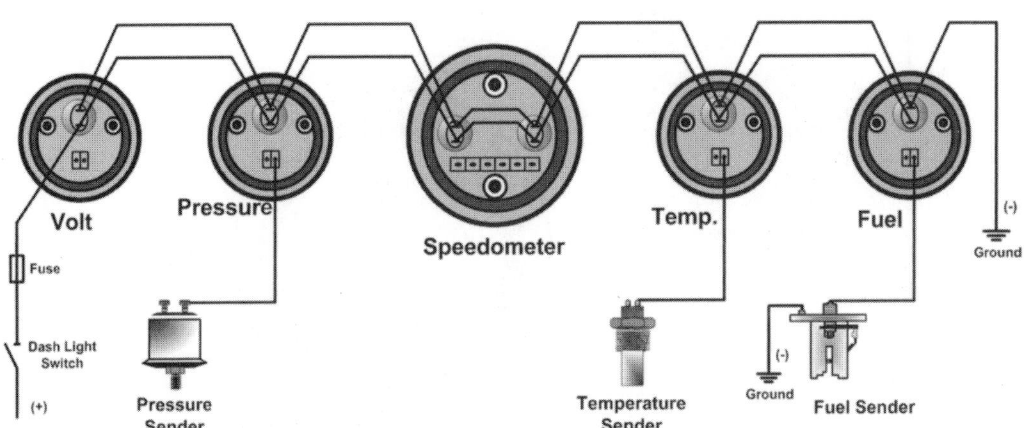

Figure 12-8 Engine Instrumentation Systems

12.13 Pressure Monitoring. The monitoring of pressures is fundamental to the proper operation of any engine. This includes lubricating oil and filter differential pressures, fuel and filter differential pressures, coolants (both seawater and fresh water), turbocharger charging air pressure and air inlet pressures, gearbox and transmission oil pressures and engine crankcase pressures.

Oil Pressure Monitoring. The oil pressure sensor unit is a variable resistance device that responds to pressure changes. It is very common to assume that the meter or alarm is wrong. Oil pressure sensor units should be removed every year and any oil sludge cleaned out of the fitting as this can clog up causing inaccurate or no readings. Low oil pressure readings are caused by low lube oil level, or a clogged oil filter creating a lowering in oil pressure. A faulty oil pump also can cause a lowering in pressure; a rise in oil temperature due to an increase in engine temperature, or an oil cooler can also cause a problem. When sensor units are poorly grounded or Teflon tape is improperly applied to threads to make a high resistance contact, a problem can occur.

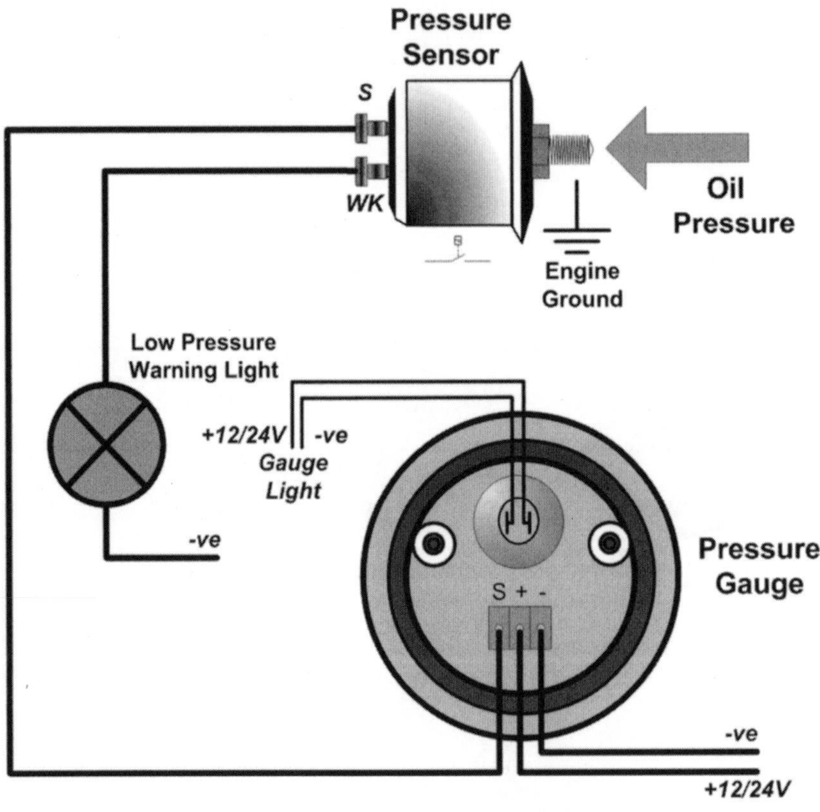

Figure 12-9 Oil Pressure Monitoring

12.14 Temperature Monitoring. The main temperature monitoring points utilizing the same sensor types include lubricating oil, transmission oil, coolants (seawater and fresh water), fuel temperature, after-cooler and turbocharger inlet air.

Water and Oil Temperature Gauges. The monitoring of water temperature is essential to the safe operation of the engine. Temperature extremes can cause serious engine damage or failure. Sender units are resistive and give a resistance proportional to temperature in a non-linear curve. If the gauge readings are not correct and a gauge test shows it to be good, check the sensor. Before you check the sensor unit, you must assume that the main causes of high temperatures are: the loss of fresh water-cooling caused by faulty water pump impeller, a loose rubber drive belt, low water levels, fouled coolers and increases in combustion temperatures. Loss of salt water cooling caused by blocked intake or strainer, faulty water pump impeller, clogged cooler or aeration caused by leak in suction side of pump. Increased engine loadings caused by adverse tidal and current flows, or overloading. Sensor units are poorly grounded or Teflon tape is improperly applied to threads to make a high resistance contact.

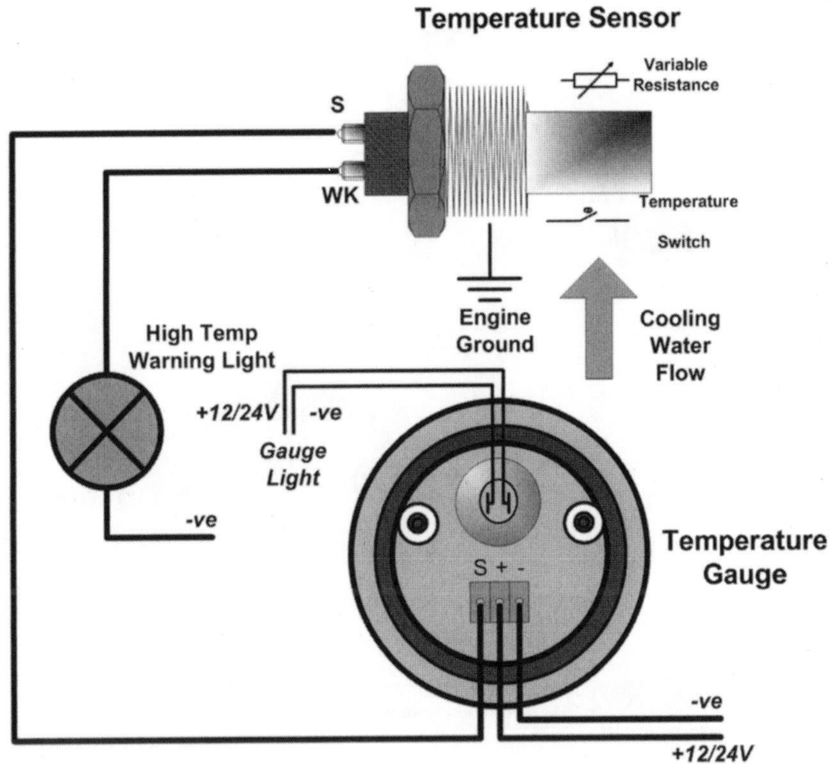

Figure 12-10 Water Temperature Monitoring

12.15 Exhaust Gas Temperature Monitoring. Exhaust gas temperature monitoring is used in commercial ships and is recommended on all yachts. Engine problems are easier and faster to identify than water temperature and oil pressure monitoring. These can be problems within the cooling water system; increased engine loads that may be caused by adverse tidal and current flows; air intake obstructions caused by clogged air filters, or where installed, blocked air coolers; combustion chamber problems caused by defective injectors, valves etc. Larger engine boats will also have cylinder monitoring, and this allows identification of problems specific to cylinders to be identified and monitored. Smaller engines may have a sensor installed on the main exhaust manifold. Pyrometer compensating leads and wiring should be routed clear of other cables to avoid induction and inaccurate readings.

Operating Principle. Exhaust temperature sensors are called thermocouples, or pyrometers. These sensors consist of two dissimilar metals (iron/constantine; copper/nickel; platinum/rhodium. nicrosil/nisil, nickel/aluminum), which at the junction will generate a small voltage proportional to the heat applied to the sensor. The voltage is measured in millivolts (mV). The typical thermocouple consists of a sensing junction, and a reference junction. The open circuit voltage is measured with a high impedance voltmeter and is the temperature difference between the sensing junction and the reference junction. The thermocouple junction is also called the "hot junction." The compensating cables between the junction and the measurement meter are electrically matched to maintain accuracy. They are polarity sensitive, so must be connected positive to positive.

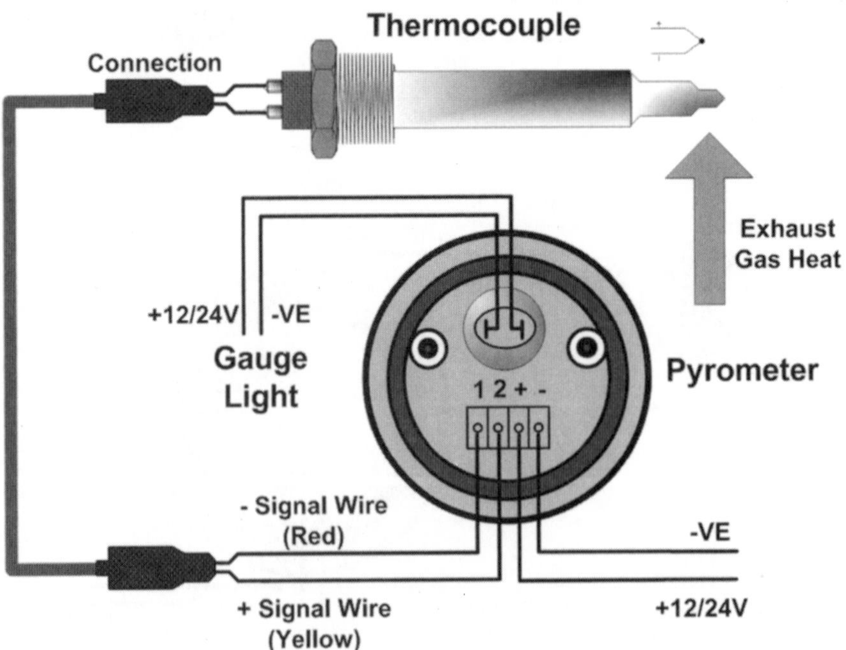

Figure 12-11 Exhaust Gas Temperature Monitoring

12.16 Engine Tachometers. The tachometer is used to monitor engine speed, differential or synchronization, shaft revolutions and turbocharger speed.

a. **Generator Tachometer.** This tachometer type receives a signal from a mechanically driven generator unit. The generator produces an AC voltage proportional in amplitude to the speed and this is decoded by the tachometer. Variations in speed give a proportional change in output voltage which changes the meter reading. The most common fault on these units is drive shaft mechanism damage.

b. **Inductive Tachometer.** These tachometer types have an inductive magnetic sensor. This sensor detects changes in magnetic flux as the teeth on a flywheel move past. This sends a series of on/off pulses to the meter where it is counted and displayed on the tachometer. Ensure that the sensor unit is properly fastened. A common cause of failure is damage to the sensor head by striking the flywheel when adjusted too close.

c. **Alternator Tachometer.** This type of tachometer derives a pulse from the alternator AC winding, typically marked 'W'. The alternator output signal frequency is directly proportional to engine speed. The pick-up is taken from the star point or one of the unrectified phases. Typical connections for VDO tachometers are illustrated. If the alternator is faulty, there is no reading. There are a number of different alternator terminal designations used by various manufacturers and the main ones are W, STA, AC, STY, SINUS. If there is no output terminal a connection will have to be made as shown in the illustration and an alternator tachometer installed.

d. **Synchronization Tachometers.** The synchronizer or differential tachometer is used to show precise speed difference between each engine in a twin-engine installation. Use of the meter allows balancing to be carried out.

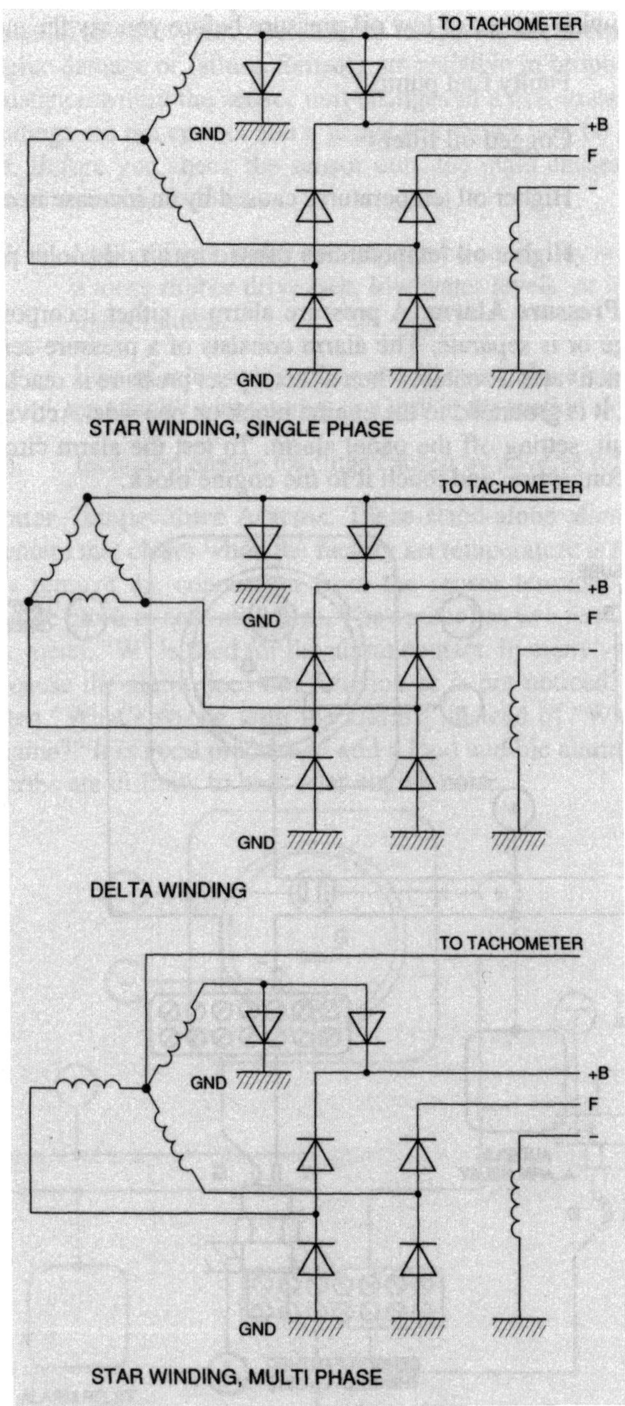

Figure 12-12 Alternator Tachometer Circuits

12.17 Bilge and Tank Level Monitoring. The monitoring of on-board fuel, water and bilge levels is an essential task. A simple electrical gauge can be installed that provides the necessary information.

a. **Resistance Sensor.** The majority of tank sensors in use operate by varying a resistance proportional to tank level. The two basic sensor types are as follows:

(1) **Immersion Pipe Type.** This sensor type consists of a damping tube, with an internal float that moves up and down along two wires. These units are only suitable for fuel tanks. The big advantage with these sensor types is that they are well damped, virtually eliminating fluctuating readings.

(2) **Lever Type.** The lever type system consists of a sensor head located on the end of an adjustable leg. The sensor head comprises a variable resistance and float arm pivot. As the float and arm moves relative to fluid levels, the resistance alters and the meter reading changes, typical resistance readings are in the range 10–180 ohms. Lever type units should be installed longitudinally, as athwartships orientation can cause serious problems with the vessel rolling. Where these units are used for fuel or water, the primary difference between them is that for water sensor units, the variable resistance is located outside of the tank to avoid water problems, while the fuel has a resistance unit in the tank.

b. **Capacitive Sensors.** This type of transducer operates on the principle that the value of a capacitor depends on the dielectric between plates. The sensor unit measures the capacitance difference between air and the liquid.

(1) **Output Values.** The sensing circuit outputs a voltage proportional to the level in the typical range of 0 to 5 volts.

(2) **Faults.** The most common fault in these systems is water damage to the circuit board, usually because of tank condensation.

c. **Pressure Sensors.** These sensors are considerably more expensive, but they are very accurate and less prone to damage. The transducers are either placed at the bottom of the tank, or on a pipe that is taken out of the side at the tank bottom.

(1) **Output Values.** The sensors output either a 4–20 milliamp or 0.6–2.6 volts proportional to the pressure of the fluid in the tank. The pressure value is proportional to the tank volume.

(2) **Faults.** If the sensor is located on a small pipe it may become clogged.

d. **Air Sensors.** Air operated bilge switches such as the Jabsco Hydro Air use an air column to pressure activate a remote mounted switch. They can switch 20 amps, and are ignition protected. One advantage is they are less prone to jamming than float switches.

12.18 Electrical System Monitoring. There are a number of parameters for monitoring electrical systems and methods for installing instruments:

a. **Charging Voltmeters.** Many instrument panels incorporate a voltmeter to indicate the state of the charging. Voltmeters are fairly coarse, and only partially useful in precisely assessing battery voltage states but they are a useful indicator on the charging system. Many voltmeters have a colored scale to enable rapid recognition of battery condition, red for under or overcharge and green for proper range.

b. **Charging Ammeters.** Charging ammeters are reasonably popular and an easy guide to the level of charge current from the alternator. There are basically two types of ammeter:

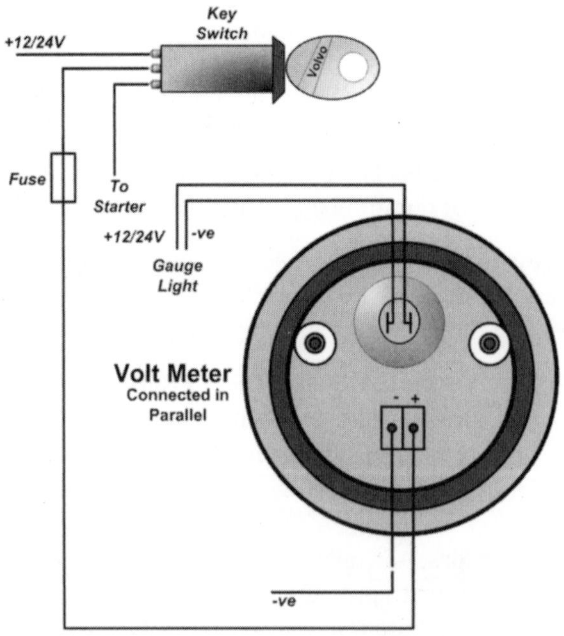

Figure 12-13 Voltmeter Connections

(1) **In-line Ammeter.** This ammeter type has the main charge alternator output cable running through it. In many cases, the long run to a meter causes unacceptable charging system voltage drops and undercharging. An additional problem with installing such ammeters on switch panels is that the charge cables are invariably run with other cables and cause radio interference. If you are going to install this type of ammeter, make sure that the meter is mounted as close as possible to the alternator. If these ammeters start fluctuating at maximum alternator and rated outputs, this is generally due to voltage drops within the meter and cable. The under rating of connectors is a major cause of problems.

(2) **Shunt Ammeter.** The shunt ammeter overcomes the voltage drop problem. The shunt is essentially a resistance inserted in the line. Sense cables (16 AWG twisted pair) are connected across it and can be run to any meter location without voltage drop problems as the output is in millivolts. The ammeter must always be rated for the maximum alternator output. Many installations do not do this; the shunt or meter is often damaged, and there are big voltage drops in the charging line.

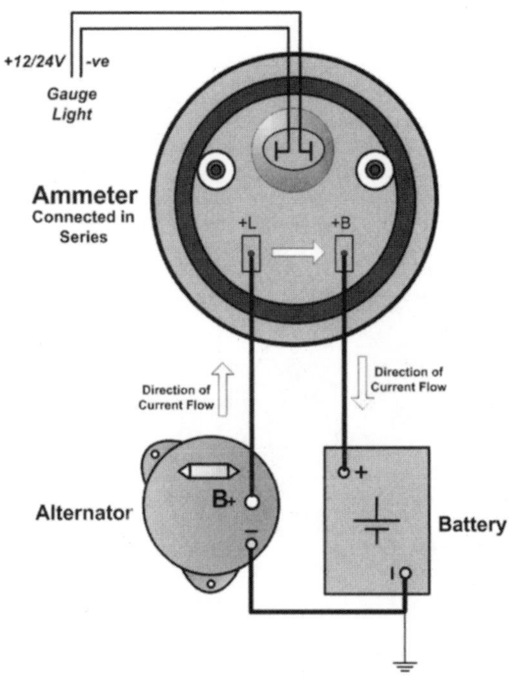

Figure 12-14 Ammeter Configurations

12.19 Hour Counters and Clocks. An hour counter is essential for keeping a record of maintenance intervals. Essentially it is a clock that is activated only when the engine is operating. There are a number of methods of activating hour counters:

a. **Ignition Switch.** This is the easiest and most practical method. The meter is simply connected across the ignition positive and a negative so that it operates when the engine is running.

b. **Oil Pressure Switch.** This is not a common method. Some installations activate through the switch, so that it operates only when the engine is running.

c. **Alternator.** In many installations the counter is activated from the alternator auxiliary terminal D+ or 61.

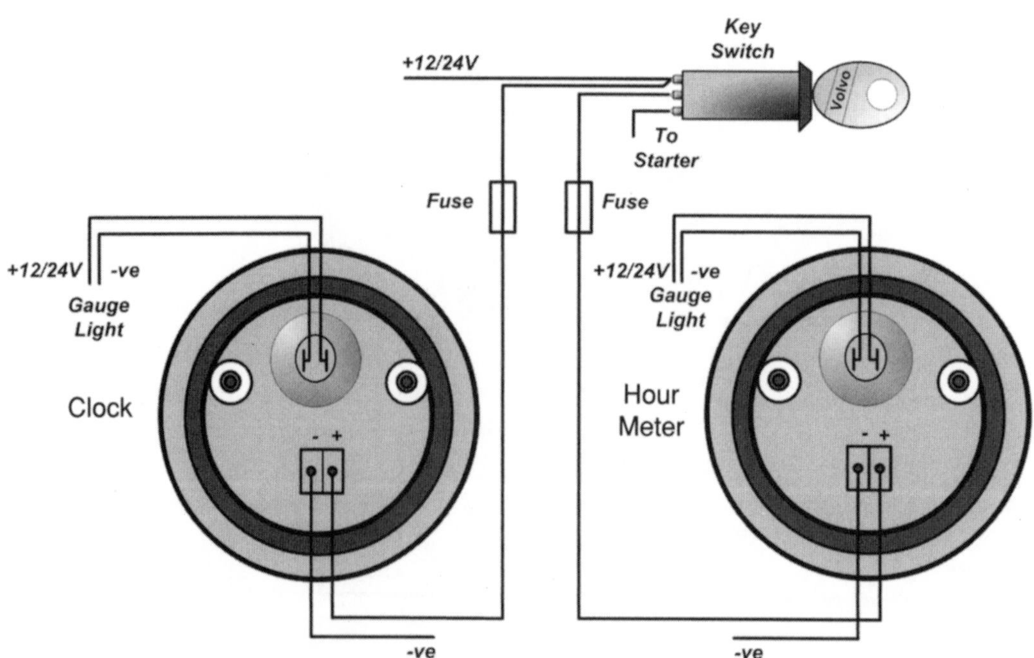

Figure 12-15 Hour Meter and Clock Connections

12.20 Acoustic Alarm Systems. Acoustic alarms are generally connected to warning light circuits, and the buzzer is activated by a relay. Acoustic warnings are activated along with the lamp from sensor contact "W". The acoustic alarm should be activated through a relay, not through the sensor contact, which is not rated for such loads.

a. **Buzzer Test.** Using a lead connect a positive supply to the buzzer positive terminal, and check that a negative one is also connected. If the buzzer operates remove the bridges. Ideally a test function should be inserted into the circuit so that the alarm function can be verified.

b. **Operating Test.** With alarm lights on, put a bridge from negative to the buzzer negative, as sometimes a "lost" negative is the problem. Connect a positive supply to the relay positive, typically numbered 86. If the relay does not operate and the buzzer is working then the relay is suspect. Verify this after removal using the same procedure. Note that sometimes a relay may sound like it is operating but the contacts may be damaged and open circuited. If a buzzer is not operating along with lights either a cable or connection is faulty, or the operating relay is defective.

c. **Mute Function.** On many home-built engine panels, it is essential to silence the alarm, and this entails placing a switch in line with the buzzer. The lamp will remain illuminated to indicate the alarm status.

d. **Time Delays.** During engine start up, a time delay is necessary to prevent alarm activation until oil pressure has reached normal operating level. Time delays are typically in the range of 15 to 30 seconds.

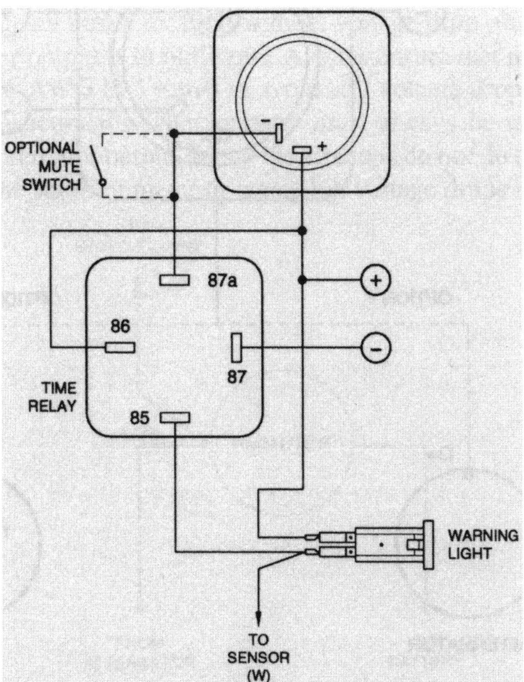

Figure 12-16 Acoustic Alarm System

12.21 Instrumentation Maintenance. Maintaining instruments is relatively simple.

a. **Electrical Connections.** A regular check of all sensor unit terminals and connections, along with a test of alarm functions, is all that is required.

b. **Oil Pressure Sensors.** Oil pressure sensors should be removed every year and any oil sludge cleaned out of the fitting. Sludge-clogged sensors may be inaccurate or show no reading.

12.22 Gauge Testing. Use the following procedure on suspected gauge faults.

a. **Open Sensor Test.** Remove the sensor lead marked "G" from the back of the gauge. Switch on meter supply voltage. The gauge needle should now be in the following positions:

(1) Temperature gauge: left-hand hard-over position.

(2) Pressure gauge: right-hand hard-over position.

(3) Tank gauge: right-hand hard-over position.

b. **Sensor Ground Test.** This test involves the bridging of sensor input terminal "G" to negative. The sensor lead must be removed and the meter supply on. The gauge needle should now be in the following positions:

(1) Temperature gauge: right-hand hard-over position.

(2) Pressure gauge: right-hand hard-over position.

(3) Tank gauge: left-hand hard-over position.

12.23 Sensor Testing (VDO). Disconnect the cables, and using a multimeter (digital or analog), set the resistance (ohms) range to approximately 200 ohms. Place the positive (red) meter probe on the terminal marked "G" on sensor, if it has a dual alarm and sensor output, the alarm output is marked "W". Place the negative (black) meter probe on the sensor thread.

a. **Temperature Sensors.** Readings should be as follows:

(1) 40°C = 200–300 ohms.

(2) 120°C = 20–40 ohms

b. **Pressure Sensors.** Readings should be as follows:

(1) High pressure (engine off) = 10 ohms

(2) Low pressure (engine running) = 40psi:105 ohms, 60psi:152 ohms.

c. **Fuel Tank Sensors.** Reading should be as follows:

(1) Tank Empty = 10 ohms.

(2) Tank Full = 180 ohms.

Table 12-9 Instrument Troubleshooting

Symptom	Probable Fault
Gauge does not operate	Power off Gauge supply cable off
Temperature gauge needle hard over	Sensor fault Cable fault
Pressure/tank gauge needle hard over	Sensor fault Cable fault
Alternator tachometer no reading	Alternator fault Lead off alternator terminal Alternator not "kicked" in
Generator tachometer no reading	Meter fault Broken drive mechanism Meter fault Generator fault Cable fault
Inductive tachometer no reading	Sensor mechanically damaged Sensor clearance excessive Sensor fault Meter fault
Low gauge readings	Negative connections to engine block and sensors degraded
Oil pressure alarm activated	Low oil pressure (oil pump fault) Low oil level High oil temperature (cooling fault) Blocked sender unit Sender fault Cable fault
Water temperature alarm	High water temperature Low cooling water level Salt water cooling inlet blocked Cooling water pump fault Loose drive belt Sensor fault
No audible alarm	Relay fault Audible alarm fault Connection fault Lamp failure Lamp connection fault Alarm circuit board fault

AC Power Systems

13.0 AC Power Safety. AC is potentially lethal, and systems and equipment must be correctly selected, installed and maintained. The following safety precautions must be undertaken at all times when carrying out work.

WARNING

a. Never work on "live" equipment. Always isolate and lock out equipment before opening. Attach a Danger Tag.

b. Always remove the shore power plug and isolate local main switch.

c. With an inverter, or inverter/charger unit, always isolate the DC input.

d. Always make sure the circuit is dead or de-energized before starting work.

e. Never work on AC equipment alone; always have someone ready to assist if you accidentally receive a shock.

f. Learn artificial respiration and CPR techniques.

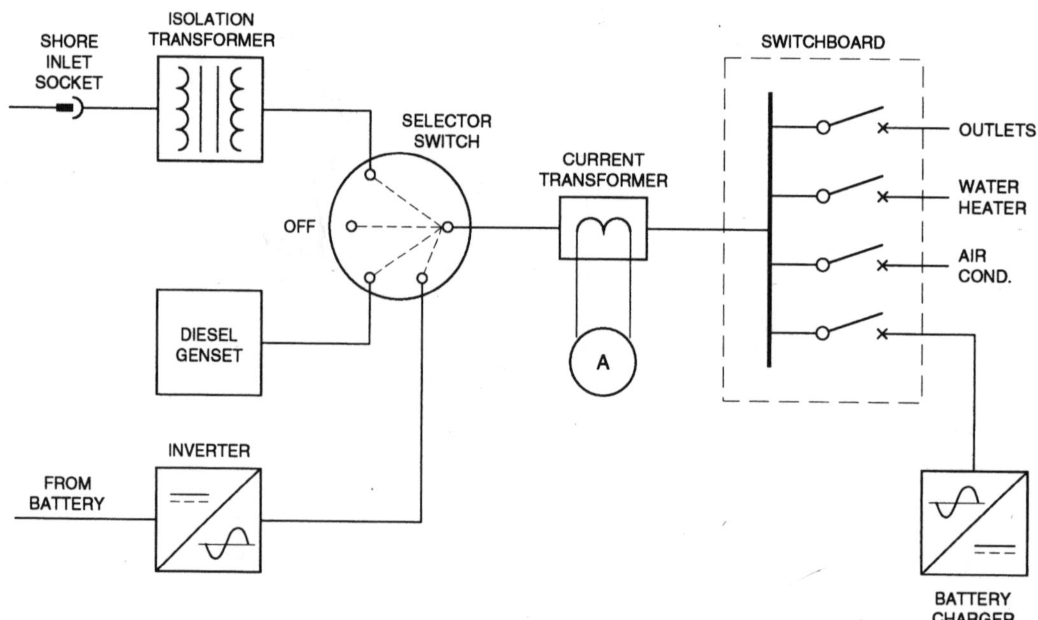

Figure 13-1 AC Power Systems

13.1 Shore Power Systems. The proliferation of appliances such as microwaves, air conditioning, washing machines, stove, hot water, TV, power tools and other devices has increased the requirement for power both at and away from the marina. The AC power systems on many sailing yachts now consist of several elements that include shore power installations, inverters, generators, AC motors and starters, and AC installations with grounding and circuit protection. The following basic recommendations should be used in conjunction with the relevant national or other standard.

Any boat that is connected to a marina or any other shore power circuit is generally required to comply with the relevant provisions of local and national electrical codes. Recommendations contained within this section are advisory only, and do not override the legal responsibilities of boat owners to meet specific requirements.

Connection of the vessel to a marina power system imposes certain obligations on the boat owner. A vessel must comply with national or other electrical standards. Some acceptable standards are as follows:

(1) ABYC. E-11 - AC & DC Electrical Systems on Boats. These are recommendations only. (Prior to July 2003 this was E-8) The United States Coast Guard has mandatory requirements for electrical systems in Title 33, CFR 183 Subpart I, Section 183.

(2) US National Electrical Code (NEC).

(3) NFPA 302, Fire Protection Standard for Pleasure and Commercial Motor Craft.

(4) Regulations for the electrical and electronic equipment of ships (UK Institution of Electrical Engineers) BS7671:2001.

(5) Lloyd's Register of Shipping (LR). Rules and Regulations for the Classification of Ships.

All boat AC electrical systems should be installed and tested by an AC qualified and appropriately licensed electrician.

Where possible, always use an AC qualified and licensed marine electrician, or a shore-based licensed industrial electrician. An accident caused by a fault in the system due to incorrect wiring may void your insurance policy and expose you to criminal and civil liability. Many marine electricians have automotive backgrounds and are not AC qualified, so check up first.

Shore power inlet sockets should be of the self-closing type and rated to IP56. All inlet sockets should have a means for locking in the plug when inserted.

Inlet sockets should be weatherproof in accordance with international protection standards, IP56 or equivalent NEMA rating, which require protection against heavy seas. Inlet sockets must have spring-loaded, self-closing and locking covers. The inlet sockets must also be of a male type connector only. Do not use inlet sockets that are designed for caravans, trailers or recreational vehicles, as they are not of the standard required. The plug should have a screw locking-ring that prevents the plug from being pulled out with boat movement. Suppliers such as Marinco (www.marinco.com) and Hubble (www.hubble.com) can supply good quality equipment. Ratings of inlet sockets and plugs are 15 amps for 240-volt systems, and where domestic extension cables are used these are often 10 amps only. 30 amps is the nominal rating for 115 volts systems. Check that the O ring seals are always in good condition as the seal will also fail.

Shore power inlet sockets should be mounted at a position and height that prevents mechanical damage and immersion.

The sockets should be in an accessible location and as high as possible above the deck line. They must be located so that there is no mechanical damage risk. They should also be shielded from rain so that driving rain at 45° angle will not enter the plug and socket.

The vessel should not be connected to any marina power supply outlet that is damaged.

Inspect the conditions at the marina before you connect your boat. Look for damage to outlets and tracking on pins, wires and damaged conduits etc. If the marina systems are in a degraded condition it represents a risk to you, your crew and boat, and possible nuisance tripping of GFCI units. Check the dock supply voltages with a multimeter. If they are low (more than 5%) this is not good for on board electric motors. Some battery chargers will also have lower outputs. Start your largest appliance, plus operate all the systems proposed and see whether the circuit will withstand it, in many cases it does not. You must check and report deficiencies, as you may become liable. If they are not reported, they do not get fixed.

Shore power leads should be weatherproof, and have a high visibility outer sheath.

Shore cables should be suitable for outdoor use. The cable must be heavy duty rated which is typical of most outdoor rated extension cables, and special shore power lead suppliers. High visibility outer sheaths are typically yellow or orange. These outer sheaths are also UV stabilized unlike standard duty domestic leads, which should not be used. It is good practice to also have a strain relief grip over the cable to prevent unnecessary strain on the cable, particularly where the plug into the supply pedestal has a screw locking ring fitting. The flexible cable should be arranged so as to permit normal movement of the vessel without stress in the full predicted tidal range. It also must prevent water travelling along the cable to the inlet receptacle and be secured so that immersion in the water is not possible. Additionally, provision must be made to prevent the plug falling into the water if it is accidentally disconnected. Plugs should be double insulated and be made of impact resistant material. If a plug has accidentally been immersed in water, do not use it. Disconnect and dismantle for cleaning and drying first. Cables should be regularly inspected, and the outer insulation checked for cuts or other damage. Do not repair or join a damaged cable, it must be replaced. Cables should always be run from the pedestal to minimize exposure to damage, or present tripping hazards to other people on the marina.

No shore power ground conductor should be disconnected as a method for reducing corrosion.

Some boat owners disconnect the shore power ground connection either in the plug or at the socket inlet to open the ground path ashore. This should never be done; in many cases it removes the safety ground of equipment, and creates serious electric shock risks.

The AC power panel should have a polarity indicator and polarity reversal switch.

Switchboards should incorporate a reverse polarity indicator, and have a changeover switch. It is quite common to find marina supplies with the neutral and active (hot or live) conductors reversed. This condition is indicated and the switch simply reverses to the correct polarity. Most marina supply outlets are protected by a circuit breaker or a GFCI, which may trip on connection of the boat if ground and active are wrongly connected.

The AC power panel should have power input isolation, power source selection and short circuit protection.

The switch panel should have input isolation and short circuit protection. The input isolation can be a trip free circuit breaker rated at the maximum power input of the panel. The selection switch is for switching of each input source (shore power, inverter, generator) without paralleling. Parallel connection of an inverter will destroy it.

13.2 115VAC 60Hz Single Phase System. This voltage (also stated as 117 or 120 V) is primarily used in the United States and Canada, Brazil and most South American countries. Normally a 30-amp supply requires a supply cable of 4.0 mm^2. The following circuit illustrates a typical shore power system. In this configuration, only one wire in the power inlet is "hot" or energized.

(1) Black wire is "hot".

(2) White wire is neutral.

(3) Green wire is ground or earth.

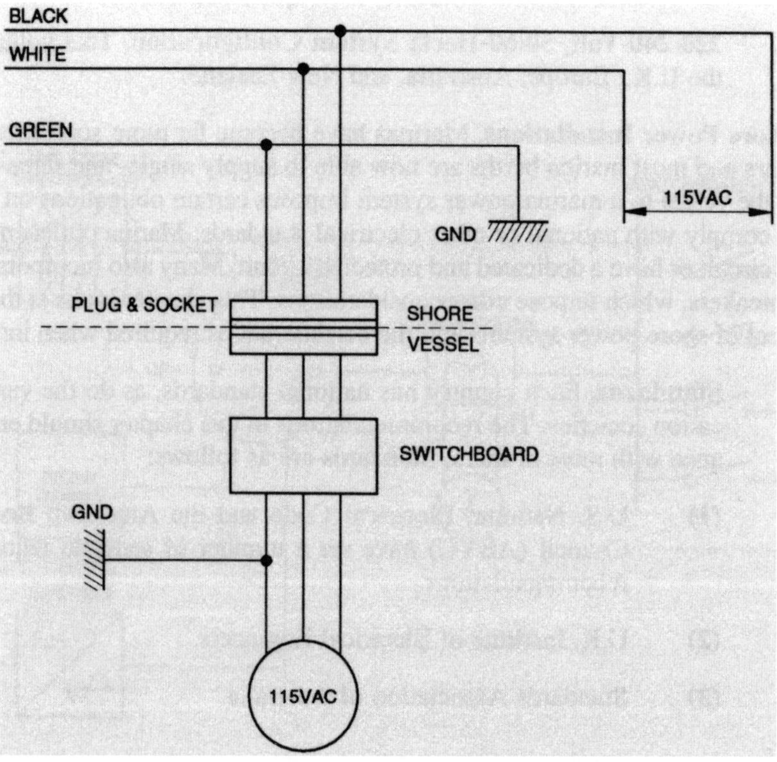

Figure 13-2 115 VAC Shore Power Systems

13.3 115/230 Volt Systems. The following circuit diagrams and color codes are for typical American dual voltage shore supply systems. Observe the following:

(1) Red wire is "hot".

(2) Black wire is "hot".

(3) White wire is neutral.

(4) Green wire is ground or earth.

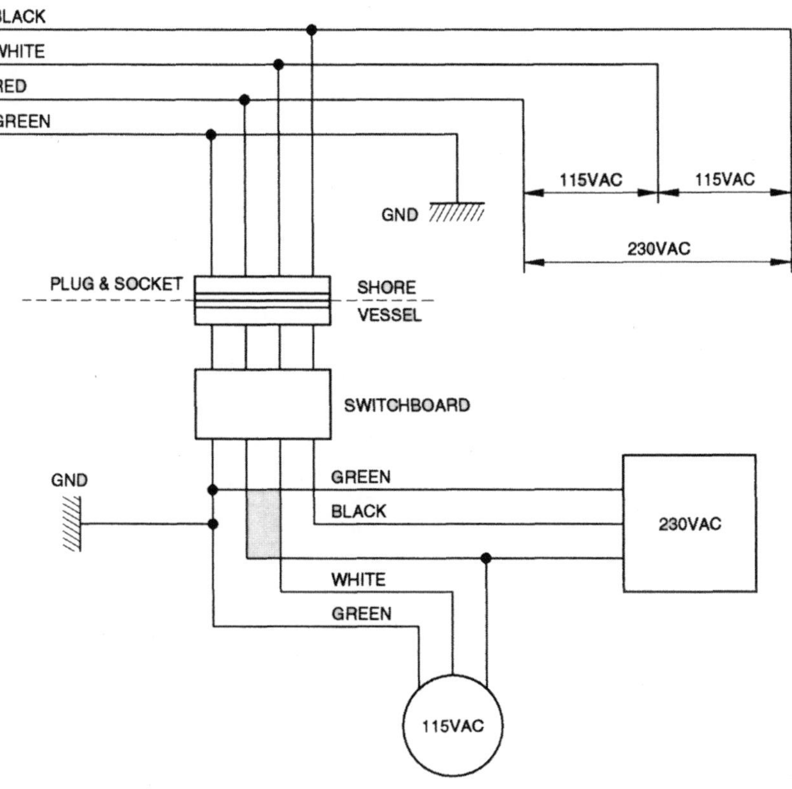

Figure 13-3 115/220 VAC Shore Power Systems

13.4 220/240 Volt Systems. 220/230/240VAC 50Hz Single Phase System. This voltage is the IEC standard and used in the UK, Europe, Australia, New Zealand, South Africa, Fiji, New Caledonia and most other countries. The IEC is standardizing most countries at 230 VAC. The following circuit diagrams are for typical systems using IEC standard color codes, and incorporating an isolation transformer. Normally a minimum 15-amp supply rating requires a supply cable of 2.5 mm^2. Many marina supplies only have a 10-amp supply. The cable should be a length of approximately 45 feet and anything over this will start to introduce volt drop problems at rated load.

(1) Brown wire is "hot" (used to be red).

(2) Blue wire is neutral (used to be black).

(3) Green/yellow stripe wire is ground or earth (used to be green).

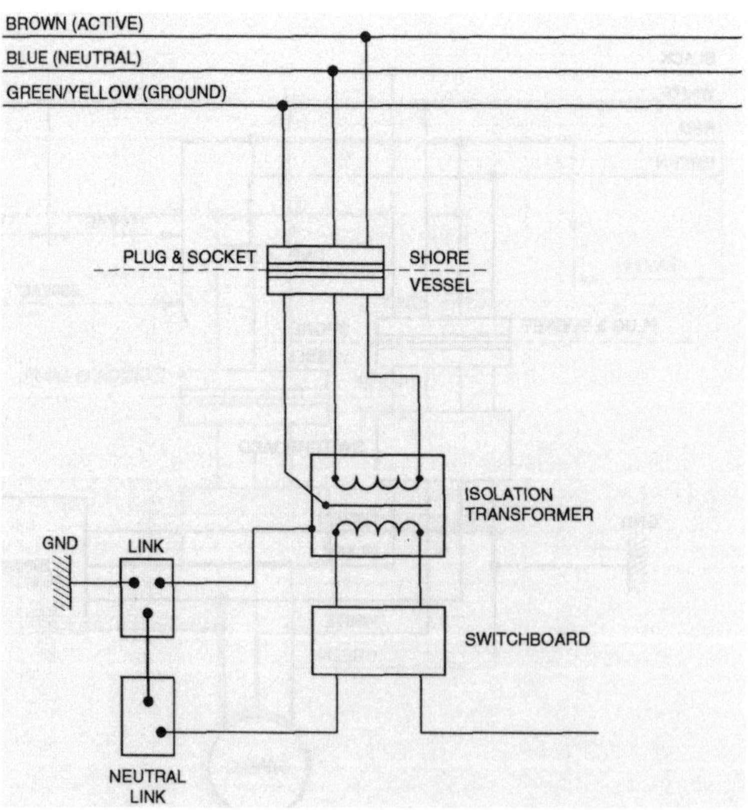

Figure 13-4 220/240 VAC Shore Power Isolation Systems

13.5 440/460VAC 60Hz 3 Phase System. This arrangement is more commonly used on large vessels, super yachts and larger sailing yachts. In larger yachts having 3-phase power supplies, connection to shore power is less common or unavailable. Few marinas are able to offer power supplies. 3-phase offers equipment 20% lighter and smaller than single-phase, and 150% greater efficiency. In a 3-phase vessel power is usually generated in a three phase, star or wye configuration. This means that the center point is grounded at the neutral. In a 208 VAC 3-phase wye system, you get 208 V between phases, and 120 VAC phase to ground. Connected loads are usually symmetrical with electric motors. Where loads are connected to each phase, unacceptable imbalances can occur. Single-phase loads must be evenly distributed across phases. In some instances 115/220V transformers are used to power lower voltage systems.

a. **Phase Marking.** Circuit designations are R, S and T for each phase, also it may be R, Y and B to represent the phase cable colors. In the US this may be L1, L2 and L3. Secondary motor circuits may be also called U,V and W or U1, V1 and W1.

b. **Phase Colors.** The phase colors are black, white and red in the US. In most other parts of the world they are red, yellow and blue, although yellow may be replaced by white.

c. **Phase Sequencing.** Phases must be connected in the correct sequence for a motor to run in the correct direction. If a motor is running in reverse rotation, reverse any two phase cables to reverse the direction.

d. **Phase Current Loads.** In any 3-phase motor, each phase should have the same current. If there is an imbalance, this indicates a high resistance in the connections or in some cases a failing motor winding.

13.6 Transformers. Transformers are used to raise or lower voltages, in electronics equipment power supplies, in starters for control voltages, and in isolation transformers for shore power supplies. Larger boats transform 3-phase from 480VAC to 208/120VAC for power and lighting circuits. It consists of a primary winding and a secondary winding on an iron core. Power applied to the primary magnetizes the core and induces a voltage into the secondary winding. The most common type is the isolation transformer, which electrically separates input and output. Autotransformers are lighter and cheaper, with connected primary and secondary windings. Maintenance is a simple cleaning of dust and an insulation test every year. Visually check windings and wire insulation for heat damage. The cycle "hum" can cause vibration, which can loosen screws and all connections should be tightened. At switch on there are high inrush currents, up to 25 times the rated current, and these must be considered when selecting protection devices. Booster transformers are available that also act as isolation devices, and are used to boost lower dockside supply voltages.

13.7 Domestic Plugs. The following is additional information on power plugs, as the various plug types required when travelling can be confusing. You should acquire an international adapter kit.

(1) US type: 2 parallel pins. (US, Canada and South America).

(2) Australian type: 2 and 3 pins (Australia, Fiji, New Zealand).

(3) European type: 2 pins round (all of Europe, South Africa).

(4) UK type: 3 pins square (UK, South Africa, Portugal).

Isolation transformers should be rated for the maximum current rating of the shore supply inlet.

By eliminating the ground path, ashore isolation transformers isolate galvanically the vessel from the shore power system, reducing electrolytic (stray current) corrosion. Most marina outlets rarely exceed a maximum of 15 amps/240-220 volts or 30 amps/115 volts. Some transformers are dual input with inputs of 120/240 or 480/240 and this gives some flexibility. Transformers do not alter frequency. Frequency may be different in some places, and this affects motor speeds (50 Hz is slower than 60Hz); for battery charging, domestic appliances and resistive loads there is little problem. Victron have also incorporated a Soft Start unit, which manages high start currents that would otherwise trip the marina supply breaker, and short circuit protection.

Table 13-1 Isolation Transformer Rating Table (0.8 pf)

Output (kW)	Output (kVA)	Current 120 V	Current 240 V
3	3.74	34 Amps	17.0 Amps
4	5.00	45 Amps	22.7 Amps
5	6.25	57 Amps	28.4 Amps
7	8.75	80 Amps	39.8 Amps

13.8 Shore Power Inverters. A relatively new development from Atlas Energy, Mastervolt and Asea Power, shore power inverters use frequency conversion and control technology to output a stable AC voltage to the boat. In many marinas and other AC power supplies, the shore voltage is low, fluctuating in voltage levels or generally unstable. The shore inverter accepts all input ranges and frequencies, and will always output a stable waveform and voltage level irrespective of the input variations. The units operate similarly to UPS (uninterruptible power supply) units and use inverters and pulse-width-modulation techniques that allow wide input voltage ranges and ensure stable, good waveform outputs. These units will suit the larger boats as they are relatively expensive, but for boats at the dock for long periods, they will solve many power problems of computers and other sensitive equipment. The systems also allow parallel connection of two shore leads in their Smart Box unit. Atlas also have the ShorPOWER converter supply onboard 3-phase power from a single-phase power inlet.

13.9 Power Management. There are several power management systems on the market that combine a transfer switch and switchboard. The electrical loads are grouped into heavy, medium and light. The power input groups such as shore power, generator and inverter are supplied to the panel. When the major power supply source such as the generator is supplying the board all load groups are available. If a limited shore power source is connected, heavy load groups are inhibited from operation. As the system has input voltage monitoring, when the voltage drops, the system will switch off equipment to prevent damage and overloads. The system transfer switch operates in 200ms. Another automatic device from Victron is the Filax transfer switch unit. This allows changeover from shore, generator or inverter in just 20ms, which stops computers rebooting, video and microwave clocks resetting.

13.10 AC Switchboards and Panels. The following basic guidelines should be observed:

The AC switchboard should be of the dead front type, and constructed of non-hygroscopic and fireproof material.

Open dead front type switchboards must be of a non-conductive and non-hygroscopic material. If protective and isolation devices are not to be integrated into the main electrical panel, consider an industrial or domestic consumer distribution panel or module as an alternative. These panels are made of plastic with a splash-proof cover, and have all earth and neutral conductors, main switch or RCD and MCBs within one compact unit.

The switchboard should be located in a position to minimize exposure to spray or water.

Locate the switch panel or distribution unit in a dry place such as a cupboard or other suitable and safe area. Many choose to locate them at the navigation station, and while it looks good, it does expose the board to damage.

AC systems should not be located or installed adjacent to DC systems. Where DC and AC circuits share the same switchboard, they should be physically segregated and partitioned to prevent any accidental contact with the AC section. The AC section must be clearly marked with Danger labels.

Where DC and AC are installed on the same panel, which is common on many yachts, cover the exposed AC connections at the rear of the panel, or preferably enclose them in a separate compartment.

13.11 AC Circuit Protection Principles. In any AC system, there are several control and protection devices. It is normal to have these co-ordinated and this is called discrimination or selectivity. The principle is that a fault appearing in any part of the system is cleared or disconnected by the protection device immediately upstream of the fault. Discrimination ideally should be full, but in most cases is only partial. Partial discrimination may be achieved up to a specific current level, and any fault current levels above that cause all breakers to trip. This is achieved by having different current settings, with a motor thermal overload set, a switchboard distribution breaker, and generator supply breaker all set at different levels. Tripping times are dependent on the rate of current increase. If a fault such as an overload rises up to a level of 0.75 above normal current, the thermal relay will trip. If the current continues to rise such as in an impedant short-circuit, the magnetic circuit breaker supplying the circuit should trip. With a full short circuit current the supply breaker also will operate very fast. If there is no discrimination, all circuit breakers will open up to the generator supply.

13.12 AC Short Circuit. Short circuits are relatively rare in yacht electrical systems. A short circuit is where two points of different electrical potential are connected, that is live to neutral, live to ground, phase to phase, and phase to ground. A short circuit causes a very rapid rise in current which can reach several hundred times the nominal value in milliseconds, and causes high thermal and mechanical stresses which can destroy the cables, and the busbars which feed the fault. There are two thresholds in short circuits. The first is where electrical arcs commence, insulating materials start to breakdown and parts start to deform. The second level is where contacts melt and weld together, electrical arcs continue and insulators start to carbonize. Short circuits can be either a lower level condition called an impedant or intermediate short-circuit, usually caused by deterioration in insulation. A full short-circuit is usually caused by directly connecting two phases or a similar connection fault.

13.13 Causes of a Short Circuit. Typically short circuits are caused in order of probability by loose connections, insulation damage or failure, metallic bodies or conductive deposits on terminals within junction boxes and motor terminals, broken conductors within a cable, dust and moisture gradually tracking and burning across to create a short, and after new installations or repairs there are connection or crossover errors.

13.14 Short Circuit Calculations. Prospective short circuit current (Isc) can be calculated for circuits although this is often only done in larger yachts with large power systems and built to survey or Class. This is defined as the calculated RMS value of short circuit fault current at any point in the system, should a conductor of negligible impedance replace the protection device. Values depend on the supply voltage and line impedance.

13.15 Short Circuit Protective Device (SCPD). Typically this is either a fuse or a circuit breaker. Fuses are devices whose conductors melt to break the fault current. Circuit breakers detect short circuit current and open the poles to clear the fault. An SCPD must detect and break high fault current levels quickly before reaching peak current values. Given the high rise times and magnitude of currents this must be fast to be effective. Speed of the SCPD is determined by the peak current and breaking capacity of contactor thermal device to withstand overload. A DOL (Direct On Line) starter has a start curve of approximately 7.2 x current for 10 seconds. Thermals must protect against low-level faults but allow for machine start-up times and current. An MCB combines the action of an isolation switch, overload protection, and short circuit protection.

a. **Overload Protection.** This function is a thermally operated one. Tripping values are normally to hold 110% of rated value and trip at 137% of rated value at 25°C.

b. **Short Circuit Protection.** This function is a magnetic one. A solenoid coil within the breaker trips when the factory set short circuit current value is reached. Under short circuit fault conditions a large arc can be generated, and breakers use the generated magnetic field to direct and quench the arc in a chute.

13.16 Selecting Protective Equipment. Equipment must be selected properly if it is to perform properly under fault conditions.

The main switch or protection circuit breaker should be rated for the maximum current capacity of the circuit.

The MCB must be selected to protect the cable, not the equipment. The MCB must be rated to hold at the maximum demand, such as motor starting loads, which can be 4 to 6 times rated load. The device voltage must suit the system voltage. The interrupting capacity must be able to cope with any prospective fault current levels. The MCB current rating must hold at 100% of operating current, and trip at 125% at 40°C.

13.17 Cable Installation. The cabling installation requirements for AC systems are virtually the same as those defined for DC systems. These are briefly redefined.

Conductors are to be selected based on the maximum current demand of the circuit. Ambient temperatures exceeding the rated temperature of the cable should be de-rated.

Conductors should be selected based on the maximum current of the connected load. Where cables are to be installed in hot machinery spaces consideration should be given to de-rating if the ambient temperature exceeds the cable rating temperature.

Conductor size should be selected with a maximum allowable voltage drop of 5% for all circuits.

The maximum acceptable cable voltage drop is 5% in all circuits. The voltage drop problem can be prevalent in high starting current equipment, and where equipment may be running at maximum load a larger cable size may be a good option.

Conductors should be stranded, insulated and sheathed (double insulated) and where possible of shipwiring standard. All conductors should have a minimum cross-sectional area of $1.0mm^2$.

Cable ratings and insulation materials should conform to recognized national standards. Where possible cables classed as "Shipwiring Cables" should be used. 3 Core (2 core and ground) cables should be used and typical ratings are given in Table 14-2. Shipwiring cable is expensive. If you do not want to pay the costs of such cable, and do not have to meet Class requirements, install 15/30-amp heavy duty outdoor rated, orange or yellow-sheathed extension cable. Do not install domestic triplex single strand type cable under any circumstances.

Conductor color codes should conform to either IEC or US standard codes.

IEC: Brown wire is active "live or hot"; blue wire is neutral; green/yellow stripe wire is ground or earth. Three phase systems should consist of the primary color red, yellow and blue for each phase, and in some cases white is used for the yellow phase. Switching and control circuit wires may be any other color that cannot be confused or mistaken.

US: Red wire is hot "active"; black wire is hot "active"; white wire is neutral; ground is green wire.

Cable runs should be installed as straight as practicable. Cable bend radii should be a minimum of 6 x cable diameter.

Cables should be installed in as straight a run as practicable. Tight bends should be avoided to reduce unnecessary strain on conductors and insulation. The minimum cable bend radii is 6 x the diameter.

Cables should be accessible for inspection and maintenance.

All cables must be accessible for maintenance and inspection. All cables, in particular those entering transits should be capable of access for routine inspection, and adding other circuits.

Cables should be protected from mechanical damage, either where exposed, or installed within compartments.

All cables should be installed so as to prevent any accidental damage to the insulation or application of excessive stress on the cable.

Cables passing through bulkheads or decks should be protected from damage using a suitable non-corrosive gland, bushing or cable transit. Cables transiting decks or watertight bulkheads should preserve or maintain the watertight integrity.

Cable glands are designed to prevent cable damage and ensure a waterproof transit through a bulkhead or deck. They should be installed properly and not degrade watertight bulkhead integrity.

Cables should be supported at maximum intervals of 8 inches (200mm). Supports and saddles are to be of a non-corrosive material. Where used in engine compartments or machinery spaces, these should be metallic and coated to prevent chafe to the cable insulation. Cable saddles should fit neatly, without excessive force onto the cables, or cable looms, and not deform the insulation.

Cables should be neatly loomed together and secured with PVC or stainless saddles to prevent cable loom sagging and movement during service. In machinery spaces metal saddles are often used; however they should have a plastic sleeve placed on them to prevent the sharp edges chafing the cable insulation.

Cables should be run as far as practicable from DC power cables, network, data and signal cables.

Cables should be run separately from DC cables and should never be within the same loom or cable bundle. They should also be run clear of any data cables, network or signal cables.

All neutral conductors terminated in a neutral link should be identified to correspond to the marked circuit number.

When connecting AC circuits the neutral should be marked and connected to the same numbered terminal in the neutral terminal block. Similarly if an earth or ground terminal block is used to collect the grounds, the ground for the circuit should have the same terminal number. For example circuit 1 has neutral 1 and ground number.1.

The ground conductor should not have any switch, fuse or other device installed.

The ground conductor should not have anything connected in line to it that may cause opening of the circuit. See notes on galvanic isolators.

All ground connections should be mechanically secured, and be protected against mechanical damage or corrosion.

Ground connections must be mechanically secured so that they cannot come loose. The ground connection point should be protected from mechanical damage or any accidental disconnection. The connection should be clear of water or moisture so that corrosion cannot occur. The connection should be coated for protection.

Table 13-2 AC System Cable Ratings

Cable Size	PVC	Butyl Rubber	EPR
1.0 mm^2	8 amps	12 amps	13 amps
1.5 mm^2	11 amps	16 amps	17 amps
2.5 mm^2	14 amps	22 amps	23 amps
4.0 mm^2	19 amps	31 amps	32 amps
6.0 mm^2	25 amps	39 amps	40 amps
10.0 mm^2	35 amps	53 amps	57 amps
16.0 mm^2	40 amps	70 amps	76 amps
25.0 mm^2	60 amps	93 amps	102 amps
35.0 mm^2	73 amps	115 amps	120 amps
50.0 mm^2	85 amps	135 amps	155 amps

Equipment should be grounded and the maximum resistance between any ground point and the boat ground shall be 1 ohm.

The maximum resistance of the grounding (earthing) system must be 1 ohm between the main ground terminal block and the boat ground. It should also be less than 1 ohm between any grounded point and the boat ground. Any fault arising on an ungrounded or inadequately grounded item of equipment may cause exposed metal to be "alive" up to rated voltage. Accidental touching and grounding by a person may cause serious electric shock, injury and death. Grounding provides a low impedance, low resistance path for any fault arising on exposed and bonded metal. During fault conditions, extremely large current levels of several hundred to thousands of amps may flow. This high current usually ruptures fuses or trips circuit breakers. Improper or degraded grounding or a high resistance ground may cause circuit conductor heating and fire. Corrosion protection is an important consideration, but the safety factor is more important. In multiple grounded systems ashore any current flowing in the grounded neutral conductor will also cause a voltage rise on that conductor above the mains supply voltage. This will also be impressed on the grounding system, and this is typically a few volts above ground potential. In a fault on the mains, the supply neutral voltage on the installation grounding system will rise in proportion to the load, and the value will depend on the impedance of the main ground system. At all times, it is recommended that an AC licensed electrician be used to ensure that installations are done correctly.

13.18 Ground (Earth) Leakage Protection. The most reliable and accepted method for personal protection is the installation of earth leakage protection devices. Many marinas now install these on each circuit. The new devices are now considerably more advanced; they include residual current devices (RCD) or ground fault current interrupters (GFCI) or residual current circuit breakers (RCCB).

Ground fault protection devices should be installed on all boat power outlet circuits.

GFCI/RCCB/RCD units should protect all power outlets on a boat. This may be on the entire circuit or integral to the power outlet. They are not required on water heating or electric stoves as element leakages will cause nuisance tripping. This also can occur in a clothes washer and dishwashers with water heating elements.

Where earth leakage protection devices are installed such as RCD (Residual Current Devices) and GFCI (Ground Fault Current Interrupters), they should be tested monthly to verify operation.

All RCD/GFCI units should be tested every month using the integral test facility. The GFCI has an integral test button that simulates a ground fault to test the tripping function, and a reset button. Devices also require checking with a special test unit, as the self-test button is not always reliable.

a. **Installation Requirements.** The selection and installation of an GFCI is based on the tripping values, and therefore the level of protection. The values are as follows and tripping times are around 25ms:

   (1) **5-30 mA Value.** These values are used for quick tripping and protection against personal shock. Many GFCI are set at 5mA.

   (2) **100 mA Value.** This level is designed to give fire protection.

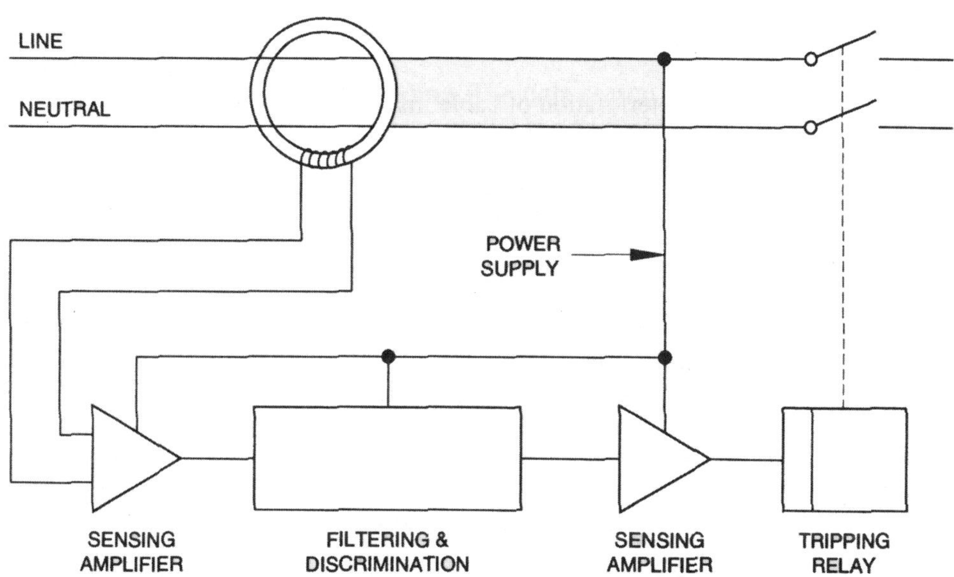

Figure 13-5 Residual Current Protection Devices

b. **GFCI Tripping.** Due to the environment, earth leakages are commonplace, and nuisance tripping is common at marina berths. Not all trips are nuisance; they indicate a problem that must be investigated and rectified. The principal causes of trips are as follows:

(1) Connection of a neutral and ground (earth) connection downstream of a GFCI.

(2) A crossed neutral between protected and unprotected circuits.

(3) Deterioration in cable insulation.

(4) Water and moisture in terminal boxes, and cumulative leakages from a number of sources of small leakage paths.

(5) Absorption of moisture into heating elements including steam irons, refrigeration defrost elements, stove and hot water elements and electric kettles. This problem disappears if element operates for half an hour or more.

(6) Tracking across dirty surfaces to ground.

(7) Intermittent arcing internally in appliances.

(8) High voltage impulses caused by switching off inductive motor loads.

(9) High current impulses caused by capacitor start motors.

c. **GFCI (RCD) Operation.** The GFCI/RCD units work on an electromagnetic principle as illustrated below:

(1) A toroidal transformer detects magnetic fields created by current flow in the active and neutral conductor of the protected circuit.

(2) Under normal conditions, the vector sum of the currents, known as residual current, is effectively zero and the magnetic fields cancel.

(3) If a condition arises where current flows from active or neutral to ground, the residual current will not be zero and the magnetic field will establish a tripping signal to the protected circuit.

d. **Installation Checks.** Installation should be performed by an AC licensed electrician and tested using test equipment made for the purpose. The following tests must be performed using a 500-volt (Megger):

(1) Disconnect supply, neutral and earth. Test between active and earth. On new installations readings must exceed 1 meg ohm, and a minimum of 250k ohms on existing systems.

(2) Test between neutral and earth. Readings must be a minimum of 40k ohms.

13.19 Circuit Testing. The following *must* be tested before putting any circuit into service, and as part of routine inspection and testing:

The insulation resistance between all circuit insulated poles and ground, and between poles ground must be greater than 1 meg ohm.

A 500-volt DC insulation Megger tester should be used. Disconnect all electronics and appliances, turn power off and disconnect the main grounding conductor. All switches should be in the on position. Insulation resistance between ground and live conductors must be a minimum of 1 meg ohm. Water heater elements must be at least 10k ohms. An ohmmeter should be used to check between all active and neutral poles on each circuit to ensure that only load resistances are present, and with all switches on there is no short circuit through either cable damage or incorrect equipment connection.

The insulation resistance between all switchboard busbars, and between busbar and ground must be greater than 1 meg ohm.

A 500-volt DC insulation Megger tester should be used. Disconnect all electronics and appliances, turn the power off and disconnect the main grounding conductor. All switches should be in the on position. Insulation resistance between busbars and ground must be a minimum of 1 meg ohm.

The insulation resistance of all generator and motors, cables, windings and control gear and ground must be greater than 1 meg ohm.

A 500-volt DC insulation Megger tester should be used. Insulation resistance between all parts and ground must be a minimum of 1 meg ohm. Where possible the tests should be made on hot machines.

There must be no transposition of active and neutral conductors.

All switches, circuit breakers, outlet live pins, equipment terminals must be checked and be of the same polarity. No transposition of neutral and actives (crossed connections) is allowable.

There must be no transposition of ground and neutral conductors.

All equipment and outlets must be checked to ensure that there are no crossed connections.

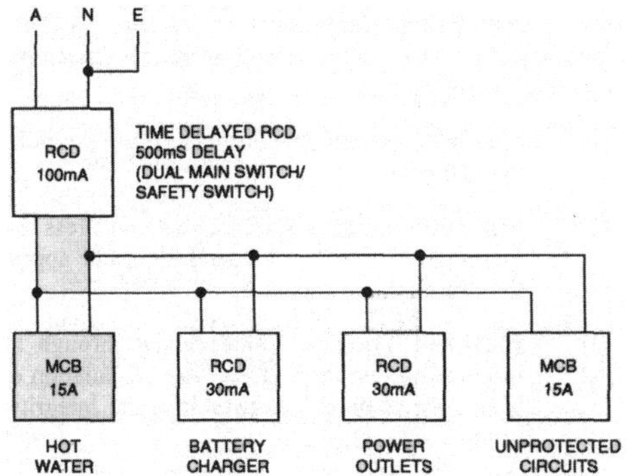

Figure 13-6 AC Circuit Protection

13.20 Generators. The majority of generators are single phase, with three phase machines being used on larger vessels. The majority of units come complete with sound shields and only require external connection of cooling water, fuel, electrical and exhaust systems. Factors that are important in generators are the stability of the output voltage, the stability of the frequency, the quality of the output waveform, and the ability to withstand high starting currents. Most manufacturers have significantly reduced weight, physical size, noise emission and vibration levels. Diesel generators generally have the same principles and requirements as those for main propulsion diesels and in practice they operate longer hours than main engines. A number of alternators driven off main propulsion engines have been developed, and are common on canal boats and barges which have long motoring periods when in transit. Typical of these are the Auto-Gen unit which has a sine wave output unit of 4.5 KW, (19.6 A). This type of unit is belt coupled to a main crankshaft pulley, and requires up to 9 HP at full rated output. The units use a clutch and control system to compensate for engine speed variations and ensure frequency stability. Other types use an alternator to feed an inverter system with fixed stable frequency. Earlier units had a modified sine wave output. A Northern Lights generator is illustrated below with various systems.

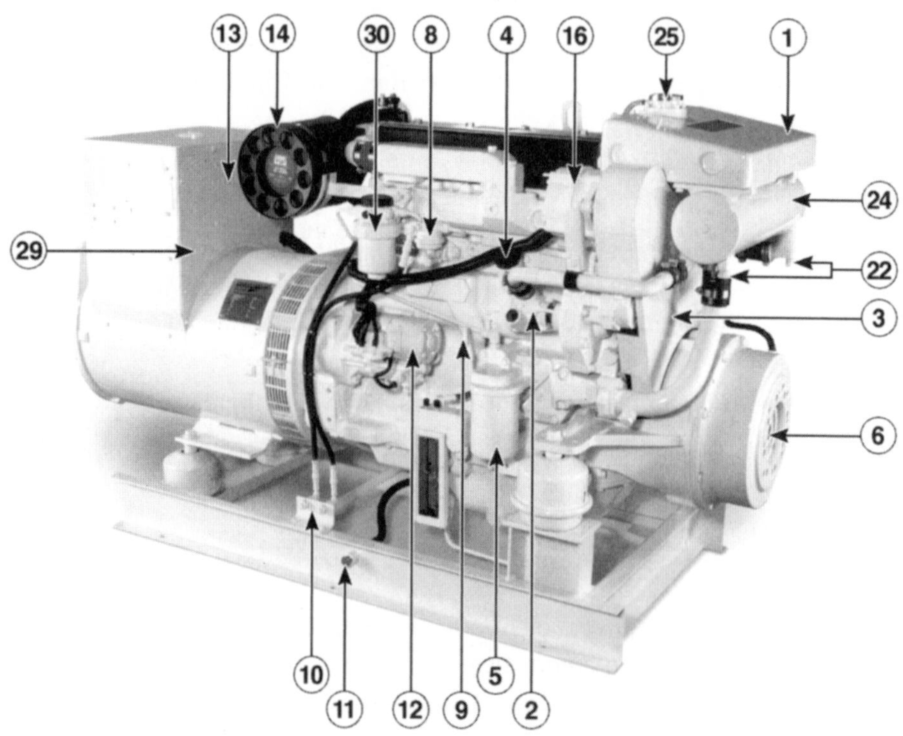

1: Expansion Tank; 2: Injection Pump Drive; 3: Drive Belt Cover; 4: Lub Oil Filler; 4: Lub Oil Filter; 6: Power Take Off; 8: Fuel Lift Pump; 9/10: Fuel Manifold; 11: Lub Oil Drain; 12: Starter; 13: DC Circuit Breaker; 14: Air Cleaner; 16: DC Alternator; 22: Heat Exchanger Zincs; 24: Heat Exchanger; 25: Coolant Filler; 29: AVR Fuse; 30: Fuel Filter

Figure 13-7 Northern Lights Generator *(Courtesy Northern Lights-Lugger)*

a. **Installation.** The basic installation factors are as follows:

(1) **Fuel Systems.** Clean fuel is essential and the installation of a separator system such as a Racor is essential. Check them regularly and drain off any accumulated water.

(2) **Exhaust System.** Improperly installed exhaust systems are a major cause of failure. The seawater inlet should not be fitted with a scoop type inlet. It can pressurize the water and subsequently force water past the pump impeller, which can cause the muffler to fill and in worst cases into the exhaust manifold and engine cylinders. The exhaust outlet should be installed above the loaded waterline; the transom is the best location. Ensure that the vented loop (siphon break) is installed where required and regularly check that it operates. Ensure that the exhaust lines are properly installed with respect to loops, and slope, and that there are no points where water can be trapped. As with all engines use caution when cranking over the generator as water can fill the muffler and back up into the engine. When bleeding the fuel system, Northern Lights recommends closing the seacocks, and removing the water pump impeller during the process. An innovation from Northern Lights is the Gen-Sep which separates the water from the exhaust gas and drains it via a through-hull fitting.

(3) **Seawater Systems.** Ideally a separate water system supply and strainer should be used. Ensure that the strainer is cleaned regularly, as it is often forgotten until a high temperature occurs. Where the generator has anodes check these regularly.

b. **Control Systems.** Generators are generally remote start and stop systems with automatic engine protection systems. In most generators, all shutdown functions are suppressed for 10–20 seconds at start-up, and this may include low water pressure and low oil pressure. Some larger generators on large yachts and super yachts will have generator units that have a pre-lubrication pump operating when the engine is stopped, and a failure in pre-lube pressure will inhibit the start. On vessels with multiple generators and automatic stop and start functions, all systems must be selected to Auto. In any system it is important to know what interlocks are in the start ready chain. Most generators will trip automatically on overspeed, high crankcase pressure, high temperature, low oil pressure, and low oil levels.

c. **Governor.** The mechanical governor controls the fuel rack and maintains constant speed. On starting the engine it runs up to nominal speed and then the governor maintains it. Many now have electronic governors.

13.21 Generator Fuel Consumption. Table 13-3 gives approximate fuel consumption rates at full rated load for a number of engine (not electrical) output ratings. These will act as a general guide in working out similar on-board consumption values.

Table 13-3 Generator Fuel Consumption

Cylinders	Capacity	Speed	Output	Fuel Rate
2	0.5 liter	3000	7.5 kW	1.5 l/hr
3	1.0	1500	7.7	2.5
4	1.3	3000	9.8	3.2
4	1.3	1500	19.4	5.9
4	1.5	1500	11.9	2.6
4	1.8	1500	13.0	4.5
3	2.5	1500	21.0	6.1
4	3.9	1500	34.0	10.2

13.22 AC Alternators. Alternators are generally robust and constructed to marine standards. The alternator consists of the main stator winding, exciter stator winding, the main wound rotor, the rotor exciter winding, the cooling fan, terminal box, the bearings and the rotating rectifier.

Single Phase Alternators. The single-phase 2 or 4 pole alternator is the most common configuration and may be either brushless and self-excited or brush with sliprings and externally excited. Alternators may be single bearing and directly coupled to the engine, or be dual bearing machines driven by coupling.

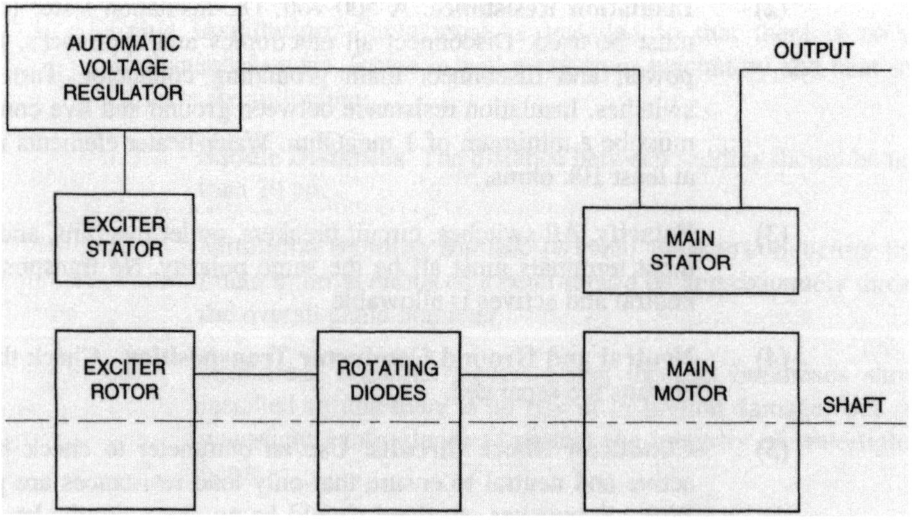

Figure 13-8 Single Phase Alternator Diagram

(1) **Operation.** At initial start-up, there is sufficient residual (remanent) voltage left in the machine to establish the main field. Once rated speed and output is reached the automatic voltage regulator (AVR) controls the output voltage in response to system variations. Frequency is a function of speed and the engine governor maintains speed and therefore frequency stability.

(2) **AVR Operation.** The AVR controls the excitation voltage level. The control voltage is applied through the brushes and sliprings to the rotor-mounted excitation winding and the diode rectifier. The rectifier DC output then goes to the main rotor excitation winding rotating field and controls the field strength.

13.23 AC Alternator Parameters. The alternator has the following parameters:

a. **Voltage.** The typical rated output voltages are 115/230, or 220/240 volts for single-phase machines, and 440/480 volts for three phase machines. The automatic voltage regulator (AVR) maintains nominal output. The AVR is an electronic regulator that senses output terminal voltage and varies the field strength to maintain correct value. Regulation is typically within 2% of nominal rating. The AVR must be able to control the output rapidly in response to large load fluctuations, and recovery in good machines is typically 3% of rated output within 0.25 seconds when full load is applied. Voltage is not a function of speed or frequency when an alternator is running at or near rated speed.

b. **Frequency.** Frequency is specified in hertz, the number of alternating cycles per second. Output frequency is a direct function of speed and varies in response to speed fluctuations. Stability is dependent on the ability of the machine to maintain nominal frequency over the complete power output range, and this is typically within 1%. The engine governor controls engine speed. When a large load is applied, such as a motor starting, the generator loads the engine, causing it to slow momentarily. The governor reacts by increasing fuel flow and speeding the engine up. When the load is removed, the reverse occurs. Stability depends on response time, and governors are factory set. A small time lag is inherent in the system and helps minimize hunting, caused by continual alterations based on small load fluctuations. Frequency and specified engine speed depend on the number of poles within the alternator. Two pole machines generate one cycle per revolution, and require an engine speed of 3000 rev/min for 50 hertz. Four pole machines generate two cycles per revolution, and only require 1500 rev/min for 50 hertz.

c. **Power.** Power output is stated in either kVA or kW ratings. These are defined as follows:

(1) **kVA Rating.** The KVA rating is simply the power output, which is the current multiplied by voltage to give volt amps, and divided by 1000 to give a KiloVoltAmp rating.

(2) **kW Rating.** Kilowatt rating is the kVA rating which is then multiplied by the power factor, typically 0.8. This is the actual power output.

13.24 AC Alternator Rating Selection. Rating selection must consider a number of factors. A total expected load analysis must also be undertaken to calculate the peak loads that might be encountered:

a. **Starting Currents.** Starting current values may be as high as 5–9 times that of actual normal running current. The in-rush current at starting causes these high currents, and the energy required to overcome bearing and load inertia. Duration of the peaks are typically less than one second, and most alternators can withstand 250% overloads for up to 10 seconds.

b. **Power Factor.** In simplified terms, Power Factor (PF) is the ratio of useful power in watts, to the apparent power (volt amps) of the circuit. Power (watts) = Volts x Amps x Power Factor. In a purely resistive circuit such as a heater, the alternating current and voltage are said to be in phase. The average power over a complete cycle is the product of the voltage and current in volt amps. When reactance is introduced into the circuit, the voltage and current become out of phase, so that during any cycle the current is negative, and the voltage positive. The resultant value is less than the volt amp value. Inductive reactance causes current to lag the voltage and this will be an electrical angle between 0 and 90°. Resistive loads are said to be in phase, with no angle of difference and these are termed as unity power factor or one. In electrical circuits, capacitive reactance cancels out inductive reactance. The use of capacitors can improve low power factors, and this is generally limited to fluorescent lighting systems. Most machinery nameplates specify power factor ratings. Available alternator output power decreases with any reduction in system power factor values so the higher the better.

13.25 Generator Rating Calculations. From a load analysis the following calculations can be performed to estimate required minimum alternator size:

a. Generator to run at 80% maximum load with 20% reserve.

Largest Single Load Value = 2400 watts

Max. Start Current (Ir) = $\frac{\text{Power (watts)}}{\text{Volts x PF}} = \frac{2400}{240 \times 0.8} = 12.50$ Amp

Max. Start Current (Is) = Ir x 4 = 50 Amps

b. If alternator can withstand overloads of 250% for 10 seconds, the starting current (Is) must be divided by 2.5 = 20 Amps.

c. Rating is therefore 20 x 240 = 4.8 kVA or 0.8 x 22.8 x 240 = 3.8kW. Adding a 20% minimum margin to operate at 80% = 4.5kW.

d. In selecting the margin to the calculated required output rating, an estimate must be made of the maximum load likely to be applied. Some loads are resistive and do not have large starting currents, such as kettles and heaters. If the device having the startup current is to operate simultaneously with other equipment, the other loads need to be added. So if an air conditioner with 3.8 kW rating is to be run with hot water, then add 2.4kW, and for a kettle 1.8 kW then a figure of 7 kW, plus a margin for other small loads such as lights, TV etc you require 8.4kW. The higher the rating the higher the initial capital cost, the greater the weight and space required. A decision must be made as to what equipment usage regime will be followed in order to reduce the generator to the lowest suitable size. A generator should be loaded to at least 35% and ideally operate at 75–80%. The running of gensets diesels on light loads causing cylinder glazing will increase maintenance costs.

Table 13-4 AC Load Analysis Table

Equipment	Typical Rating	Actual Rating
Inductive Loads		
Air Conditioner	3600	________
Washing Machine	1800	________
Hydraulic Pump	3600	________
Water Pump	2400	________
Fluoro Lights	40	________
Pressure Cleaner	2400	________
Refrigerator	500	________
Washing Machine	1200	________
Hand Tools	650	________
Microwave	1200	________
Food Processor	1200	________
Battery Charger	1200	________
Dive Compressor	3600	________
Resistive Loads		________
Toaster	1000 Watts	________
Kettle	1200	________
Incandescent Light	60	________
Water Heater	2400	________
Television	100	________
Video Player	200	________
Hair Dryer	500	________
Iron	1200	________
Fan Heater	2400	________
Low Energy Light	60	________
Coffee Maker	750	________W
TOTAL LOAD	________W	

13.26 AC Equipment Ratings. Many of the generator over rating problems can be resolved by carefully choosing appliances for use on board. Inductive loads such as coils, solenoids, motors and fans have "inrush" currents when starting, several times the normal running current. Resistive loads such as toasters and kettles do not. Many fast boil kettles have an element rated at 2.4 kW (240 VAC). On a 4 kW generator that is more than half load. Buy a lower rated kettle at around 1200 watts. Most automatic toasters have large current consumptions. The older type fold down side toasters have a lower current draw, and are more reliable. Microwaves should be a simple and compact unit with a relatively low power rating. Some regular size microwaves on high have ratings of around 1600 watts. Many fan heaters on high setting are also rated at 2.4 kW, and place a significant load on a generator. Hot water heating elements should be 1.2 kW instead of a 2.4 kW unit. High current AC motors can be fitted with soft-start starters that reduce or limit starting currents. This will depend on the load attached.

13.27 Generator Systems. The following are details of some well known generators. Most have optional sound shield options, with very low noise and low vibration levels. Also all use good design practices with respect to boats by having items needing servicing, such as oil filters, all accessible on one side.

a. **Northern Lights.** (www.northern-lights.com). Northern Lights generators have always had a reputation as being rugged and commercial grade. These generators operate at a lower speed of 1800 rpm with 4-pole brush and brushless synchronous wound rotor alternators, with Class H rated winding insulation. The generator prime mover is a Lugger diesel engine. Most generators have an alternator rated at up to 40 amps that can be used as house battery charge source in addition to the AC charger. Voltage regulation is rated at just +/- 1% RMS for the 8–99 kW range, while units in the 104 to 520 kW range are rated at +/- 0.5%. The 5 and 6 kW machines are rated at +/- 5%, and have a special dedicated auxiliary winding to power the AVR which improves starting performance. Excitation uses shaft-mounted diodes, which are sensibly made accessible although failures are very rare. Fuel consumption for the 32kW units is typically 2.6 gph (9.8l/hr) at full load. The units also have zinc anode corrosion protection, which must be checked. Control systems use relays and the engine stop is an electric operated solenoid.

b. **Onan.** The range has units in the 4 to 95kW range. Latest e-QD systems include digital controls with integral diagnostics. Speed control has an electronic governor with close isochronous regulation to maintain output frequency stability under heavy loads typical of high current equipment starting. The engine has an automatic glowplug pre-heat system to ensure good starting.

c. **Kohler.** The units are available in the 4–65 kW range. Kohler use their own AVR called PowerBoost for very good regulation characteristics, and the alternators are brushless.

d. **Fisher Panda Generators.** These are regulated asynchronous generators running at 3600 rpm with 2 poles. An asynchronous alternator has a rotor of highly magnetic material without windings. This allows the smaller sizes to be realized; however the higher speeds also emit more noise, and efficient sound shields are essential. This retains greater heat and water cooling is used. Regulation is achieved using a voltage control system (VCS) which governs engine speed of the diesel with respect to load in the range +/- 3 volts at 80% of maximum load. The system also incorporates an integrated start booster (ASB) that minimizes the droop caused by heavy current consumers such as electric motors at start up and can withstand 300% overloads. In conventional machines the voltage level drops below rated value. The asynchronous generator is different in that it does not have rotating windings or diodes, brushes, and or a wound rotor. One model has both a three-phase and single-phase windings for dual operations, which has advantages where single-phase loads would create phase load imbalances. The units also have another feature with automatic negative pole uncoupling via a relay, which minimizes corrosion, and all sensors are isolated as well. These units also have water-cooled alternators via a copper/nickel/bronze heat exchanger that minimizes corrosion. Diesel units in use are Farymann and Kubota units on smaller units and Mercedes, MTU, MAN, Yanmar etc on larger units. One additional useful feature is the ability to install on the Panda 10 HTG-Duo model, a high output 230 amp DC alternator system, and where AC is not required to also then run at lower speeds.

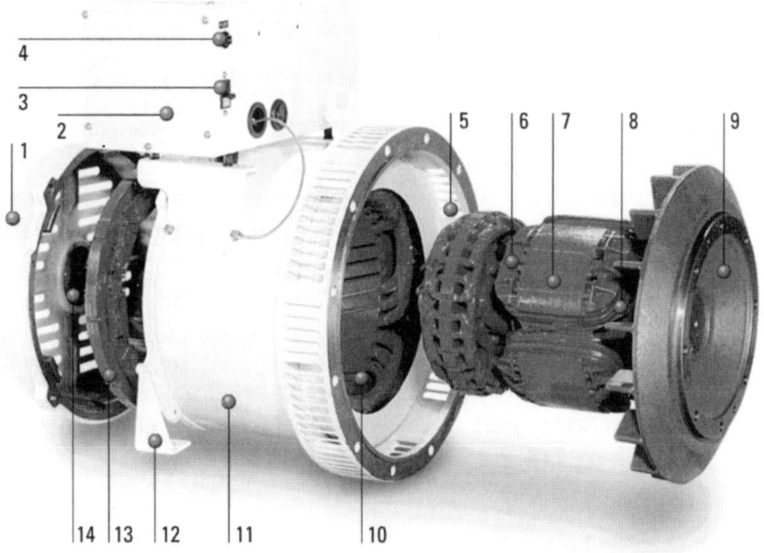

1: Bearing Housing; 2: Junction Box; 3: AVR AC Circuit Breaker; 4: DC System Circuit Breaker; 5: Cast Iron Housing; 6: Exciter Rotor; 7: Main Rotor; 8: Cooling Fan; 9: Mounting Ring; 10: Skewed Stator; 11: Frame; 12: Mounting Foot; 13: Exciter Stator Bolts; 14: Bearing Carrier

Figure 13-9 Northern Lights Alternator *(Courtesy Northern Lights-Lugger)*

13.28 Generator Protection. Diesel generators have several protection systems.

a. **Low Oil Pressure Shutdown.** Most generators have automatic shutdowns on low oil pressure to protect the engine. This is usually a pressure switch activated function.

b. **Low Oil Level.** This activates when oil level in the sump falls; it is a level switch.

c. **High Water Temperature Shutdown.** Generators will shutdown on detection of a high jacket cooling water temperature.

d. **Low Water Level Shutdown.** This is activated when expansion tank level falls.

e. **Low Seawater Flow Shutdown.** Loss of seawater cooling, either due to plugged strainer or impeller failure, will initiate an automatic shutdown.

f. **High Exhaust Temperature.** A high exhaust temperature will initiate an automatic shutdown.

g. **Over-speed Shutdown.** If the generator over-speeds an automatic shutdown is initiated. This can be caused by governor failure.

h. **Over-voltage Shutdown.** If the AVR malfunctions and the voltage fails to high the generator will be shutdown.

13.29 Alternator Protection. The alternator system requires the following protection:

a. **Protection.** Protection of generator electrical circuits consists of one or more of the following, which may be integral to the main circuit breakers for larger vessels, or be part of a separate protection system:

(1) **Overload.** Some gensets have an overload circuit breaker fitted at the genset control box. Reset if tripped, if repeatedly tripping, fault find the system and remove the cause of the overload. In many cases the problem may simply be too many appliances operating and overloading. Overloads are characterized by time delays between reset and tripping. If a major motor is installed and it is seized or has a locked rotor, tripping may be immediate due to high current.

(2) **Short Circuit.** This generally is a circuit breaker, which is mounted at the genset control box. If it trips immediately after initially resetting, trooubleshoot and correct the fault first.

(3) **Reverse Current.** This protection is generally only seen on larger installations or where two units are parallelled. If the load sharing function fails, or when manually taking a generator off the board, the load is taken to zero, this will trip.

(4) **Low Frequency.** Not all generators have this protection, adjust only according to manufacturer's instructions.

(5) **Undervoltage.** Undervoltage trip relays are used in larger installations and are normally interlocked with main circuit breakers. It usually indicates an AVR fault.

b. **Grounding.** All exposed metal capable of carrying a voltage under operating or faulty conditions must be grounded to an equipotential point. The generator frame should be securely connected to the boat ground system. In most generators the starter negative is also bonded locally to the AC ground. This will require bonding to the main boat ground. In single-phase installations, the neutral is connected to the distribution system neutral at the main switchboard, and not frame. Connection should be in accordance with installation instructions.

c. **Parallelling of Machines.** In larger vessels with more than one generator, parallelling of the units onto the main switchboard busbar may be necessary. To do this a frequency meter will be required to correctly synchronize the machines, or lights are used. The engine speed is controlled using the manual speed control. When the synchroscope is moving very slowly clockwise and nearly stopped, and is approaching the 11 o'clock position, close in the circuit breaker. Some vessels will have an automatic phasing and synchronizing module.

13.30 Alternator Maintenance. Alternator maintenance is simple and easily carried out.

a. **Alternator Inspection.** Carry out the following tasks:

(1) Remove the alternator access covers, and check the gaskets and seals. Connection boxes should be clean and dry.

(2) Inspect the interior for dirt, dust, oil, and water. Clean as required with a vacuum cleaner all accessible surfaces and windings.

(3) If the generator has a brush excitation system check the excitation brushes for breaks, chips and make sure they move freely in the holders. Check brush springs and shunts.

(4) Make sure the sliprings are smooth, have no scoring and have a shiny surface patina. Do not polish or use emery paper.

(5) Check for loose electrical and mechanical connections.

b. **Insulation Test.** Use a 500-volt Megger to test all active conductors to ground. Measure and record the temperature of the alternator stator circuit insulation resistance in the boat maintenance log. Isolate the exciter and measure the rotor insulation resistance and record. Any reading less than 1 meg ohm should be rectified.

13.31 Generator Mechanical Systems. The alternator prime mover gives the most in-service troubles. The principles involved are identical to the main propulsion engine. If the lubrication, cooling, and fuel and air quality are maintained, long-term, trouble-free operation is assured.

a. **Coolants.** Genset diesels may be cooled by sea water or have a heat exchanger with a closed-circuit cooling system. The coolant provides the medium for transferring engine heat to the primary sea water coolant and for controlling the overall operating temperature of the engine. It is essential to maintain adequate heat transfer. The coolant must remain free of salt water contamination to prevent corrosion or the formation of sludge and scale that may impede coolant flow or block coolers. A coolant which has no inhibitors, incorrect inhibitors, or improper concentrations of inhibitors will ultimately cause problems with rust, sludge, fatigued water-pump seals, and reduced heat transfer rates as engine block water passages become coated with insulating layer of scale. This will gradually result in overheating and all the damage that goes with it.

(1) **Additives.** A number of additives are available that improve the performance of coolants, including sulfates, chlorides, dissolved solids, and calcium. Coolant should also have an antifreeze additive to prevent freezing and engine damage in cold climates. Most ethylene-glycol-based antifreeze solutions contain the inhibitors required for normal operation.

(2) **Corrosion Inhibitors.** This is generally a water-soluble chemical compound that protects the metal surfaces within the system against corrosion. Compounds can include borates, chromates and nitrites. Inhibitors with soluble oils should never be used as a corrosion inhibitor.

b. **Lubricating Oil.** Lubrication has the dual function of reducing friction between moving parts and taking away some of the heat generated during combustion. It is essential to use the correct grades of oil for the prevailing temperature conditions, and to change the filter regularly along with oil. Oil viscosity must be maintained if correct lubrication is to be achieved, and this depends on the engine remaining within proper operating temperature range. Lubricating problems include:

(1) **Fuel in Oil.** Fuel in the oil creates the risk of a crankcase explosion and is characterized by low lube oil viscosity.

(2) **Water in Oil.** Water in the oil causes emulsification, which destroys the lubricating properties. After repairing a leak, completely flush out the system. No moisture must remain.

(3) **Microbe Growth.** Moisture in the system can encourage microbial growth in the oil. Once the system is infected, considerable flushing is required if it is to be eliminated.

c. **Fuel.** Uncontaminated fuel is essential to good combustion and efficient engine operation. There has been a history of water-contaminated fuels supplied to unsuspecting yachts from bunker barges. Observe the following installation precautions:

(1) **Filters.** Install a filter and water separator (Racor), preferably with a water-in-fuel alarm. In most cases, the generator takes fuel from the same tanks as the main engine, which should have similar protection.

(2) **Purge Fuel Tanks.** Microbial growth can occur in water-contaminated fuels. Tanks must be regularly purged of water and kept topped up to avoid condensation.

d. **Electrical System.** The electrical and monitoring system is similar to the main engine system. It consists of the following:

(1) **Starting System.** Always install a separate starting battery for the generator. Start batteries for a typical 4–6 kVA genset are recommended at 70AH/325CCA. This provides backup power for the main engine in an emergency, or alternatively ensures that the genset can be started if the main engine is out of service. Some generators have an interlock in the starting system that prevents starting if the generator is running.

(2) **Charging System.** The alternator charging system on most gensets is very small, typically around 25 amps maximum. Normally, it is only used to charge the start battery, although alterations can be made to charge main engine or house batteries. In some cases, the alternator can be uprated to 55 amps or even 80 amps, which when connected up to a cycle regulator, provides the main battery charging source, limiting the run time requirements of the main engine. Care must be taken as output shafts cannot always cope with large mechanical side loads.

(3) **Monitoring System.** Most generator units have basic control panels with alarms only for high water temperature or low oil pressure. Many newer units also incorporate an hour meter, pilot light, overload alarm, and a water warning alarm. It is generally easy to install gauges which give a clearer indication of performance.

(4) **Preheat Systems.** Many generators have a preheating system which uses a traditional cylinder glowplug in the combustion chamber. Some engines also have preheaters in the air intakes, referred to as a cold start aid.

13.32 Generator Maintenance. The maintenance tasks for the main engine are also valid for generators.

a. **Fuel System.** Renew and clean filters every 1000 operating hours.

b. **Lube Oil System.** Replace oil and filters every 500 operating hours. Always check oil viscosity, and for signs of water, fuel or microbial growth that may affect viscosity and quality.

c. **Air System.** Replace filter element every 1500 operating hours.

d. **Coolant System.** Check coolant levels monthly. Periodically check inhibitor concentrations with test kit. Check thermostat.

e. **Cleaning.** Keep engine clean of oil and dirt.

f. **Belts.** Check and tighten alternator and water/fuel pump drive belts monthly.

g. **Charging System.** Using voltmeter, check that the charge voltage is approximately 13.5-14.5 volts.

h. **Mountings.** Check rubber mountings for cracks and fatigue.

i. **Electrical.** Check that battery and starter connections are clean and tight.

j. **Anodes.** Sacrificial zinc anodes in the cooling system should be checked every 6 months and renewed if corroded.

13.33 Generator Operation. Consider the following when operating the genset:

a. **Starting.** After starting, always check that sea water coolant is discharging overboard to ensure that coolant is passing through the engine. Don't wait for a high-temperature alarm to warn you of possible engine damage.

b. **Operating Temperatures.** Run the genset for 3–5 minutes before putting a load on it. This gives the engine a chance to increase to a normal operating temperature.

c. **Loading.** Do not let the generator run on light or no load for extended periods. This will cause cylinder glazing and deposits within the engine which will increase maintenance costs. If you have an electric water heater, put it on to increase load and make the most of the available energy.

13.34 Gasoline Gensets. Portable gasoline gensets typically have ratings up to around 3 kilowatts. They are designed primarily for land-based applications, not for use in marine environments. There are many in use on cruising yachts, but they have significant disadvantages.

a. **Explosive Fuel.** Gasoline is extremely volatile and good ventilation is essential to avoid potentially hazardous vapor concentrations.

b. **Exhaust Emissions.** Gasoline gensets emit toxic carbon monoxide; belowdeck concentrations can be extremely hazardous. Good ventilation is essential.

c. **Electric Shock.** As components are not designed for marine environments, the risk of electric shock does exist. Grounding is also more difficult. Grounding should be to the vessel AC ground source. A prominent boating journalist once recommended that grounding on fiberglass vessels was unnecessary due to the insulating nature of the deck it sat on. The premise was that it was difficult for anyone to ground out a fault on the unit and hence electrocute themselves. Not true! Don't risk it!

d. **Installation.** The lubrication system is not designed to operate at any angle of heel, so the units must be run in a near level attitude. Because of this, multihulls may have more use for portable gensets, and since a number of fairly large multihulls are propelled by outboards, the fuel will be identical.

Table 13-5 Generator Troubleshooting

Symptom	Probable Fault
Generator will not start	Fuel supply valve closed
	Fuel filter clogged
	Flat battery
	Control power failure
	Stop solenoid jammed
	Fuel lift pump fault
Low cranking speed	Low battery voltage
	Battery terminal loose
	Starter motor fault
Will not hold load	Fuel filter clogged
	Air filter clogged
	Air in fuel system
	Governor fault
	Voltage regulator fault
High temperature	Air cleaner clogged
	Injector pump fault
	Injector atomizer fault
	Thermostat fault
	Heat exchanger clogged
	Water pump fault
	Water pump belt loose
	Low coolant level
	Seawater strainer clogged
	Seawater pump fault
Low oil pressure	Low oil level
	Oil filter clogged
	Oil cooler clogged
	Oil pump fault
	Bearing problem
	Pressure gauge fault
Exhaust smoke black	Air cleaner clogged
(incomplete combustion)	Low compression
	Injector pump fault
	Restricted exhaust
	Head gasket leak
	Engine cold
	Valve stuck
Exhaust smoke white	Low compression
(unburnt fuel vapor)	Head gasket leak
	Cold start
Generator starts and stops	Fuel filter clogged
	Air in fuel system
	Fuel supply valve closed
Generator misfiring	Injector pump fault
	Overheating

13.35 Generator Electrical Troubleshooting. Faults within generators are normally confined to bad connections at the alternator. In rare cases cables may chafe and insulation is damaged. The main faults that may arise on the alternator are:

a. **Over Voltage.** This condition arises where the voltage regulator is faulty.

b. **Under Voltage.** This condition arises where the voltage regulator or excitation circuit is faulty.

c. **Voltage Fluctuation.** This occurs when a voltage regulator is faulty. In brush machines this occurs when a brush sticks, or the sliprings are dirty and there is arcing.

d. **Current Pulsation.** This is caused by diesel problems, injector failures etc. It is seen on the ammeter as a needle oscillation.

e. **Unable to Sustain Load.** This fault typically occurs with a faulty regulator, or where a rotor diode is breaking down or has failed. In units with capacitors, faulty capacitors, or transient suppressors can also cause this.

13.36 AC Machinery. The most common type of equipment is the electric motor. Motors are very robust and generally give years of trouble-free service. The principal problem with AC motors is the starting currents, which can be two to five times their rated load. Most starters have a Direct-On-Line (DOL) starting system. On-board repairs are generally limited to bearing replacement. Rewinding and similar repairs will be undertaken by a shore repair facility. On many occasions, an AC motor may have a very low winding insulation value, due to moist air or being flooded. To undertake repairs, the following should be undertaken to get up and running again:

a. **Winding Drying.** Dismantle the motor completely. If the motor has been immersed, replace the bearings and do the following:

(1) Wash the motor stator with fresh water.

(2) Place the stator in the oven at approximately 70°C for at least four hours.

(3) Check insulation value to case with 500-volt (megger) tester. Reading should be at least 1 meg ohm.

(4) Recheck insulation reading after 4 hours to ensure that the reading remains high.

b. **Maintenance.** The maintenance requirements of AC motors are minimal and consist of the following:

(1) **Insulation Testing.** Every 6 months, the winding-to-ground insulation resistance should be tested using a 500-volt (megger) tester. Readings should be a minimum of 1 meg ohm.

(2) **Terminals.** Connections should be checked and tightened.

(3) **Bearings.** If bearings are not sealed, they should be repacked every two years, depending on run times. If a motor is stationary for much of the time, maintain the bearings by manually turning the shaft at least once a month.

13.37 Bearing Replacement. Removing bearings and pulleys is a task that requires both care and skill. More damage is done to motors and machinery because of improperly installed bearings than nearly any other cause. A good puller set is very useful. I have a very good Proto set that caters to most pulling tasks.

Table 13-6 AC Motor Troubleshooting

Symptom	Probable Fault
Ground fault	Insulation resistance broken down (moisture, overheating, aging or mechanical damage to winding)
Motor overheating	Ground fault Mechanical overload Bearings seizing
Circuit breaker tripping	Ground fault Mechanical overload Winding intercoil short circuit Terminal box cable fault
Motor overload	Stalled load Seized bearings Terminal connection loose
Bearings hot	Bearing lubrication failure Drive belts overtensioned Bearings worn
Vibration	Coupling misaligned Bearing failure Loose frame holding bolt

13.38 Inverters. Inverter technology has advanced considerably in the last few years. Inverters range from small portable units of just 150W up to large fixed systems of 5kw with some units that can be parallelled with an automatic synchronization module. The combination charger/inverter is also very common. The latest units use IGBT (Isolated Gate Bipolar Transistors) which provide precise output control and waveforms, over older MOSFET type systems. Transformers have been reduced in size and weight as conversion is done at high frequencies. Reliability on most systems is greatly increased over earlier generation units that unfairly gave inverters a bad reputation. Units use fans to assist in heat dissipation.

a. **Output Waveforms.** Output waveforms are an important consideration when looking at proposed applications:

(1) **Trapezoidal.** The majority of inverters have a trapezoidal waveform. This is suitable for most equipment, but microwaves and some inductive loads do not operate at full output, dropping efficiency by some 20% or more in some cases. Fluorescent lights may also be less efficient at starting and a 3-microfarad capacitor across input will improve starting characteristics. Interference is possible on these waveforms.

(2) **Modified Sine Wave.** Some units have what is called a modified or quasi sine wave output, which closely resembles pure sine wave. It is not and offers slightly less performance. Interference is possible on these waveforms. Appliances and equipment such as microwaves and VCR units with clocks often run either slow or fast. Battery chargers for cordless portable drills are susceptible to early failure. If a charger gets excessively warm or hot turn it off as some incompatibility may exist. There is a ferro-resonant line conditioner called a Line Tamer, Model PCLC which can improve performance. These are available from Shape Electronics in Illinois (tel. 708-629-8394).

(3) **Sine Wave.** Sine wave output units are the ideal and offer quality better than shore mains power but are more expensive. If you are using sensitive equipment sine wave should be used. They are available in most ratings up to 1800W in 12-volt systems, with outputs at 230V RMS +/- 5%. Frequency stability is typically +/- 0.05%. Quality units have low harmonic distortion, typically less than 3%, and low EMI (Electromagnetic Interference) levels.

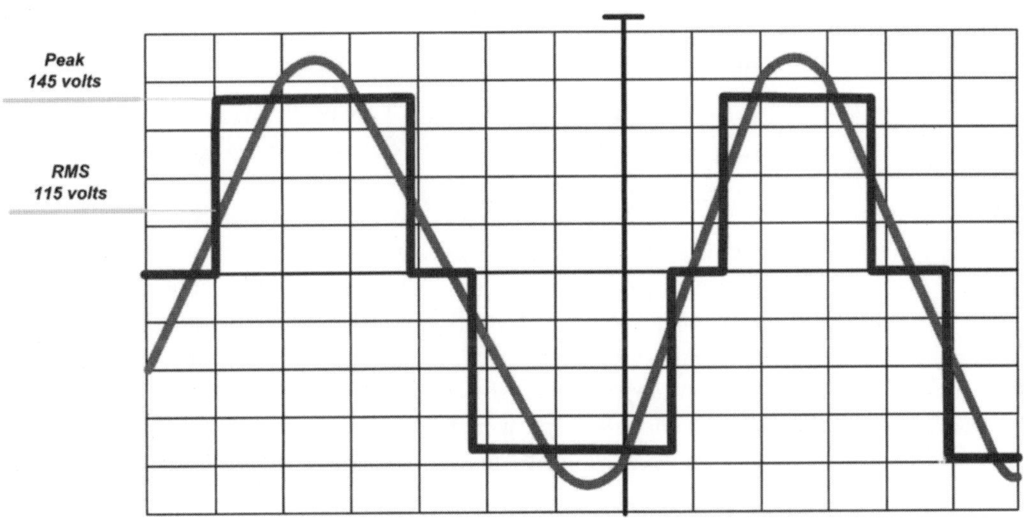

Figure 13-10 Inverter Output Waveforms

b. **DC Loads.** The typical inverter is capable of drawing large DC current loads from the battery. It is essential for battery capacity to be able to support these loads without affecting the existing electrical system and connected loads, particularly electronics equipment. A simple method of calculation is to divide the power in watts by 10 for 12-volt systems or 20 in 24-volt systems. It should be noted that some manufacturers offer inverters for 12, 24 and 48 volts inputs depending on system voltage. The minimum battery capacity required for an inverter is 20% of the inverter capacity, ie 2000 watts 400 Ah capacity.

Table 13-7 Inverter 12 VDC/220 VAC Current Loads

AC Load	DC Current Draw	Peak Overload
200 W (0.8 amps)	18 amps	42 amps
400 W (1.6 amps)	37 amps	100 amps
600 W (2.4 amps)	56 amps	145 amps
900 W (3.6 amps)	84 amps	180 amps
1200 W (4.8 amps)	120 amps	290 amps
2400 W (9.6 amps)	240 amps	580 amps
3000 W (12 amps)	300 amps	750 amps

c. **Transfer Systems.** If another AC power source is connected across the inverter output, the inverter electronics will be seriously damaged. Shore power supplies, generator and inverter outputs must not at any time be paralleled.

(1) **Rotary Switches.** The normal selection system is a rotary cam switch. The switch should be a center off type with the inverter to the side opposite to generator and shore power.

(2) **Automatic Transfer Systems.** Many units are also able to automatically switch the supply from shore power or generator to the inverter on supply loss giving the same features as a UPS. Freedom series inverter/chargers have an automatic "load sharing" facility that offers protection to connected AC loads from variations in shore power or generator power. When the shore or generator power is connected the unit transfers the power to any connected AC loads and some power is used to charge the batteries. When in the charge/transfer mode the incoming AC power is monitored and when a voltage drop (or increases) longer than 8 milliseconds occurs charging ceases, load transfer ceases and inverter mode starts.

d. **Efficiency.** The typical inverter is now approximately in the range of 85% to 95% efficient at rated output. This efficiency level has evolved with new electronic switching technologies such as MOSFET and IGBT.

e. **Combination (Combi) Units.** Most of these units have a modified sine wave output. The chargers can be set for various battery types such as flooded cell, Gel or AGM batteries, and those from Heart use a 3-stage microprocessor charging process. An automatic transfer switch and relay transfers between charger mode and inverter mode when AC power is off. As a note of caution, always remember that combi units may automatically supply AC so proper isolation from the battery source is essential before opening or working with the unit.

f. **Input Cables.** Always install the maximum size cables possible as considerable volt drop problems are possible at peak loads. Ensure cable connections are tight.

g. **Inverter Grounding.** Inverters must be grounded. There are documented cases of fatalities where this has not been done. Grounding introduces a number of factors that require consideration. The corrosion aspects must never compromise the safety requirements. Most standards or recommendations (ISO, ABYC, USCG) specify that the AC ground be connected to the DC negative. This requirement has raised considerable controversy in shore power installations as well as inverters. Many do not do this to reduce the risks of galvanic corrosion. The ABYC recommends that the inverter chassis be bonded to the DC negative, although this is doubling up the grounding as the AC ground and DC are already bonded.

h. **Ground Fault (GFCI) Protection.** It is normal to install a GFCI on the AC output of the inverter. Heart interface uses the Pass & Seymour/Legrand unit, although any make will be satisfactory. As indicated the GFCI must be tested regularly. The GFCI should be function tested when inverting or when transferring generator or shore power supplies. Do not test when the inverter is in idle mode, as the unit may not trip, and the GFCI electronics circuits could be damaged by the inverter idle mode sense pulses. Nuisance tripping does occur on inverters, generally caused by neutral to ground leakages, usually from surge suppression circuits which contain capacitors connected across active and ground or neutral and ground. This is attributed to waveform harmonics on modified sine wave outputs. Another cause of GFCI tripping is due to the improper connection of the inverter AC output neutral to the main neutral bus. Inverters such these from Heart ground the output neutral when in the OFF mode or in inverting mode. If the output neutral is connected to the main neutral, the main neutral will also be grounded. The GFCI will detect this and trip out before the inverter is able to disconnect the ground from the output neutral during shore power and generator transfers.

i. **Interference.** The control electronics such as logic circuits and memory circuits in inverters can be corrupted. This may be from lightning strike surges, onboard electrical power system surges, voltage dips and spikes. In many cases the unit may simply require resetting if no signs of catastrophic failure have been detected such as smoke or burning smells. Try the following procedure first before calling in technical assistance.

(1) Switch off power at remote panels if fitted.

(2) Switch off inverter main power switch.

(3) Disconnect the AC input power source.

(4) Disconnect the DC negative cable for at least 5 minutes.

(5) Reconnect the DC negative cable. Sometimes a small spark will be seen as filter capacitors start charging.

(6) Switch on the inverter power switch and switch on an AC load to ensure inverter has a load. Check that inverter supplies AC power.

(7) Reconnect the AC input power source and check that automatic transfer functions.

(8) If this does not restore operation, check fuses have not blown or circuit breakers have tripped. Check all connections are secure.

j. **Auto Start.** Most units have an auto start capability. This means they remain in a standby or idle mode until a load is switched on.

(1) **Idle Mode.** Input current in idle mode is typically around 10–50 mA (1.5W). This load value must always be included in DC load calculations.

(2) **Activation Load.** The load required to activate most inverters is approximately 6–10 VA or greater. If the vessel is to be left unattended for an extended period the DC supply should be switched off. In some cases loads such as fluorescent lights or electronics equipment may not activate, and another load should be momentarily switched on to cut the inverter in.

k. **Protection.** Most inverters have an under voltage cutout that is typically set at around 10.5 V. Units also have overload protection and a high voltage cutout, thermal overload protection that will shut down the inverter if an overtemperature condition is reached. Reverse polarity indication and protection, short circuit protection of the output and input voltage ripple too high, and some units have AC backfeed protection and a GFCI on the output.

l. **Ventilation.** Good ventilation is essential to reliable operation and full rated outputs. The unit should be installed in a dry, and well-ventilated area. Sufficient vertical clearance should be allowed for natural convection of heat from unit. De-rating factors are illustrated in Table 13-8.

Table 13-8 Temperature De-rating Factors

Temperature	Output Rating
+40°C to +50°C	80% rated output
–10°C to +40°C	100% rated output
–10°C to –20°C	140% rated output

m. **Ratings.** Units generally have an output rating based on a resistive load and for a nominal period, typically 30 minutes (eg. 1600 watts). The continuous rating is the normal continuous operation rating (eg. 1000 watts). The peak or maximum rating is the maximum short duration load that the inverter can withstand (eg. 3000 watts). Most units are capable of withstanding the short duration and intermittent overloads that are required, especially with motor starts. The surge rating enables them to withstand short time overloads of up to approximately 200% over the continuous rating for 5 seconds.

13.39 Microwave Ovens. The number of onboard microwave ovens has increased rapidly over recent years. Coupled with good freezer capacity, they make meal preparation much easier, especially in bad weather. Additionally, they conserve cooking gas. Microwaves convert the AC input voltage into very high frequency energy using a magnetron. They are around 50% efficient, so a microwave rated at 650 watts, for example, would actually consume around 1300 watts at maximum output. An inverter would have to be able to supply that value. If operating off a square wave inverter, efficiency could drop by up to 30% due to the power supply waveform. These domestic appliances should be permanently installed to prevent moist, salt air and condensation from getting into the magnetron, which may cause corrosion and premature failure. It is a good idea to put a bag of silica gel inside the case.

SECTION TWO

ELECTRONIC SYSTEMS

NAVIGATION STATION DESIGN

14.0 Navigation Station Design. Before you start installing navigation equipment, especially if you are fitting out a new vessel, consider carefully the following requirements that ensure reliable performance:

a. **Aesthetic Considerations.** There is a certain amount of satisfaction in having a good looking navigation station. It will attest to a seamanlike attitude that is not lost on your guests, and it is nice to show off. The trap is that a nicely presented navigation station is worthless if the equipment malfunctions or is unreliable because of a lack of planning or a failure to consider the technical requirements of the equipment. Make it look good, but above all make sure it all works.

b. **Location.** The nav station is invariably located at the bottom of the companionway steps where it is easily accessible. In many cases, the electronics are exposed to spray or even solid water if the washboards are carelessly left out in the event of a knockdown. Many problems are associated with this exposure. Precautions should include the following:

(1) **Equipment Selection.** Select equipment rated as splashproof so it can withstand exposed positions and intermittent spray.

(2) **Waterproofing.** Mount instruments in a panel that prevents water from getting behind to connectors and power connections. In exposed positions, protect the instruments with clear perspex or plastic sheeting if possible.

c. **Ergonomics.** Instruments should be positioned so they are easy to operate and monitor. Instruments should be grouped into functional blocks where possible. Keep communications equipment in one block, position fixing equipment in another. Equipment must be fitted so that access is unobstructed. Many nav stations are a jumble of systems thoughtlessly crammed into any available space. Important considerations are as follows:

(1) **Display Visibility.** Position displays at an angle that is normal to observation. Many instruments are mounted vertically for observation when sitting down, but when at sea in normal operation, they are generally monitored when standing up. Difficulties are often especially pronounced with LCD displays.

(2) **Accessibility.** Make sure you can easily reach and operate controls. On some badly designed stations, you either have to stretch awkwardly or a knob is placed in such a tight corner you can't get to it.

(3) **Lighting.** Make sure that there is adequate lighting with a good deckhead light above or at the chart table.

d. **Electrical Factors.** Consider the effects the instruments can or may have on one another.

(1) **Cable Routing.** Route all radio transmission cables clear of signal cables. Where cable crossovers are required, make sure they are at 90°. Properly space out and secure cables with the required separation distances. Position electronic equipment so that aerial cables and inputs exit the nav station directly. Don't route cables behind other instruments or close to other cables.

(2) **Electrical Equipment Location.** Where possible, do not locate the main electrical switchboard next to the electronic equipment. In many cases, this is nearly impossible. The trend now is to install a smaller sub-board containing circuit breakers for the electronic equipment only. This removes a great deal of interference caused by the electrical equipment.

(3) **Interference Protection.** Interference sensitive equipment such as GPS, LORAN and autopilot control units should be located in a block. Construct an aluminum housing around the section and ground it if this is a real problem although new equipment is more resilient.

(4) **Accessibility.** Make sure you have easy access to rear connections, plugs, sockets and rear fuse holders.

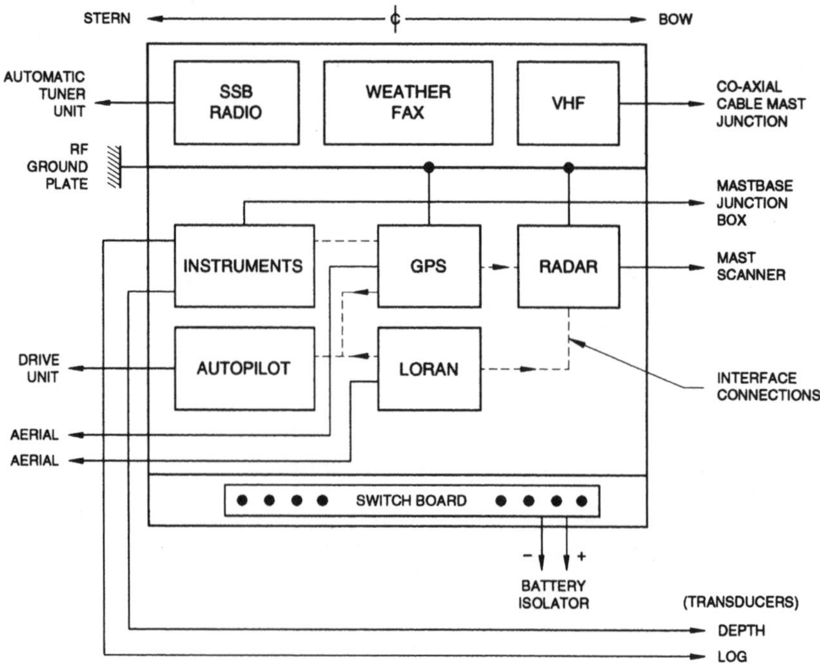

Figure 14-1 Nav Station Instrument Layout

Radar

15.0 Radar. RADAR is an acronym for RAdio Detection And Ranging. Radar is a method for locating the presence of a target, and calculating its range and angular position with respect to the radar transmitter. Good radar units make close-in navigation a lot easier for making landfalls, navigating channels, or in reduced visibility. With the advent of GPS, some mistakenly see radar as redundant, but there is no subsitute for radar as a navigational aid. Radar indicates where things are; GPS indicates where you are. Many times I have wished for radar when closing on a shore or port in bad visibility at night. Radar offers many very useful functions:

- Position fixing from geographical points.
- Positions of other vessels.
- Positions of buoys.
- Rain and squall locations.
- Land formations when making a landfall in poor visibility.
- Collision avoidance at night and in poor visibility.

15.1 Radar Theory. Radar transmits a pulse of radio frequency (RF) energy and on small boat radars the IF frequency is 60 MHz. This is radiated from a highly directional rotating transmitter called the scanner. Any reflected energy is then received and processed to form an image. The time interval between transmission of the signal and reception of reflected energy can be calculated to give target distance. The subject of radar reflection theory is complex and is covered extensively in Chapter 16. It is essential to understand how radar signals behave on various target materials if radar is to be fully utilized.

15.2 Radar Scanners. A typical scanner revolves at around 24 rpm. In practice, the larger the scanner, the narrower the beam width, and the better the target discrimination. Of the two main scanner types, the beam widths of enclosed scanners are always larger than the open array types. This factor is one of the trade-offs that has to be considered when selecting a radar scanner unit. If it can be accommodated, an open array scanner performs far better. A typical 2kW, 18 inch radome has a power consumption of 28 watts and 9 watts on standby, and 4kW, 24 inch radome consumes 34 watts and 10 watts on standby. Typical warm up times are around 60–70 seconds. The two scanner types in common use are as follows:

a. **Enclosed Scanners.** The enclosed array scanners are commonly used when mast mounted on yachts and an 18-inch radome weighs in at 7.5 kg (16.5 lbs). There are two basic types of antenna elements in use:

(1) **Printed Circuit Board.** Printed circuit board, phased antenna arrays are commonly fitted to enclosed scanners. The antenna is on a circuit board instead of the more expensive slotted waveguides.

(2) **Slotted Waveguide.** Center-fed, slotted waveguide arrays are normally used on open array antennas and larger range radomes.

b. **Open Array Scanners.** An open array scanner has a beam width nearly half that of enclosed units, which gives far better target discrimination. If you can tolerate an open scanner, such as on a sternpost, the improved performance is worth it. The downside is that power consumption is greater. A 4-foot array weighs in at 30 kg (66lbs) so they tend to be more suited to power vessels.

c. **Sidelobe Attenuation.** Beam widths are not precisely cut off. There are zones outside the main beam where power is wasted and dissipated. End slotted waveguides are often used in new radars to suppress sidelobes, which generate false echoes. False echoes are more pronounced on short ranges at increased sensitivity.

d. **Frequencies.** All small boat radars operate on microwave frequencies in what is termed the X band. Frequency ranges are 9200 to 9500, a wavelength of around 3 cm.

e. **Output Power.** Peak power ratings are given for the actual microwave output power. A 24 nm radar is typically around 2.0 kilowatts and a 48 nm unit is 4 kilowatts. A kitchen microwave operates on the same principle. Given the effect microwaves have on food, always follow the warnings on eye protection. It is quite common on naval vessels with high power radars to incinerate any birdlife in the rigging at start-up.

f. **Range Discrimination.** Range discrimination or resolution is a function of transmission pulse length. When the distance between two targets on the same bearing is longer than the pulse length, they are shown as separate. When the distance between targets is less than the pulse length, they appear as one target. Most radar sets automatically alter pulse length with a change in range settings.

g. **Beam Angles.** Radar transmissions are similar to the light beam from a lighthouse in that a radar beam has a defined angle in both vertical and horizontal planes. The beam width is normally defined as the angle over which the power is at least half of maximum output. Usually this is rated at -3dB.

(1) **Horizontal.** Enclosed radome type scanners have a horizontal beam width of around 25° for all radomes scanners. Open array scanners have a beam angle as low as 1.85° for 4-foot open array scanners and 1.15° for 6-foot open array scanners.

(2) **Vertical.** Enclosed radome type scanners have a vertical beam width of around 5.2° for 18-inch radomes and 3.9° for 24-inch radomes. Open array scanners have a beam angle of around 25° for all open array scanners.

(3) **Heel Angles.** The heel of a vessel, and therefore of the scanner, has an adverse effect on performance. Most radars have vertical beam angles of 25°, so at a heel angle of approximately 15°, anything to windward is virtually invisible and there is a significant blind spot to leeward. That doesn't take into account the additional masking of the signal by waves. This problem is very pronounced on sternpost mounted units. One solution is to alter the attitude of the scanner using one of the several self-leveling systems available. These include gimbals systems, or manually operated hydraulic systems. The Questus system can be mounted around a backstay, mast or stay pole and has hydraulic damping on the leveling mechanism. The cheaper alternative is to level up the boat periodically and have a look, which shouldn't be a problem on a cruising yacht.

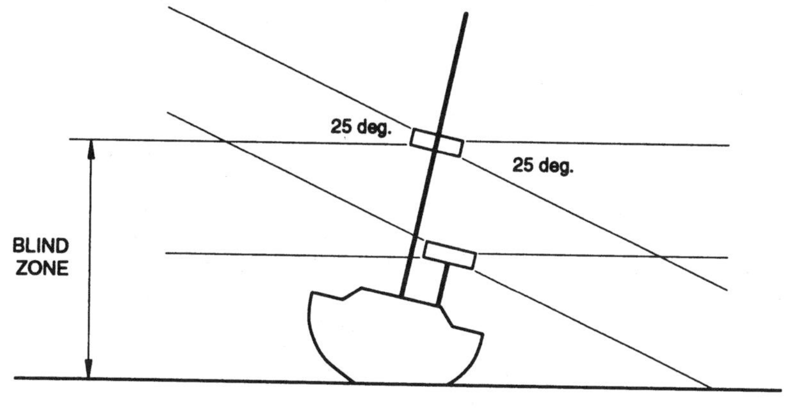

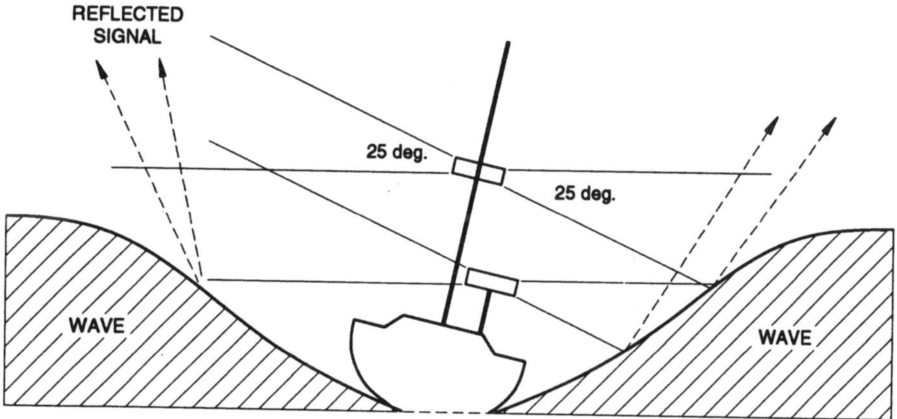

Figure 15-1 Radar Heeling Effects

h. **Target Discrimination.** Target discrimination or resolution is a function of beam width. A scanner with a narrow beam width is effectively slicing and sampling sectors of approximately 2.5° around the azimuth. Large targets will be sampled a number of times and their size quantified. A wider beam width will sample an area twice that size, but will not always discriminate between two or more targets. If a harbor entrance is narrow, the radar beam may in fact see it as part of the breakwater until the range has closed up. At longer ranges, two targets at the same distance and close together may appear as one.

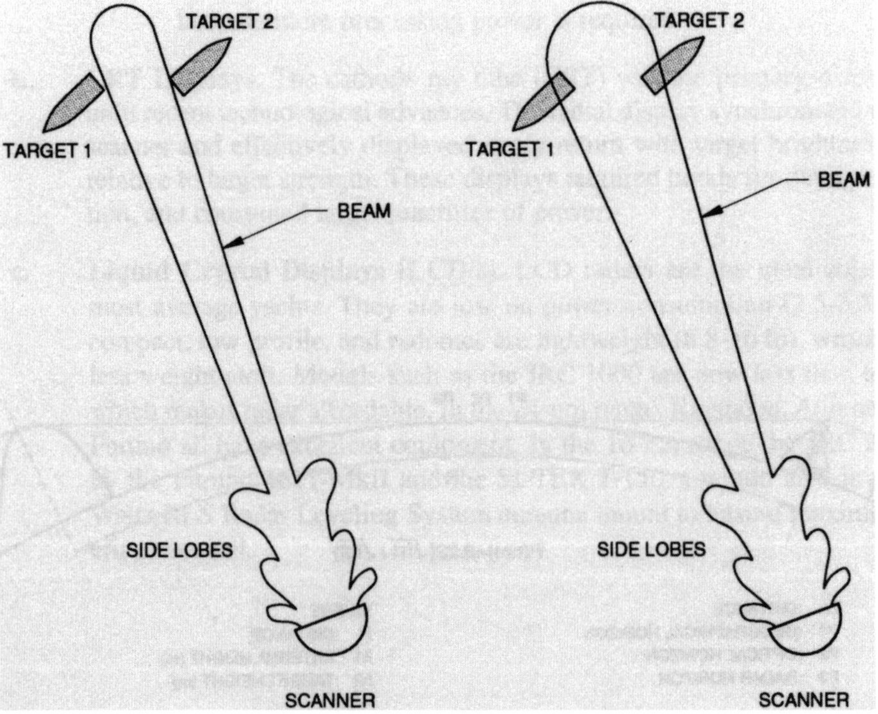

Figure 15-2 Target Discrimination

15.3 Radar Ranges. Maximum radar ranges are a function of scanner height. Table 15-1 gives the approximate horizon ranges for radar under standard conditions for targets of a known height. Conditions at sea may give better results, but it should be noted that atmospheric conditions affect the ranges as well. Radar signals travel in a straight line, but are subject to bending under normal atmospheric conditions. This bending increases the radar's horizon approximately 8% over optical horizons. Targets such as large vessels or land masses, for instance, may appear at much greater ranges, and the known height of these should be added to the scanner height. Ranges can increase or decrease depending on the prevailing atmospheric conditions.

Table 15-1 Radar Horizon Table

Target Height (meters)	Scanner Ht. 5 m	Scanner Ht. 10 m	Scanner Ht. 15 m	Scanner Ht. 20 m	Scanner Ht. 25 m
Zero	5.0 nm	7.0 nm	8.5 nm	10.0 nm	11.0 nm
5	10.0	12.0	13.5	15.0	16.0
10	12.0	14.0	15.5	17.0	18.0
15	13.5	15.5	17.3	18.5	19.8
20	14.8	17.0	18.5	19.8	21.0
25	16.0	18.2	19.8	21.0	22.3
30	17.3	19.0	20.8	22.0	23.3
35	18.0	20.0	21.8	23.0	24.3

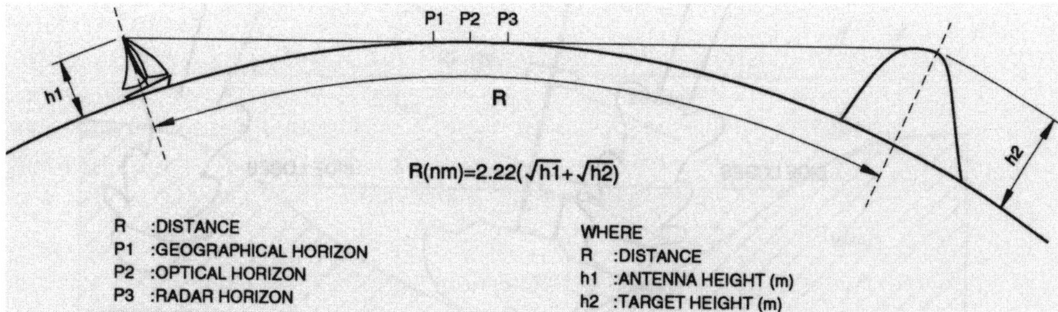

Figure 15-3 Radar Horizons

15.4 Radar Displays. The display used to be called the plan position indicator (PPI). There are now a number of display types available on various radar systems.

a. **Rasterscan Displays.** These displays use the same technology as computer monitors. Essentially, the radar screen consists of many dots called pixels. The status of the pixels installed in the memory is altered in response to signal processing changes. Resolution is quoted as the number of pixels on the screen, such as 480 x 640 pixels. Unlike the CRT display, the rasterscan display is a result of complex digital-signal processing and allows the use of numerical information on the screen. Digitally processed signals usually have to be above a minimum level to be displayed, so weak echoes are often rejected. For this reason, proper tuning and operation are essential if all targets above that threshold are to be displayed. Manufacturers have introduced a number of processing techniques to overcome these shortcomings:

   (1) **Single-Level Quantization.** This method displays all targets at the same intensity level, regardless of size or strength of return. The main problem is that targets, sea clutter, and rain do have to be distinguished.

   (2) **Multi-Level Quantization.** This method of processing assigns echoes into strength categories. The stronger echoes appear bright, while weak echoes appear dim on the screen. Inconsistent or weak echoes may not be displayed. These systems are more expensive because more processing power is required.

b. **CRT Displays.** The cathode ray tube (CRT) was the primary display type until technological advances made them redundant. The radial display synchronized with the scanner and effectively displayed every return with target brightness being relative to target strength. These displays required hoods for daylight operation, and consumed large quantities of power.

c. **Liquid Crystal Displays (LCDs).** Earlier LCD radars were the ideal solution for most yachts, and many are still in use. They are low on overall power consumption (2.5–3.5A), are compact, with low profile, lightweight radomes, which means less weight aloft. Models such as those from JRC, Furuno, and SI-Tex still continue to serve their owners well.

d. **Multi-function TFT LCD Displays.** Radar as part of an integrated system is virtually the norm now. There is a single display for all radar, chartplotter, video, instrument and fishfinder data. This data is all transferred by network. The displays are sunlight viewable with 256-color capability. Resolutions are typically 640 x 480 (VGA) and 800 x 600 (SVGA). Display power consumption is typically 9–12 watts.

e. **Heading Mode Display Orientation.** A radar display can orient to one of three configurations. These are:

(1) **North Up (N-UP).** North is always at the top of the display. Interfacing a gyro or fluxgate compass puts true north at the head of the screen. One of the advantages of this display is that both chart and display correspond and bearings are easily transferred for plotting purposes. Many plotting and navigating errors are made by incorrect transfers of screen information. As the heading changes the heading marker also changes.

(2) **Head Up (H-UP).** The top of the screen is the same as the vessel heading; all bearings are relative. You can select a ship's heading marker on screen, and as the heading changes the display also follows and rotates.

(3) **Course Up (C-UP).** The top of the screen is aligned to the selected course using an interfaced fluxgate compass. As heading changes so does the heading marker.

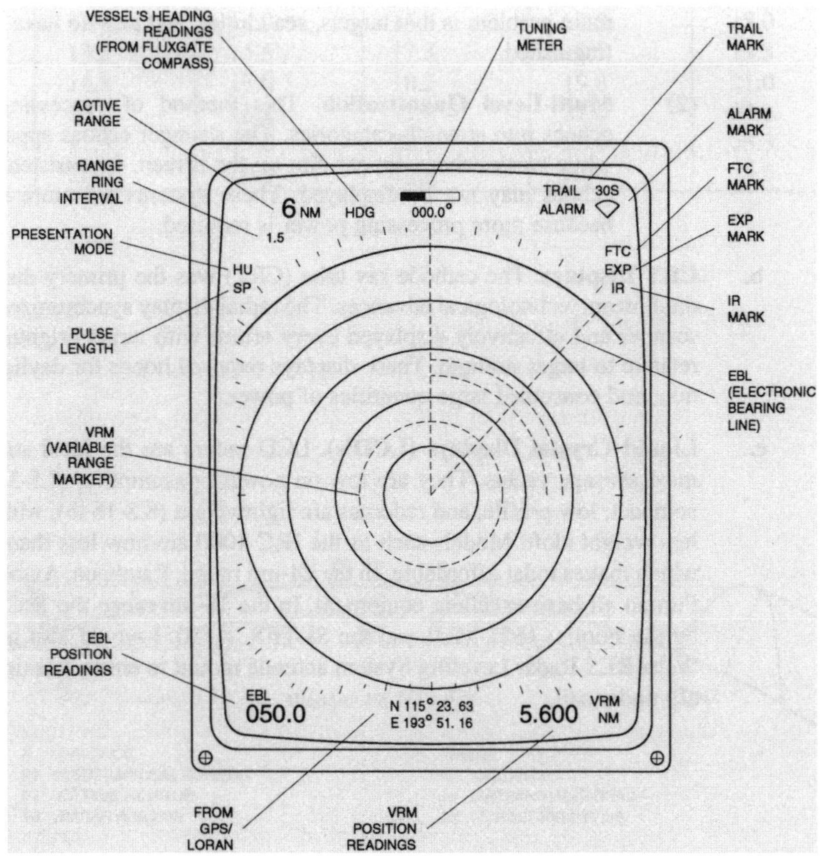

Figure 15-4 Koden 24-nm Radar Display

f. **Motion.** A radar display can be configured for true motion or relative motion. In relative motion the yacht's position remains fixed on the radar screen and all targets are relative to the yacht. In true motion all the targets are in a constant position on the display. The yacht moves across the display on the course heading and the display is similar to a chart. All other boats appear in true relationships to other targets and to displayed land masses.

g. **MARPA (Mini Automatic Radar Plotting).** The display is set up to specify target vectors, safe zones and target history. This is a function for target tracking and risk analysis. MARPA was once a big ship system. It allows up to 10 targets which are automatically tracked, and the target bearing, range, true speed and CPA (Closest Point of Approach) and TCPA (Time to Closest Point of Approach) are calculated. The result is accurate and gives a continuous evaluation of the situation. MARPA's effectiveness may be reduced or disappear when target echoes are weak, targets are close to land masses or to other large targets, the target is engaged in rapid maneuvers, rough or very choppy seastate conditions exist and the target is obscured by sea clutter or large swell conditions, the vessel is unstable in big sea conditions or the heading data is degraded.

h. **Chart Overlay.** As many new radars on multifunction displays are capable of overlaying electronic chart with the radar display, this can be selected and deselected as required. Check your manual for proper use of this function.

15.5 Radar Installation. The two most common scanner mountings are mast mounted, or sternpost mounted. Each has advantages and disadvantages.

a. **Mast Mounted.** There are a number of factors affecting the mounting of the scanner on the mast:

(1) **Radar Range.** Mast mounting increases radar range. This is clearly illustrated in the radar horizon table.

(2) **Weight and Windage.** Contrary to opinion, weight and windage are very low. Six kg for a 16-mile scanner are not really a problem in cruising yachts.

(3) **Blind Sectors.** The position of the scanner is important. Locate the bracket above or below the spreaders to minimize obstruction. There will be a small blind sector astern. Where scanners are mounted on ketch mizzen masts, you have both forward and stern shadows.

b. **Sternpost Mounted.** The sternpost mounting arrangement is more accessible than mast mounting and has become a very popular alternative in recent years. Some sternposts are hinged to allow easy lowering. I can never understand why 24-mile radars or above are used in these installations, as their range is limited by the post's height. It is preferable to use an open array antenna to improve the resolution.

(1) **Radar Range.** The radar range is reduced by a couple of miles depending on the target height. Typical height is around 3 meters, compared to around 6–8 meters for a mast mount.

(2) **Scanner Leveling.** When the boat heels, the scanner also tilts, leading to significant loss of performance and range. There are now some innovative self-leveling mountings available.

(3) **Health Risks.** There are increased health risks with stern mounted units.

c. **Safety.** There are several important safety factors that must be considered when working with radar:

(1) **Eye Damage.** Direct exposure to an operating radar transmission can permanently damage the retina or cause blindness. Safe distances are normally given as around one meter, but recent medical research has recommended an absolute minimum of two meters. In this respect, sternpost radars represent a real health hazard, especially when powerful units are installed.

(2) **Electric Shock.** Scanner units have high voltages inside. Owners should never open an operating scanner and attempt repairs or measurements. Call the authorized service person.

d. **Cables.** Cutting and rejoining of cables should be avoided, but when mast mounted on a sailing boat it is always necessary. A ferrite should be installed on each side of the break and you should consult your supplier for specific details. Wires should be reconnected correctly, and that includes the screen. In general you will need to use a junction box and there are radar specific units available. They typically have up to a 15-way terminal strip rated at 10 amps. Always keep unscreened coaxial cores to less than 30 mm to ensure EMC norm is maintained. Always ensure that radar cables are well protected from chafing where they enter the mast. Ensure that cables are not bent too tight or kinked, and minimum bend radii for small cables are 60 mm (2.5 inch) and for larger cables 82 mm (3.75 inches).

e. **Power Consumption.** Small boat radars generally have power consumptions around 4 amps. With newer technology, consumption figures are getting lower; some are nearly down to 3 amps. Open scanners typically have a power consumption 50% greater than enclosed types because they have a heavier scanner and require a more powerful motor to rotate it.

f. **Economy Mode.** This function has been incorporated into a number of radars and is very useful for power-conscious sailors. The radar can remain operating with guard zones activated and the display off to save power. If any target is detected within the guard zone, the alarm will sound and the display can be called up with one button.

g. **Grounding.** As a transmitter, radar requires proper grounding, usually at the scanner and the rear of the display unit. If a sternpost is used, ensure that a grounding cable is attached from the base to the RF ground.

15.6 Radar Operation. Correct operation of radar is essential if you are to get the maximum benefit from it. It is prudent to attend a shore-based course as well. Don't be one of the all-too-common radar assisted casualties. A radar has a bewildering array of controls, but they all have clearly defined functions which are easily learned.

a. **Power Up.** At power up, all radars have a magnetron warm-up period. When warm-up is complete, the radar always defaults to stand-by status. If operating a new set, allow it warm up for at least 30 minutes before using it or making adjustments.

b. **Range Selection.** Always set the range you wish to work on. Typically, the 12-mile range is ideal for the average yacht given the radar's horizon. On a mast mounted scanner, a greater range will enable you to detect a large vessel on or just over the horizon. Selecting a range automatically sets the appropriate range ring intervals, the pulse length, and the pulse repetition rate.

c. **Adjust Brilliance.** Adjust the brilliance control to suit your requirements. Don't make it too bright at night or so dim that targets are not clearly displayed.

d. **Adjust Gain.** The gain control removes background noise, which consist of large areas of irregular speckles on the display. Adjust the gain control so that screen speckling just starts to appear. Gain controls the signal amplification, so be very careful not to overadjust as smaller echoes can be masked, or if under the required threshold, will not appear at all. The gain is normally set high for long ranges and reduced for low ones. Menu selections are now normal and SEA reduces sea returns, RAIN reduces close rain or snow returns, FTC reduces distant rain or snow returns.

e. **Fast Time Constant (FTC).** This control reduces rain clutter. Rain clutter is proportional to the density of the rain, fog, or snow. Although the control is useful in tracking squalls and rain, caution should be used so that targets are not obscured. Heavy rain may cause total loss of target definition and cannot be adjusted for.

f. **Adjust Anti-Clutter.** This control is often referred to as the Sensitivity Time Constant control. Sea clutter, most apparent at the screen center in the region closest to the vessel, is interference caused by rough seas or wave action where some of the transmitted signal is reflected off wave faces. Most 3-cm radars transmit at a very low signal angle which grazes the water surface. On short ranges, clutter can mask targets, especially weak ones. The effect decreases at long ranges. Sea clutter always appears stronger on the lee side of the vessel because vessel heel in that direction exposes the beam to larger water areas.

g. **Tuning.** The majority of radars are self-tuning, and adjustment will be indicated on a small bar readout on the screen. Most radars can be manually tuned, but this should be done carefully and according to your manual.

h. **Pulse Length Selection.** Pulse length selection is automatic with range changes on modern small boat radars. At short ranges, pulses are at 0.05 microsecond to give better target resolution. At long ranges, they increase to 1.0 microsecond.

i. **Pulse Repetition.** Repetition rates vary across ranges from 200 to 2500 per second. Rates determine the size of the area around the vessel where there is a dead zone. At 0.05 microsecond, this is around 150 meters. At 1.0 microsecond, this reduces to 30 meters.

j. **Interference Rejection (IR).** Interference can come from a number of sources:

(1) **Other Radars.** Other radars operating in the area can cause interference on the display. This is particularly apparent near major shipping routes where powerful commercial vessel radars operate. Use the IR function to remove these unwanted signals.

(2) **Mast Clutter.** When a radar is installed, there will be a blind spot abaft the scanner due to the mast. No targets will be detected in this area at close ranges. On a sternpost or mizzen mast mounted radar, the area in front will be masked for the same reason. Caution must be exercised as this is the normal collision risk sector.

15.7 Radar Plotting. The main purpose of a radar is to detect stationary and fixed targets. A number of basic features facilitate this:

a. **Range Rings.** The concentric range rings automatically change with the selected radar range. User definable range ring settings are possible at setup.

b. **Variable Range Maker (VRM).** This function uses the range rings and the marker. The readout appears on the screen, but as with all navigation exercises, make sure you are measuring the correct target. Many errors are made this way, which is why radar should be used in conjunction with other position keeping systems, principally the charts and eyeball.

c. **Electronic Bearing Line (EBL).** The most commonly used function in conjunction with the VRM enables easy plotting of a target, but be careful, many unfortunate incident occur because a bearing was taken without checking which headup display was being used. Most radars enable selection of either actual (Magnetic/True) or relative (relative to boat heading) bearing readout in the EBL data box.

d. **Target Expansion.** This function on many radars allows short or long range contacts to be expanded. It is useful when closing on low altitude landfalls such as atolls and islands.

e. **Off Centering.** A number of radar sets have an offset function which alters the screen center (the vessel) 50% down the screen. This makes forward long range observations possible in the same radar range.

f. **Safe (Guard) Zones.** Guard zones offer real safety advantages. They can be set for complete circular coverage or for specific sectors. It is a big error to rely on this function when sailing shorthanded; proper observations still should be regularly made. On some radars, the economy mode setting saves power by letting the guard zone and alarm function operate without the screen being on.

g. **Target (Wake) Plotting.** This feature, now part of a number of radar models, allows a trail to be plotted on targets. Target plotting is time related and can be continuous or set at a number of seconds. A clear plot of the target is invaluable for ensuring that collision risks do not arise. What a change from plotting aids and chinagraph pencils!

15.8 Radar Maintenance There is not much maintenance required on a radar unit, but taking the following steps will ensure long term reliability:

a. **Connections.** Once a year, open the scanner and tighten all the terminal screws.

b. **Clean Scanners.** Clean the scanner with warm soapy water to remove salt and dirt. Do not scour or use harsh detergents.

c. **Scanner Bolts.** Check and tighten the scanner holding bolts.

d. **Gaskets.** Check that the scanner's watertight gaskets are in good condition and seal properly.

e. **Scanner Motor Brushes.** Some scanner motors have brushes. Check these every 6 months. Manufacturers sometimes provide a spare set taped to the motor.

f. **Display Unit.** Clean the screen with a clean cloth soaked in an anti-static agent. Do not use a dry cloth as this can cause static charging which attracts and accumulates dust.

15.9 Radar Troubleshooting The following table gives typical faults that can be investigated and rectified before calling a technician.

Table 15-2 Radar Troubleshooting

Symptom	Probable Fault
Scanner stopped	Motor brush stuck (if fitted) Bearing seized Scanner motor failure Scanner motor control failure
No display	Power switched off Brightness turned down Fuse failure Loose power plug Incorrectly tuned
Display on, no targets	Scanner stopped Local scanner switch off Scanner plug not plugged in
Low sensitivity	Ground connection loose Radome salt encrusted Open array salt encrusted

Radar Reflectors

16.0 Radar Reflectors. The subject of radar reflection has sparked continuing controversy over the years. There has also been a constant stream of so-called radar reflective safety devices launched upon unsuspecting yachtsmen. Not to have an effective reflector mounted at all times is, in my judgment, negligent in the extreme.

a. **Merchant Vessel Visibility.** If you have never been on the bridge of a fast merchant vessel steaming up the English Channel at 24 knots, dodging yachts that are invisible to radar, I can assure you that it is not a deck officer's favorite pastime. In deep ocean waters, there is still a requirement to be seen. While the shipping lanes may constitute areas of heavy commercial traffic, commercial vessels ply waters everywhere. The attitude commonly adopted—that no one is keeping a look-out anyway, so why bother—is fatally flawed. Most, if not all, vessels these days have the radar set with collision avoidance tracking and alarm systems, so if the vessel's radar cannot lock onto a good, consistent signal, it cannot identify and track a target. I have sailed under many flags commercially, including the much-maligned flags of convenience such as Liberia and Panama, and the officers were all qualified, contrary to popular opinion. With large and fast vessels, the earlier you are detected on radar and your course and collision risk assessed, the earlier action can be taken to change course and avoid a close quarters situation.

b. **Search and Rescue.** Besides the collision risk problem, the reflector's important role during search and rescue (SAR) operations cannot be overstated. Many SAR operations are called off at night. Much valuable air time and fuel are wasted in aerial search patterns under poor conditions and low cloud cover simply because no effective reflector is hoisted. Reaction times, rescues, and survival prospects even in spite of EPIRBs are decreased in the localization and visual identification phase of the operation.

c. **Mast Weight and Windage.** One of the main reasons given for not having a reflector hoisted is that reflectors are too bulky, cause windage, or are too heavy up the mast; yet the mast will carry the radar scanner and lights.

d. **Mast Shadowing.** Wherever you mount your reflector, there will be some shadowing from the mast. When a reflector such as a Blipper 210-7 is mounted directly in front of the mast, there is typically a 10° blind spot directly aft, the lowest collision risk sector of all. A yacht's track is far from straight, whether under autopilot or hand steering. Typically variation is in the range of 10 to 25°. Even though some reflective surface will be "seen" overhanging the mast, this movement will expose a substantial number of reflective corners, enough to offer a reasonably consistent return at a range of at least 5 miles.

16.1 Reflector Theory. To understand reflectors, a basic understanding of radar signal behavior is required.

a. **Radar Beam Behavior.** When a radar beam reaches a target, in theory it reflects back on a reciprocal course to be processed into a range and bearing for display on the screen. In practice, a beam does not simply bounce back off an object. Some materials are more reflective than others, while others absorb the signal.

b. **Reflective Materials.** The best reflective structures are made of steel and aluminum. Materials such as wood, fiberglass, and sailcloth do not reflect at all. In fact, fiberglass absorbs some 50% of a radar signal. There will almost always be some reflection, but the direction of the reflected beam will be erratic and so minimal that no consistent return can be monitored on the screen.

c. **Reflection Consistency.** Consistency is one of the major requirements of a good reflector. A good reflector consists of a metallic structure, normally aluminum, with surfaces placed at 90° to each other. If a beam is directed to the center of a re-entrant trihedral parallel to the center line, it will reflect on a reciprocal course back to the scanner. A re-entrant trihedral is simply a corner with three sides, such as the corner made up of two walls and a ceiling. The center line of the corner points in a direction approximately 36° to each of the sides making up the trihedral. The more the angle increases away from the center line from a radar beam, the less radar signal returns back. This simple fact forms the basis of radar reflectors.

d. **Radar Reflection Standards.** The basic standards include a number of specifications. Never buy a reflector that does not comply. A peak echoing area of 10 m^2 is defined as the equivalent of a metal sphere of approximately 12-feet diameter. International requirements and standards are as follows:

(1) **ISO (8729).** This is an IMO sponsored standard. It specifies an RCS of 2.5 m^2 as the minimum threshold of radar visibility.

(2) **USCG.** A standard is set for survival craft reflectors. Manufacturers are required to demonstrate a range of 4 nautical miles in a calm sea.

(3) **DOT (UK).** Set down in the Marine Radar Performance Specification, 1977, it requires that reflectors have an equivalent echoing area of at least 10 m^2.

(4) **RORC (UK).** A documented equivalent echoing area of not less than 10 m^2 is required.

(5) **AYF (Australia).** A minimum equivalent echoing area of at least 10 m^2 is required.

e. **Reflector Types.** There are a variety of reflectors on the market. See the illustrations below. The illustrations are to scale and show the various sizes of the devices, indicating relative effectiveness.

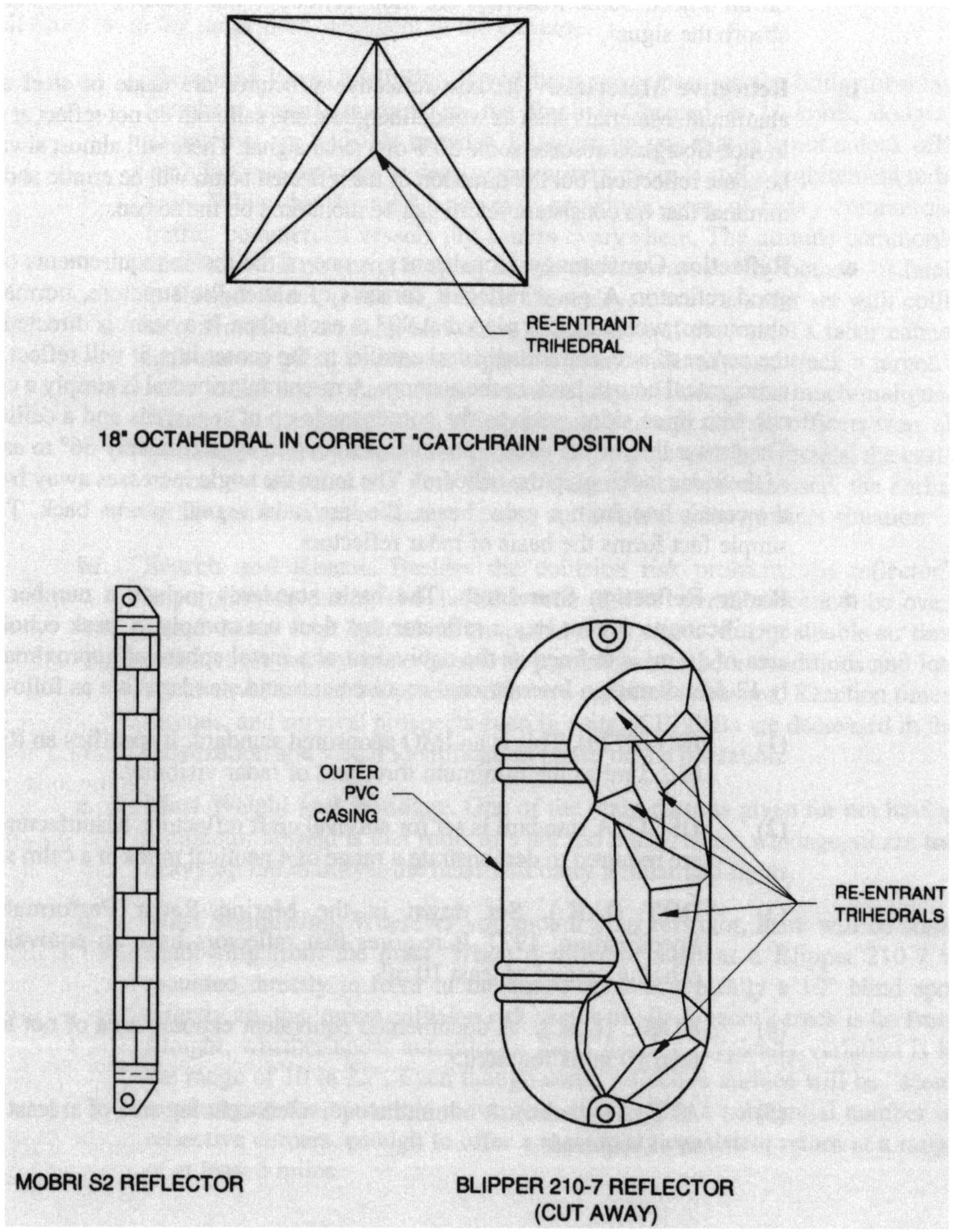

Figure 16-1 Radar Reflector Types

f. **Octahedrals.** The standard octahedral is a structure consisting of eight re-entrant trihedrals. It was developed in the early 1940s, when radar was under development. For maximum effect, the octahedrals must be mounted in the proper orientation, which is called the "catchrain" position. It is amazing how many are hoisted up by a corner; one magazine survey had a figure approaching 70%, and my survey was closer to 80%. The structure, in fact, has only 6 effective corners, pointing alternately up and down, the remaining corners being of little use. The effectiveness of the radar reflector is shown in the polar diagram. On the typical 18-inch octahedral polar diagram, the lobes where peak reflection occurs are clearly visible. The peaks clearly exceed the peak echoing area of 10 m^2. A big problem, however, is created by the large areas between the lobes, where no reflection occurs, or it is so minimal that they are under the minimum standards set down by IMO of 2.5 m2. The total number of blind spots on a correctly hoisted octahedral total nearly 120°, which is not ideal. The small peaks do not affect the result much. The bad news is that when heeled to 15°, the blind spots increase to nearly 180°. So, under sail, you have a 50-50 chance of being seen on radar, in most cases intermittently, so the radar will reject the inconsistent signal. This can be further reduced when part of the signal, after reflecting of the sea surface, cancels out another beam traveling directly to the reflector. If you are using an octahedral, anything less than 18 inches is a waste of effort.

g. **Optimized Arrays.** The Marconi-Firdell Blipper 210-7 is representative of these reflectors. The Blipper consists of an array of precisely positioned re-entrant trihedrals designed to give a consistent 360° coverage, and through heel it angles up to 30°. As a vessel moves around in a typical three-dimensional motion, each of the corners moves in and out of 'phase' to the radar signal, with one corner sending back signal directly, and others giving partial returns, resulting in a consistent return at all times. The units are rotationally molded inside a radar invisible plastic case, and the windage is only 15% of an 18-inch octahedral, and the unit weighs less than 2 kilograms. These reflectors have a reputation of meeting and exceeding all published standards. This can be seen by the numbers mounted on masts; my own survey at a major British marina was marginally over 50%. The Blipper 210-7 has been awarded a NATO stock number. The Echomax is a similar device that claims that it meets RORC and ORC as well as ISO8729 requirements.

h. **Stacked Arrays.** These are typified by tubular reflectors that resemble a fluorescent tube or rolling pin such as the Mobri and Slim Jim units. I have seen many of these taped to a backstay or stay, sometimes three or four on a yacht. They consist of an array of tiny reflectors housed in a see-through plastic case. These reflectors are purchased because they are cheap and small, not for the radar visibility factor which is the primary safety requirement. The analysis that I have looked at shows that it can only effectively return the amount of signal required in a perfectly vertical position. At any angle of heel, at 1° or more the unit return falls away to virtually nil. At best tabulated positions, at 0° azimuth, the RCS is 6.05, heeled to 1° it falls to 1.46, and at 2° to 0.18. So you can draw your own conclusions; if you have one taped to a backstay, it probably doesn't do anything.

i. **Luneberg Devices.** These devices resemble two half spheres mounted back to back; they are typified by the Visiball. They are normally fitted to the masthead in a fore-and-aft configuration. They are very heavy. The main criticism of the reflector is that the returned echo is only fore-and-aft and not athwartships, permitting a large and dangerous blind sector. More importantly, the return does not meet the minimum standards of the IMO or RORC, having an RCS of only about 0.8 m^2.

j. **Foil Devices.** I have read several articles and I have also heard many people advocating filling the mast with foil, or simply hanging a pair of stockings full of foil in the rigging. A case was heard in the UK courts regarding the loss of a catamaran in a collision with a coastal vessel during a yacht race. The skipper did not hoist a reflector because he feared windage would reduce sailing performance; he inserted instead a foil filled stocking into the mast. The Admiralty judge included the following in his decision against the catamaran skipper: "To leave an anchorage and proceed without radar into a shipping lane when the visibility is less than 75 yards, so that the navigator is blind, and without a radar reflector so that the yacht is invisible, is in my judgement seriously negligent navigation." That statement sums up the issue of radar reflectors and the necessity for having them.

k. **Cyclops.** This reflector operates on the Luneberg Lens principle. It uses concentric shells of material to reflect and refract radar signal. Two lens assemblies plus an additional two trihedrals are used to give full coverage around the azimuth and at heel. The polar diagram claims RCS of 10.5 m^2 and average of 4 m2 all round. The problem is that it is large for intended masthead mounting, and weighs some 4 kilograms. It is also expensive and has been the subject of some criticism.

l. **Radar Detectors and Transponders.** These are now becoming more affordable and there are several products on the market.

(1) **Radar Detectors.** These devices are omnidirectional units that activate an alarm when a radar signal is detected in the vicinity. Typical range is approximately 5 nm. When an alarm is activated, the units can be used as a radar direction finder and a plot of the track of a vessel can be made. The disadvantage is that with more than one fast oncoming vessel, it is difficult to plot all targets and make judgements based on the plot. Vessels may have already made collision avoidance alterations, and this only in the case that a good radar reflector is fitted so that you are radar visible.

(2) **Radar Target Enhancers (RTEs).** These are not GMDSS equipment. The reception of an incoming radar signal, the amplification of that pulse, and the retransmission of the pulse back to the radar signal source is how these devices work. This has to occur simultaneously and at the same frequency. The returned signal is displayed in enhanced form, with the relatively small return of the boat appearing significantly larger than it actually is. The McMurdo Ocean Sentry RTE claims a target enhancement factor of eight times greater than actual reflected image. This obviously has the advantage of displaying strong and consistent echoes on radar screens. Effectiveness depends on the incoming radar signal strength, height at which the RTE is installed, and the height of the other vessel's radar above sea level. The Ocean Sentry unit operates either in standby or transponder modes. In standby mode, the unit is activated only when a radar signal is present. These units operate in response to 3-cm X-band radars only, not S-band. The effective range is typically around 12 nm, but not less than around 3 nm.

16.2 Radar Reflection Polar Diagrams. Polar diagrams are the usual way manufacturers represent the performance of radar reflectors. There are two types of polar diagrams:

a. **Horizontal Polar Diagrams.** Polar diagrams are essentially signal returns plotted for all points around the azimuth for a reflector in the vertical position. This is crucial to the understanding of test claims and actual onboard performance of the reflector. The various polar diagrams for the 18-inch octahedral reflector and the Mobri tubular reflector are illustrated below. The Firdell Blipper 210-7 reflector now only has three-dimensional polar diagrams and is therefore not included.

b. **3D Polar Diagrams.** The more accurate test of a radar reflector is a three-dimensional polar diagram, which indicates performance under actual heel conditions. These are illustrated below: they are derived from computer generated results and give a close image of actual performance. The white space is the area of no radar visibility. Given that performance under heel is the critical requirement, I would caution purchasers against buying a product that cannot produce such data, or verifiable proof that it works under the normal heeled sailing conditions of a yacht.

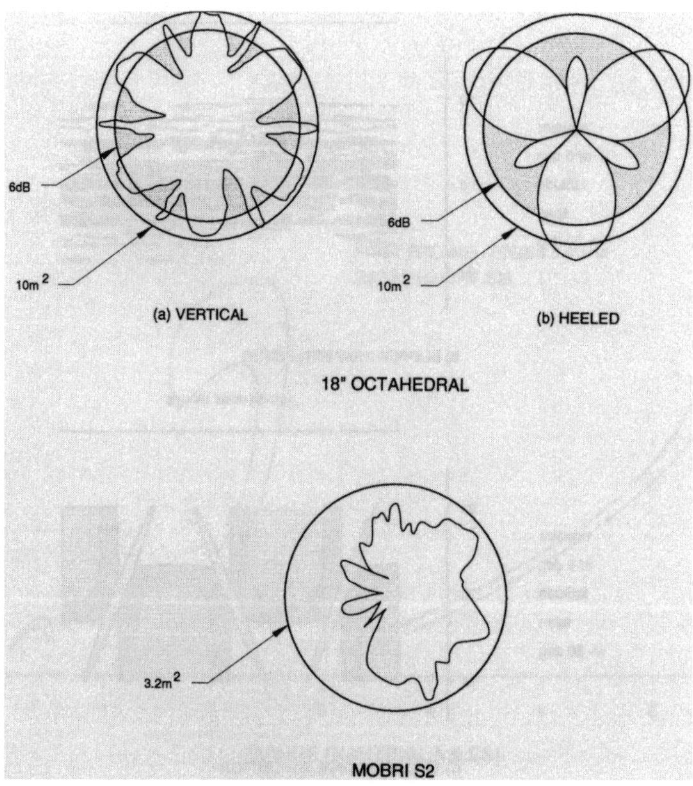

Figure 16-2 Horizontal Polar Diagrams

16.3 Radar Fresnel Zones. In some cases radar signals self-cancel, either in the transmission or return path. This problem is related to a variety of factors including radar height, target height, sea and earth surface conditions, and radar range. The regions where cancellation occurs are called fresnel or extinction zones; they can be up to a mile-wide. In such conditions the radar signal reaching the radar reflector may be relatively weak, with a weak return. The result is no return to the radar, or a return so weak that it is not processed.

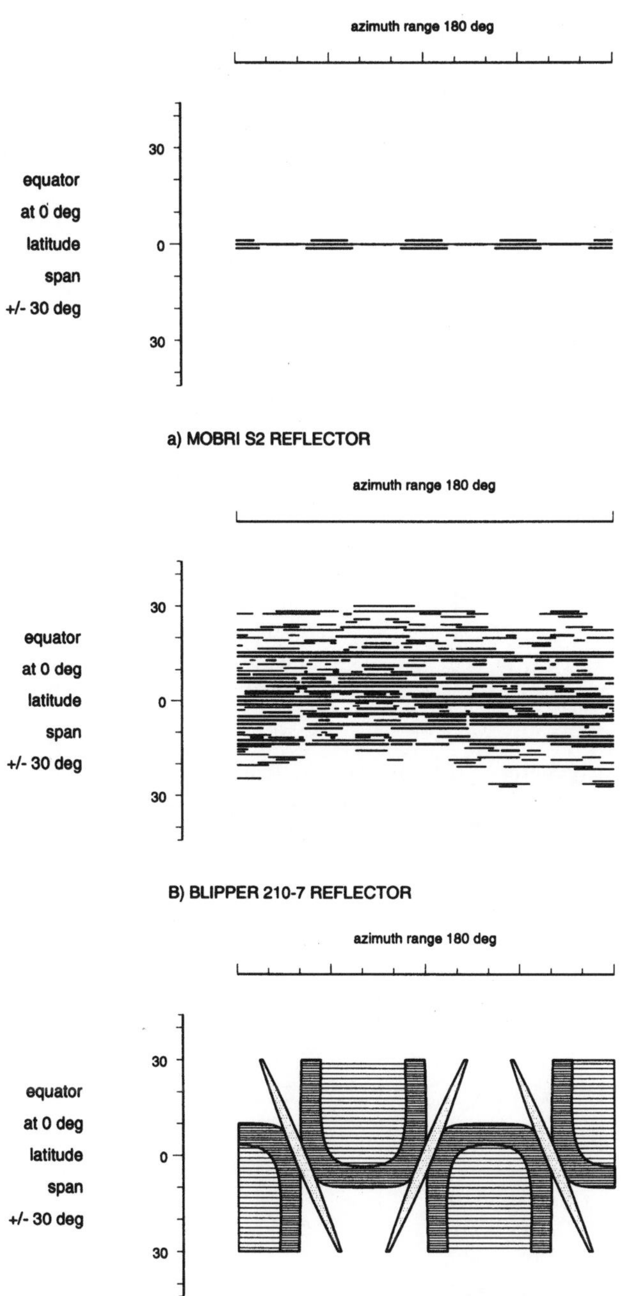

Figure 16-3 Radar Fresnel Zones

a. **Reflector Mounting.** It is apparent from the fresnel tables that the masthead is not the ideal place to put your reflector, as a relatively large cancellation zone exists. Reflectors are best mounted around the spreaders, or about 4 to 5 meters high.

b. **Fresnel Tables.** The following fresnel tables are published courtesy of Marconi-Firdell, and cover the first Fresnel Zone for 12- and 16-feet radar heights, and 4- to 22-feet target heights. The tables are based on a radar frequency of 9.4 GHz and for the range of 0.1 to 10 nm which is typical for cruising yachts.

Table 16-1 First Fresnel Zone Tables Radar Height 12 Feet

Target Height	Zone (nm)	Zone (nm)	Zone (nm)
4 m	0.140–0.457	1.536–1.578	3.626–3.692
6 m	0.212–0.481	1.692–1.741	4.086–4.166
8 m	0.288–0.487	1.814–1.868	4.440–4.529
10 m	0.393–0.453	1.906–1.965	4.708–4.805
12 m	1.978–2.041	4.915–5.016	
14 m	2.033–2.102	5.078–5.182	
16 m	2.077–2.150	5.209–5.316	
18 m	2.122–2.190	5.318–5.427	
20 m	2.141–2.224	5.410–5.521	
22 m	2.164–2.253	5.492–5.604	
24 m	2.183–2.279	5.565–5.678	
26 m	2.199–2.301	5.632–5.747	
28 m	2.211–2.321	5.695–5.811	
30 m	2.221–2.339	5.755–5.872	

Table 16-2 First Fresnel Zone Tables Radar Height 16 Feet

Target Height	Zone (nm)	Zone (nm)	Zone (nm)
4 m	0.188–0.469	1.626–1.673	3.824–3.891
6 m	0.290–0.473	1.775–1.828	4.271–4.351
8 m	1.898–1.958	4.648–4.740	
10 m	1.997–1.064	4.957–4.058	
12 m	2.077–2.150	5.209–5.316	
14 m	2.140–2.220	5.414–5.526	
16 m	2.189–2.276	5.583–5.700	
18 m	2.227–2.322	5.723–5.844	
20 m	2.256–2.359	5.842–5.966	
22 m	2.276–2.389	5.944–6.071	
24 m	2.288–2.413	6.033–6.162	
26 m	2.292–2.532	6.112–6.243	
28 m	2.288–2.447	6.183–6.316	
30 m	1.574–2.457	6.248–6.382	

Autopilots

17.0 Autopilots. The autopilot is one of the few indispensable electronic items. It is often referred to as the non-complaining, non-eating extra crewmember. The advances in autopilot technology are powerful microprocessors and equally complex software algorithms that give "intelligent" control. Raymarine call their software AST (Advanced Steering Technology) which in conjunction with a rate gyro monitors vessel pitch and roll and adapts to the boat's characteristics. Most autopilot problems occur because of incorrect installation, improper matching to the vessel, or improper operation, rather than personality conflicts. The basic function of an autopilot is to steer the vessel on a predetermined and set course, to a position or waypoint, or to wind angle. The pilot makes course corrections at an amount corresponding to the course error, usually correcting to eliminate any overshoot as the course is met. Virtually all autopilots are microprocessor based, and use the proportional rate system of operation. Correction is based on the amount of course deviation and the rate of change. Autopilots vary depending on the type of steering system used. The factors affecting autopilot selection are as follows:

a. Autopilot Selection. An autopilot is selected on the basis of a number of important criteria:

(1) The installed steering system, either tiller or wheel hydraulic, wire or direct drive.

(2) The loaded vessel displacement, as it is more valid than length, which has wide variations, and beam, draft and displacement. A common recommendation is that 20% be added to design displacement to get realistic cruising displacement.

(3) Type of sailing is also important. For cruising you must base all factors on worst weather possible, which means power ratings must be capable of coping with prevailing conditions.

b. Power Consumption. Always compare the current consumption at full-rated load, not average consumption. Many find that the pilot uses far more power than expected, although much of the heavy consumption relates to excess weather helm activity and overworking of the pilot. There is no significant difference between the average consumption of the various drive types for a specific vessel size.

c. Factors Effecting Performance. The following factors must be considered when selecting an autopilot:

- The speed of rudder travel
- The rudder size
- The required number of turns lock-to-lock
- The expected wind and sea conditions

d. **Autopilot Torque.** Torque is the force required to hold the rudder in position due to the pressure of water on the rudder, and to overcome the steering gear resistance of bearings and steering system drives. The vast majority of people underestimate this, and while the pilot works well in average conditions, it fails to keep course in bad weather.

e. **Sail Trim.** Overloading and burn-outs are almost always due to excessive weather helm. If the helm is constantly held over by the pilot, trim the sails. The following can be done to improve performance, and reduce electrical and mechanical loads:

(1) **Reduce Vessel Heel.** Minimize vessel heeling, ease the mainsheet, or traveler to leeward.

(2) **Reef Early.** It is a good idea to reef the first time you think about it. The power saving is worth the effort.

f. **Trim and Load Monitoring.** One more innovative suggestion is to insert an ammeter in the autopilot supply scaled to the full load rating. This enables the current consumption to be used as a trimming guide.

17.1 Autopilot Drive Systems. The choice of pilot is based on the steering system in use.

a. **Hydraulic.** A reversible pump must be installed in the system, and controlled by the autopilot. Hydraulic steering systems consist of a steering wheel pump and steering cylinder. The wheel pump forces oil into the cylinder from either end, depending on the direction required. The system should have a lock valve to prevent the rudder driving the wheel pump. Hydraulic systems are inherently more reliable than mechanical drives. Pump types are as follows:

(1) **Constant Running.** Pumps are usually dual speed to save power in lighter conditions. Solenoid valves control oil pressure for directional activation of the hydraulic ram.

(2) **Reversible Motor.** These pumps are the most common. They have a low overall power consumption with the pump operating under autopilot command. Typical power consumption is in the range of 2–4 amps on units for vessels up to around 45 feet but will have a maximum of approximately 20 amps. On larger vessels, this moves up to 4–8 amps. The pump unit consists of an electric permanent magnet motor, valve block, reversible gear pump, and non-return valves on the directional outlets.

b. **Hydraulic Steering Types.** There are three basic types of hydraulic steering systems.

(1) **Two Line System.** Pressurized fluid is pumped into the ram from either end, depending on the direction required.

(2) **Two Line Pressurized System.** This system has an external pressurized reservoir.

(3) **Three Line System.** Pressurized fluid flows in one direction only. A uniflow valve is installed within the system to direct all fluid back to the reservoir.

c. **Installation.** There are a number of important considerations when installing pumps:

(1) The pump must be mounted in a horizontal position. I have seen units mounted in the vertical as it was more convenient.

(2) The pump should be mounted adjacent to the steering cylinder.

(3) The pump must be securely mounted to prevent vibration.

(4) Non return valves must be fitted to the helm pump to prevent the autopilot pump from driving it instead of the ram.

d. **Pump Maintenance and Testing.** Perform maintenance and testing as follows:

(1) **Test Rudder Operation.** Drive the rudder lock-to-lock, using the pilot control unit. Ensure that the rudder moves to the same side as the required command signal. If reversed, the motor terminal connections require reversal at the autopilot control box. The oil expansion reservoir, if fitted, may require topping up. Make sure the rudder stops before reaching the mechanical stops.

(2) **Maintenance.** Dismantle the pump after 1000 hours operation. Examine oil seals and replace them (I always do this regardless of the condition). Check the motor brushes, and replace if excessively worn. Clean the brushgear with an electrical cleaner, and make sure that they move freely in the brush-holders.

e. **Hydraulic System Troubleshooting.** The following faults and symptoms are applicable to most pump systems:

(1) **Spongy Steering.** This is the most common; it is caused by air trapped in the hydraulic system. The system must be bled according to the manufacturers' instructions. When bleeding, ensure that the steering is operated stop-to-stop to expel air in the pump and pipework. This problem will greatly affect the performance of the autopilot.

(2) **System Cleanliness.** The system must be absolutely clean, and no particles of dirt should be introduced into it. This means clean hands, clean tools, and clean oil. Particles commonly lodge in check valves, causing loss of pressure and back driving of steering wheel.

Table 17-1 Autopilot Hydraulic System Troubleshooting

Symptom	Probable Fault
Excessive pump noise	Air in hydraulic system
No piston movement on command	Oil valve closed
	Pump sucking in air
	Non return valve leaking
Rudder moves back to amidships	Non return valve leaking
Wheel moves with pump operation	Lock valve leaking
Piston moves erratically	Air in hydraulic system
Rudder movement stops with increase	Pump underrated
in rudder load	Low pump motor voltage

f. **Wheel Drive.** Wheel pilots are usually located on the steering pedestal. The drive unit is mounted in line on a cockpit side; it consists of an integrated gearbox and motor, rotating the wheel via a belt. Vessel steering characteristics can be programmed into the control system, and a simple clutch lever enables instant changeover to manual steering. Belts must be correctly tensioned to avoid premature breakage or wear.

g. **Mechanical Linear Drive.** The linear drive unit is either an integrated hydraulic ram and pump system, or a motor and gearbox drive directly connected to the rudder quadrant.

(1) **Advantages.** The linear drive has a minimal effect on helm feel. It is relatively low cost, and the hydraulic units are very reliable. There is also the advantage of a backup steering if some part of the steering drive or pedestal fails.

(2) **Power Consumption.** Typical power consumption is relatively low, in the range 1.5–3 amps and 2.75–6 amps for larger vessels.

h. **Rotary Drives.** These drives are usually fitted on vessels where linear drives cannot be installed, where there are space restrictions, or an inaccessible or small quadrant cannot accommodate any other drive. The motors on these systems consist of an electric motor coupled to a precision manufactured epicyclic gearbox. Power consumption is typically in the range 2–4 amps, and 3–8 amps for larger vessels.

17.2 Autopilot Installation. There are a few fundamental points to observe when installing autopilots. Appraisals and post-mortems of many offshore races revealed that many problems were directly attributable to improperly installed autopilots. The following factors should be considered, as they are the major causes of problems:

a. **Anchoring.** Always ensure that the drive units are mounted and anchored securely. It is sensible to mount a strong pad at anchoring points, as it is quite common on fiberglass vessels to see the hull flexing because the inadequate mounting points are unable to take the applied loads.

b. **Wiring.** There are a number of important points to consider:

(1) **Power Cables.** Make sure that power cables to drive units are rated for maximum current demand and voltage drops, as cable runs are normally long. As standard, I install a minimum 6 mm^2 twin tinned copper cable to the motor and computer unit.

(2) **Radio Cables.** Make sure that all wiring is routed well away from radio aerial cables since interference is a major cause of problems during radio transmission. Ensure that a ground cable is run from the computer unit to your RF ground. In rare cases you may have to put on a foil shield to SSB tuner unit interconnecting cables as well.

c. **Fluxgate Installation.** There are a number of important points to remember:

(1) **Location in Fiberglass and Wooden Boats.** The fluxgate compass should be installed in an area of least magnetic influence, and close to the center of the boat's roll to minimize heeling error. Turning errors can arise if the compass is not properly compensated. The southerly and northerly turning errors increase as distance from the Equator increases. This causes slow wandering and slow correction; normally compensation reduces this problem.

(2) **Location in Steel Vessels.** Steel vessels pose problems due to the inherent magnetic field in the hull. Mount the fluxgate sensor at a minimum of 5 feet above the deck. Note that, as this is often on the mast, it may become disturbed when radar or radio cables passing through the mast are carrying current or signal.

(3) **Cables.** Ensure that the compass is mounted clear of any cable looms or any other metallic equipment. As fluxgates are invariably installed under saloon bunks, do not store any metallic items such as tool boxes or spares there, as often happens.

d. **Course Computer Location.** This should be located clear of magnetic influences and away from radio aerial cables. While older units were prone to induced interference, newer units are generally made to strict international noise emission standards.

17.3 Autopilot Controls. Many adjustments can be made to achieve optimum autopilot operation. As a note of caution, do not use in any channels, confined areas or heavy traffic zones, as VHF and SSB operation can cause sudden course changes. Wireless remote controls are also now available for autopilot control, and this makes quick course changes easy from anywhere on deck. The various controls are as follows:

a. **Deadband.** This is the area in which the heading may deviate before the pilot initiates a correction.

b. **Rudder Gain.** This relates to the amount of rudder to be applied for the detected heading error, and must be calibrated under sail. It is inextricably linked to proper compass set-up and damping.

c. **Rudder Feedback.** Rudder feedback provides instantaneously the precise rudder position information to the pilot. It is essential that the feedback potentiometer is properly aligned. Most new pilots have a high resolution potentiometer that offers more precise feedback than earlier and coarser units.

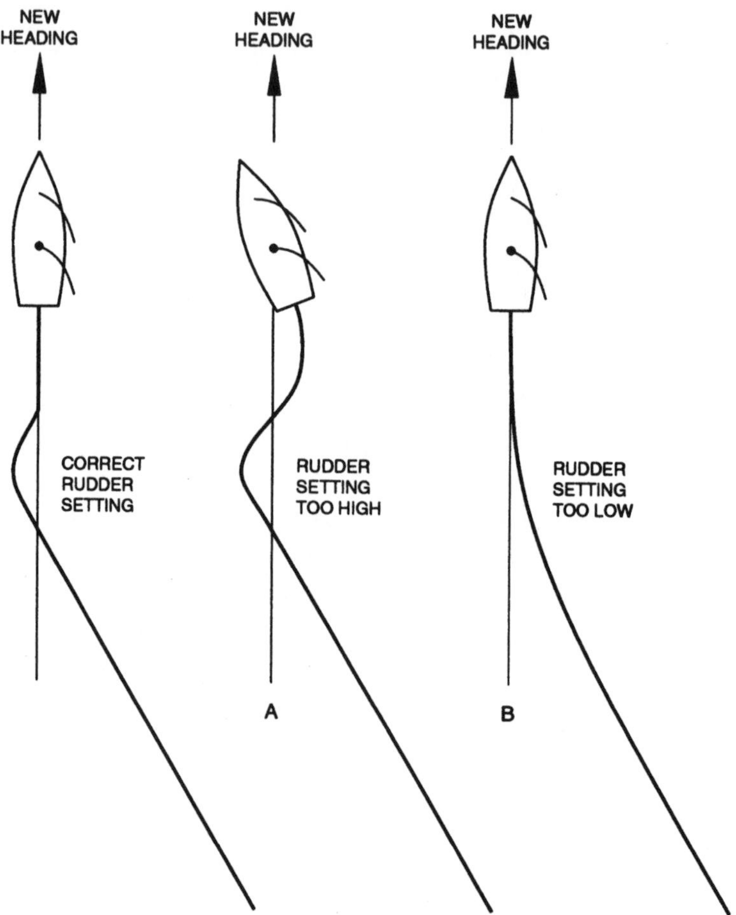

Figure 17-1 Rudder Gain Effect

d. **Rudder Limits.** This controls the limit of rudder travel. The autopilot must stop before reaching the mechanical stops or serious damage may result.

e. **Rudder Damping.** This calibration is used where a feedback transducer is installed and minimizes hunting when the pilot is trying to position the rudder.

f. **Rate of Turn.** The rate of turn limitation is typically 2° per second.

g. **Auto Tack and Gybe Function.** Automatic tack and gybe functions are ideal for shorthanded sailing. This is a user programmable turn through to the same apparent wind angle on the opposite tack.

h. **Dodge Function.** This function allows a manual course change to avoid debris, etc. and then automatic return to original course.

i. **Off-Course Alarms.** The off-course alarm activates when the course error exceeds typically 15° for 20 seconds. Alarm angles can be programmed in.

j. **Auto Speed Gain.** This allows adjustment of the amount of helm applied at varying boats speeds.

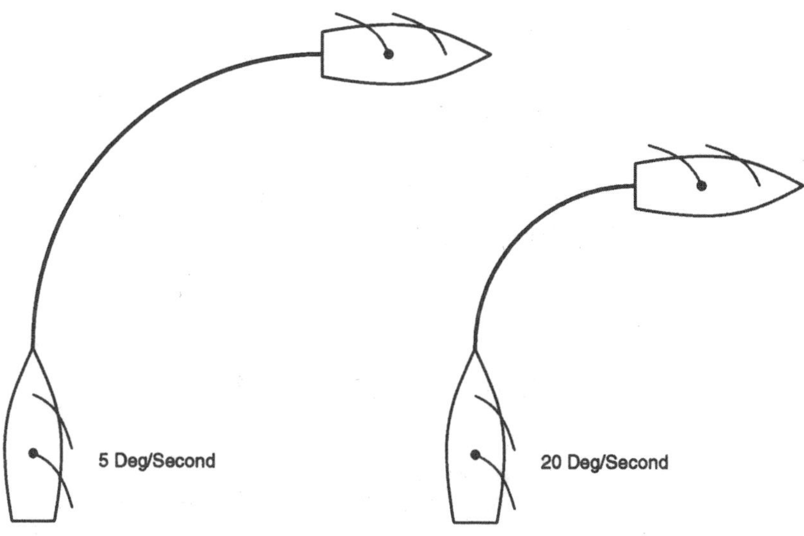

Figure 17-2 Rate of Turn

k. **AutoTrim and AutoSeastate.** These functions were pioneered by Raymarine (Autohelm) and are standard on most autopilots makes:

(1) **AutoTrim.** This function automatically compensates for alterations in weather helm, and applies the correct level of standing helm.

(2) **AutoSeastate.** This function enables the pilot to automatically adapt to changing seastate conditions and vessel responses. It alters automatically the deadband settings. It is controlled by the pilot software. The pilot does not respond to repetitive vessel movements, but only to true course variations.

l. **Magnetic Variation.** The variation must be entered into the autopilot. Newer units have automatic compass linearization to correct for compass deviation errors.

m. **Compass Damping.** The basis for good autopilot performance is the proper setting of compass damping. You should start with minimum damping and increase according to conditions. Failure to get this right will cause either lagging or overshooting as rudder is applied to maintain course. This of course has detrimental effects on power consumption rates, as well as making you sail a lot farther than you have to.

n. **Heading Error Correction.** This correction compensates for northerly and southerly heading errors. Failure to do this will cause amplification of rudder responses on northerly and southerly headings.

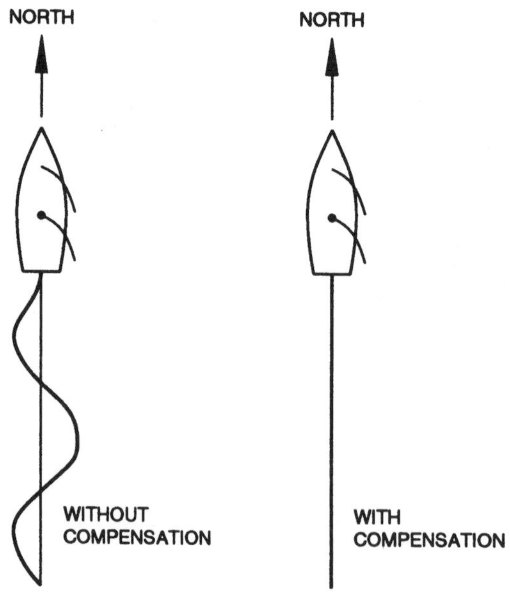

Figure 17-3 Heading Error Correction

o. **Autopilot Interfacing.** Interfacing of compasses and GPS receivers is now standard. Autopilots are now networked into the main instrument system, and take information for control as required. Raymarine uses the SeaTalk system.

(1) **GPS Receivers.** Input from the GPS allows route and waypoint navigation. Most manufacturers list the NMEA 0183 recognized sentence headers. It is important to remember that position fixing systems are subject to errors, sometimes extremely large. This will have obvious effects on the steering, so it is important to keep a regular plot as the autopilot will not be able to recognize the errors.

(2) **Fluxgate Compass.** Input from the fluxgate compass gives accurate heading data to course computer.

(3) **Rate Gyro (Raymarine).** This allows rapid real time sensing of vessel yawing prevalent in light weight vessels and multihulls in following and quartering seas. The data input supplements the fluxgate signal and allows fast correction to counter the rapid heading changes which the fluxgate cannot compensate for.

(4) **Wind Input.** This input from wind instrumentation allows steering to wind angle, either true or apparent. In vane mode Raymarine pilots take wind information and steer to wind angle and automatically steer at the optimum balanced angle for sail settings. In wind trim mode up to 9 settings allow selection of true or apparent wind to steer to.

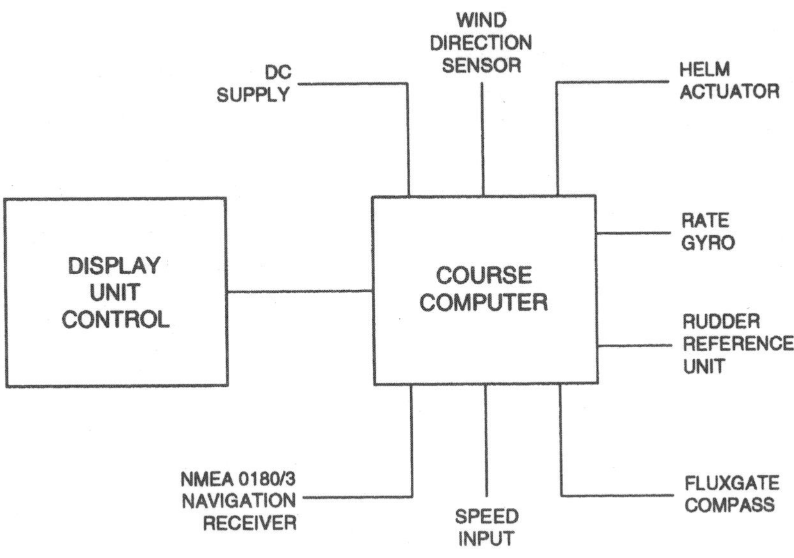

Figure 17-4 Autopilot Control System

p. Track Control. Track control enables a pilot to steer from waypoint to waypoint in conjunction with a navigation receiver. The autopilot effectively adjusts to take account of tide and leeway. To do so it takes cross track error (XTE) data and uses it to compute and initiate course changes to maintain the required track.

(1) **Limitations.** Most pilots will keep within 300 feet of desired track. Track control is less effective at lower speeds, as tidal stream effect has a greater impact. Differences are noticeable where flow speed exceeds 35% of vessel speed, and careful plotting is essential.

(2) **Waypoint Advances.** Many pilots will advance to next waypoints at a single command. This depends on reception of valid NMEA headers that are the waypoint numbers and bearing to waypoint.

(3) **Cautions.** You must be aware that, if a navigation receiver passes incorrect or corrupt position data, the pilot may alter course and put the vessel in danger. Never use unsupervised autopilot steering to position or waypoint close to the coast or in enclosed waterways. It can happen that a large error occurs on a GPS, and by the time you realize it, you are aground.

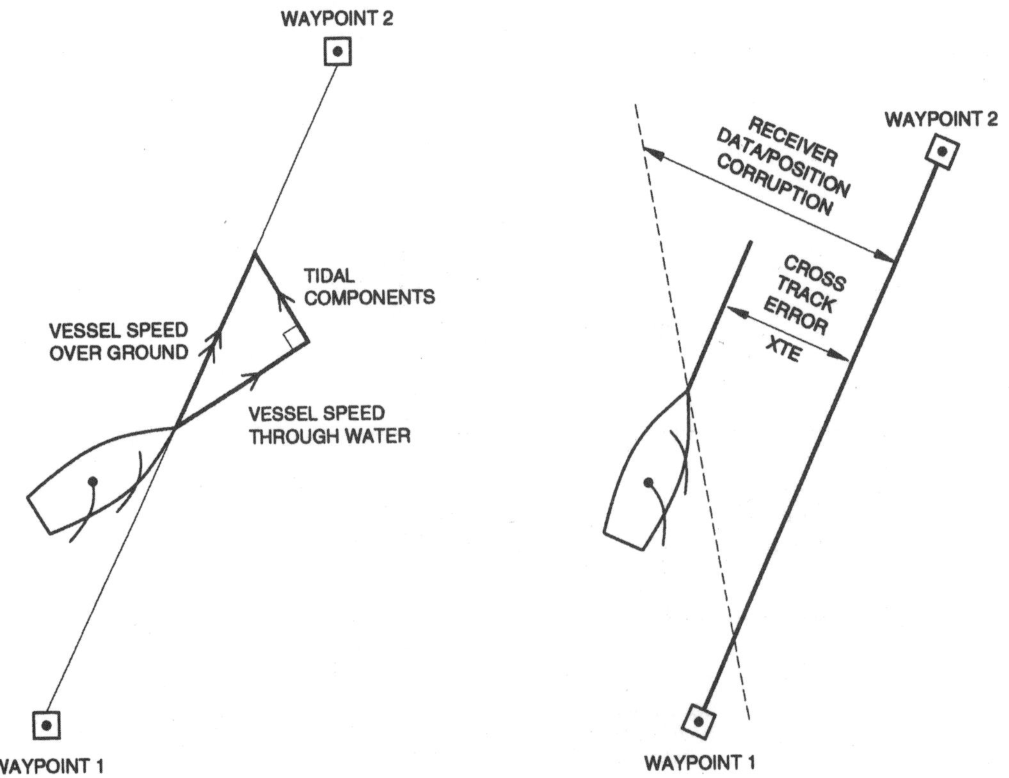

Figure 17-5 Autopilot Track Control

17.4 Autopilot Maintenance. Use the following to increase reliability.

a. **Electronics Temperature.** This applies to tiller units. Install the pilot out of direct sunshine if possible, and keep cool. Units are made of black plastic to facilitate heat transfer from internal components. While this aids heat dissipation the casing will absorb heat. To reduce this cover the unit with a light weight white sail cloth cover in hot sunny weather; use Velcro for easy removal.

b. **Corrosion Control.** Ensure that systems are not exposed to excessive salt water and that seals are intact. Exposed units will be protected by the additional cover.

c. **Plugs and Sockets.** Regularly check plugs and sockets for water and moisture. Make sure they seal properly.

d. **Cleaning.** Clean using a damp cloth. Do not use any solvents or abrasive materials. Do not use a high pressure hose.

Table 17-2 Autopilot Troubleshooting

Symptom	Probable Fault
Display is blank	Loss of power (fuse)
Display compass bearing does match magnetic	Compass not corrected for deviation
No rudder response	Loss of power
	Autopilot fuse failure
	Rudder jammed
	Plug/connection fault
	Control unit fault
Rudder drives hard over	Radio interference
	Loss of feedback signal
	Rudder limit failure
	Fluxgate compass failure
	Radio nav data corruption
	Control unit failure
	Wind data corruption
Wandering course	Calibration settings incorrect
	Overdamped compass
	Rudder gain setting incorrect
	Feedback transducer linkage loose
	Control unit fault
	Drive unit fault
Vessel slow to turn or overshoots	Gain setting incorrecty set
North/South headings unstable	Incorrect setup
Rudder angle display incorrect	Incorrect rudder offset setting
Condensation in display	Turn on illumination to dry
No position information from GPS	Check GPS output or wiring
No auto advance to next waypoint	Check GPS output or wiring

Position Fixing Systems

18.0 Position Fixing Systems. The advances in electronic position fixing systems have been nothing short of spectacular. Equally the fall in GPS prices has ensured that nearly everyone has one on board.

a. **Repeatable Accuracy.** This is defined simply as the ability to sail back to a position or waypoint previously fixed by the receiver, and is vitally important with Man Overboard (MOB) functions. If any system is placed in a static situation, and the positions plotted at intervals, there will be a wandering of position. It is important to remember that all displayed positions must be used with the understanding that errors exist. Hitting the rocks and claiming the position fixing system was at fault is not a valid defense.

b. **Predictable Accuracy.** This is less concise than repeatable accuracy. Essentially this is the difference between the position on your position fixing system and that indicated on your chart (where you are plotting your positions at regular intervals!). These errors are often induced by the vagaries of electronic fixing systems such as signal propagation problems. These errors can be attributed to datum variations, inaccuracies in the electronically derived position, etc.

c. **Chart Datum Variations.** Plotting a position on a chart has inherent errors. These errors can be caused by the GPS fix error or the transformation between GPS datum and chart datum. There may be a discrepancy that requires correction, and many charts carry appropriate notes. A wide variety of datums are used around the globe, and new charts are generally being compiled on WGS84 datum, the same datum used by GPS. Of the 3337 current British Admiralty charts, 65 datums are used, and a typical error is a 140-meter offset in Dover Strait. An official warning was issued not to rely on any position within 3 nm of land in the Caribbean. Note that Datum NAS83 on US charts is same as WGS84 (GPS) datum on UK charts.

18.1 Redundant Navigation Systems. Many boaters are curious about the status of the various systems, given the rapid integration of GPS-based systems. Satnav or the Transit Satellite Navigation System was switched off and its receivers cannot be converted for use with GPS. Decca has been shut down and its receivers cannot be converted for use with GPS. Radio Direction Finding (RDF) is still in limited use, and there has been considerable reorganization and reduction in stations and beacon frequencies, with few stations and frequencies now available. RDF is not part of GMDSS. Loran is still operational although there has been pressure to shut down the Loran system (and some chains have been shut down), but some system expansion has also occurred.

18.2 Global Positioning System (GPS). The US Dept of Defense (DOD) operates the NAVSTAR system. The system consists of 24 satellites in six polar orbits so that at least four will always be visible above the horizon at any time. Twenty-one will be in operation with three used as spares. GPS position fixing involves trilateration (often called triangulation) of position from a number of satellites, satellite ranging to measure the distance from the satellites, accurate time measurement, the location of all satellites, and correction factors for ionosphere conditions. Operation of a GPS set is as follows at power up:

a. **Initialization.** Turning the power on initializes with the closest satellite and ephemeris data (relating to the orbital parameters of the satellites) is being downloaded into memory. A period of up to 20 minutes is required to stabilize a position and verify the status of satellites, availability, etc. After a GPS is switched off, the last position is retained in memory. If your position remains within 50 nm, a position will generally be available within approximately a few minutes the next time the power is turned on.

b. **Acquisition.** The receiver collects data from other satellites in view. Based on the data, it locks on to a satellite to commence the ranging process.

c. **Position Fix.** Based on the data on position and time, the receiver triangulates the position with respect to the positions of satellites. Normally this will be displayed in two decimal places. Some units give three decimal places, but such accuracy is highly suspect and should be treated with caution. If typical accuracy is 100 meters with Selective Availability (where the accuracy of the signals is deliberately degraded), relying on a position fix with an accuracy of approximately 3 meters or less is not as accurate as you would like to believe.

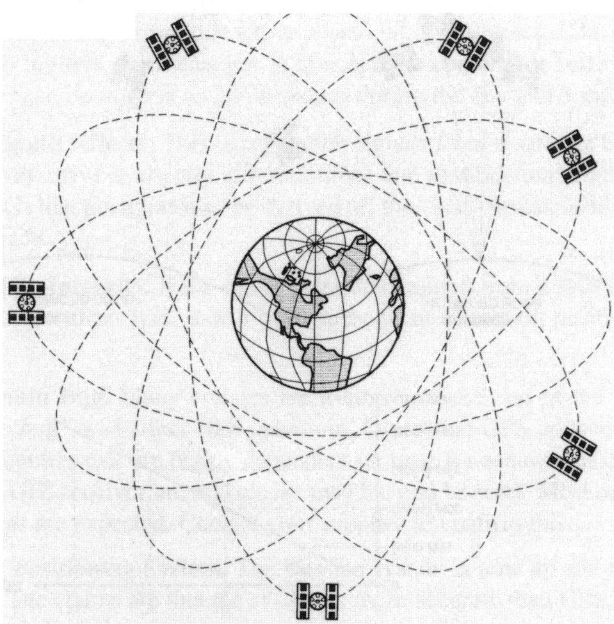

Figure 18-1 GPS Satellite Matrix

18.3 GPS Accuracy. GPS accuracy is the subject of widespread debate and controversy. The inaccuracies currently inherent in the system due to governmental policies have initiated expensive and technologically advanced solutions to improve accuracy.

a. **Precise Positioning Service (PPS).** This service is primarily for military use and is derived from the Precise (P) code. The P code is transmitted on the L1 (1575.42MHz) and L2 (1227.60MHz) frequencies. PPS fixes are generally accurate within 16 meters spherical error.

b. **Standard Positioning Service (SPS).** This service is for civilian use and is derived from the Course and Acquisition (C/A) code. Accuracy levels have been degraded to within 141 meters 95% of the time.

c. **Selective Availability (SA).** This is the process of degrading positional accuracy by altering or introducing errors in the clock data and satellite ephemeris data. SA is characterized by a wandering position, and often a course and speed over the ground of up to 1.5 knots while actually stationary.

d. **Horizontal Dilution of Position (HDOP).** Accuracy is determined by what is called (Geometric) Horizontal Dilution of Precision (HDOP), which indicates the dilution of precision in a horizontal direction. The cause is poor satellite geometry, which is due to poor satellite distribution. It is generally measured on a scale of one to 10; the higher the number, the poorer the position confidence level.

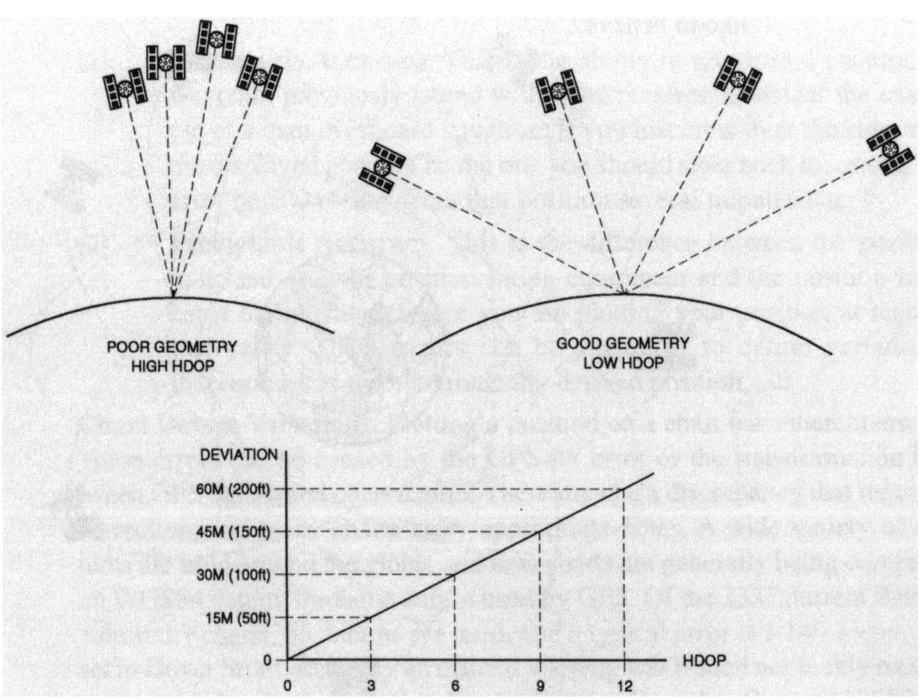

Figure 18-2 GPS HDOP

18.4 GPS Error Sources. The GPS that is considered by many cruisers to be an accurate navigation source has inherent errors that decrease accuracy. These errors are in addition to the HDOP and SA errors as described and it is important to understand them.

a. **GPS Clock Errors.** Each GPS satellite has two rubidium and two cesium atomic clocks. These clocks are monitored against terrestrial atomic clocks. Based on this information, the entire GPS system is continually calibrated against UTC.

b. **Ionosphere Effects.** Like radio signals, both ionospheric and tropospheric conditions can affect GPS accuracy. Errors occur in signal transmission times that can impose signal propagation delays. This signal refraction introduces timing errors that cause positional inaccuracies. Like radio propagation it alters given changes in atmospheric conditions, solar activity, etc. Errors can be as great as 20–30 meters during the day and 5 meters at night.

c. **Multipath Effects.** This occurs when signals from a satellite traveling to a receiver arrive at slightly different times due to reflection or alteration. The effect is that positions may be derived off the "bad" signal, resulting in inaccuracy.

d. **Satellite Integrity.** If the signal being transmitted from a satellite is corrupt due to a malfunction, it will have subsequent effects on position computations.

18.5 GLONASS Positioning System. The Russian system operates using a 24-satellite system (21 in use and 3 spares). The claims are that the system is more accurate than GPS, and this has been proven in higher latitude locations such as the UK and Europe. It is also claimed to be more reliable because it is not subject to experimental shutdowns or position degradation. At the time of publishing the constellation status was 13 satellites operating.

18.6 Differential GPS (DGPS). This system is designed to overcome the position errors with respect to Selective Availability. This system reduces errors in fixes substantially, with respect to the problems of selective availability. DGPS uses a shore based reference station located in an accurately surveyed location. The position is compared with the GPS derived position to produce an error or position offset. These errors may be due to SA or others previously covered. A correction signal to satellite range data (pseudorange differential) is broadcast by radio beacon (285-325 kHz in standard format RTCM SC104)) which is received by a radio beacon receiver, then incorporated into the vessel GPS receiver position computation to derive a final and more accurate position. The accuracy has come down to around 2 meters in some instances. DGPS services also use UHF radio correction signals as well as Spotbeam, Inmarsat, Marine Radio Beacon and IALA VHF transmitted data.

18.7 Wide Area Augmentation System (WAAS). Also known as satellite Differential GPS (SDGPS) WAAS comprises 25 ground reference stations across the US. These stations monitor GPS satellite data. Two master stations on each side of the country collect data and derive a GPS correction message. The message includes orbit and clock drift as well as signal delays. This ground base calculated ionosphere differential correction signal is then uploaded to one of two geostationary satellites and rebroadcast back again to WAAS enabled GPS receivers. Europe is developing a similar system called European Geostationary Navigation Overlay Service (EGNOS), and in Asia a system called MSAS (Multi Function Augmentation System) is under development. Accuracy is typically around 3m against 10m.

18.8 GPS Satellite Acquisition Modes. The various methods of satellite acquisition are explained below.

a. **Multiple Channel, Parallel Processing.** Multiple-channel units are now the norm with virtually all manufacturers incorporating up to 12 receivers. This powerful processing capability enables the monitoring and tracking of up to 12 satellites and the parallel processing of all those satellites in view simultaneously. These units increase position accuracy, reduce errors, and improve the HDOP. The TTFF in these units is very fast; in fact, TTFF can be achieved in several seconds. Many handheld receivers now incorporate parallel processing, and it is by far the better system to choose. In rough weather conditions, fix integrity and accuracy will generally be very high. It is important to understand that in practice the best 4 or 5 satellites are tracked and the data from 4 best are processed at the same time.

b. **Multiplex Processing.** Multiplex systems were the norm on early GPS units. They use one or two channels to sequentially handle satellites at high processing speeds. They are sometimes referred to as pseudo-multichannel systems because performance under ideal conditions was nearly as fast and accurate as that of true multiple-channel systems. The high speed sampling and processing of ephemeris occurs concurrently with the ranging function.

18.9 Space Weather and GPS Effects. The ionosphere is well known for the effects it has on HF and Ham radio. Much less known are the effects on GPS. It is an important source of range and range rate errors for users of GPS satellites where high accuracy is required. Ionosphere range error can vary from a few meters to tens of meters, with troposphere range error at a peak up to 2–3 meters. The ionosphere has a dispersive effect. It can alter rapidly in value, changing significantly over one day. In practice the troposphere range error does not alter more the + or – 10% over long periods. GPS signals pass through the ionosphere but suffer propagation delays. Ranging errors of tens of meters can occur in extreme ionosphere conditions, and typically it is 5–10 meters. These generally equatorial events are often associated with plasma bubbles that characterize the unstable state of the equatorial ionosphere at night.

a. **Plasma Bubbles.** Ionosphere plasma bubbles are a natural phenomenon consisting of large regions within the atmosphere where there are large depletions of the ionosphere plasma. They were first detected in Brazil in 1976 and continue to be a major problem within their offshore oil industry, and are subject to much research. Plasma bubbles are known to interfere with satellite communications in the frequency range VHF to 6GHz, and are known to interfere with GPS causing position errors. The plasma bubbles are closely aligned with the earth's geomagnetic field lines along which they may extend thousands of kilometers and across geomagnetic field lines. They occur after sunset and exist at night only, and activity generally is prevalent during periods of maximum solar activity.

b. **Scintillation.** Irregularities in the ionosphere produce diffraction and refraction effects, causing short-term signal fading, which can severely stress the tracking capabilities of the GPS receiver. Signal enhancements also occur, but the GPS user cannot get any benefit from brief periods of strong signal. Fading can be so severe that the signal level will drop completely below the receiver lock threshold and must be continually re-acquired. The effects are called ionospheric scintillations, and the region can cover up to 50% of the earth in varying degrees. Strong scintillation effects in near equatorial regions are observed generally 1 hour after sunset to midnight. Precise measurement using GPS should be avoided if possible from 7 to 12 pm local time during periods of high solar activity and during months of normal high scintillation activity. There are also seasonal and solar cycle effects that also reduce chances of encountering scintillation in near equatorial regions. In the period April to August, the chances are small of significant scintillation in the American, African and Indian regions. In the Pacific scintillation effects maximize during these months. From September to March the situation reverses. The regions where the strongest scintillation effects are observed are Kwajalein Island in the Pacific and Ascension Island in the South Atlantic. The occurrence of strong amplitude scintillation is also closely correlated with the sunspot number; in years with near minimum solar activity, there are little if any strong scintillation effects on GPS. Where GPS is used for autopilot waypoint steering in these regions, skippers should be alert for course changes, or simple unexplainable and short term periods of inaccuracy.

18.10 GPS Installation and Troubleshooting. The reliability and accuracy of your GPS system depends on a proper installation. Now that most sailing yachts have GPS as their primary navigation source, it is essential they be properly installed.

a. **Aerial Installation.** Aerials should be sited so that they are clear of spars, deck equipment and other radio aerials. Where possible the aerial should have as wide a field of view as practicable, while being located as low as possible. In installations that utilize a targa or sternpost with mounted radar, ensure that the GPS aerial is not within the beam spread of the radar antenna. Ensure that the location is not prone to fouling by ropes and other equipment that may damage the aerial.

b. **Cabling.** Many GPS problems are a result of cabling problems. Power supply cables should be routed as far as practicable from equipment cables carrying high currents. Aerial cables should also be routed well clear. It is extremely important for the aerial cable not to be kinked, bent, or placed in any tight radius. This has the effect of narrowing the dielectric gap within the coaxial cable, which may cause signal problems. Ensure that all through-deck glands are high quality in that they properly protect the cable and keep water from going below. Thrudex (Index) makes cable glands that enable the plug to be passed through along with the cable. Do not shorten or lengthen an aerial cable unless your manufacturer approves it.

c. **Connectors.** Ensure that all connectors are properly inserted into the GPS receiver. Ensure that screw-retaining rings are tight, because plugs can work loose and cause intermittent contact. The coaxial connector from the aerial into the receiver should be rotated properly so that it is locked in. External aerial connections should be made water resistant where possible. Self-amalgamating tape is a useful method for doing this. If you have to remove and refit an aerial connector, ensure that you use considerable care and assemble the connector in accordance with the manufacturer's instructions. Use a multimeter on the resistance range, and check the center pin to shield resistance. Low resistance generally means a shorted shield strand. Resistance is typically 50–150 ohms.

d. **Grounding.** The ground connection provided with the system must be connected to the RF ground system or negative supply polarity depending on manufacturer's recommendations.

e. **Power Supplies.** A clean power supply is essential to proper operation. Use either an in-line filter or install suppressors across "noisy" motors and alternator. The power supply should not come from a battery used for engine starting, or used with any high current equipment such as an anchor windlass or electric toilet. Note that many cheaper unsuppressed fluorescent lights also cause interference that may create data corruption.

f. **GPS Maintenance.** Perform the following routine maintenance checks. Many problems can be identified and rectified before the system fails.

(1) Check the aerial to make sure the connections are tight and the plugs in good condition. Ensure that it is mounted vertically and has not been pushed over, and this is a common problem.

(2) Ensure that all connectors are properly inserted. In particular, examine the external aerial connector for signs of corrosion, especially the outer shield braiding.

(3) Many GPS units have internal lithium batteries with a life span of only around 3 years so ensure that the battery is renewed prior to any voyage.

(4) Make a hard copy list of all waypoints for reference and reprogramming if required.

g. **GPS Troubleshooting.** You should attempt some basic troubleshooting before you call a technician or remove a GPS unit for repair by the manufacturer. Many problems are related to peripheral equipment rather than the unit, and simple checks may save considerable sums of money.

(1) **Large Fix Error.** The GPS system may be down, or a satellite may be shut down. Check your NAVTEX transmissions or other navigation information source for news of outages. SA may be activated, or the HDOP may simply be excessive due to poor satellite geometry in your location. With older multiplex receivers, loss of signal may be a problem in heavy sea states.

(2) **Small Fix Error.** Errors that are not significantly large but consistently outside normal accuracy levels are attributable to a number of sources. The signal may be subject to an excessive amount of atmospheric disturbances, such as periods of extensive solar flare activity. This may be confirmed by similar HF reception difficulties, which also suffer propagation problems. The aerial connections and part of the installation may have degraded, so check the entire system. Make sure aerial orientation is vertical and not partially pushed over. Check that some aerial shadowing has not been introduced.

(3) **No Fix.** This is often caused in sequential receivers by loss of a satellite view or when a satellite goes out of service. Another common cause is the aerial being pushed over to horizontal, so check that it is vertical. Aerial damage from having been struck by equipment is another major cause of a sudden fix loss. Check all cables and connections. If these show no defects, a check of all initialization parameters may be necessary; if those check out, the receiver and aerial may require shore servicing.

(4) **Data Corruption.** This error is often caused by power supply problems. Check whether the incident coincides with engine or machinery run periods. Radiated interference is also a possibility, often from radio equipment. A lightning strike with resultant electromagnetic pulse can also cause similar problems. Another quite common cause of data corruption is that caused by "fingers." Has another person unfamiliar with operating the GPS altered configuration parameters such as time settings or altitude?

Table 18-1 - GPS Troubleshooting

Symptom	Probable Fault
Large fix error	GPS satellite system down
	Selective availability switched on
	High HDOP
	Severe atmospheric problem
	Satellite acquisition loss (heavy weather)
Small fix error	Atmospheric propagation problem
	Aerial shadowing
No fix	GPS satellite system down
	Aerial fault
	Aerial cable fault
	Aerial pushed over to horizontal
	Aerial "view" obstructed
Data corruption	Power supply interference
	Radiated interference

18.11 Loran-C. Loran-C (Long Range Aid to Navigation) is a pulsed, low-frequency hyperbolic radio-navigation aid. 24 U.S. LORAN-C stations operate in partnership with Canadian and Russian stations to provide navigation coverage in Canadian waters and in the Bering Sea. The LORAN-C system provides 0.25 nautical mile absolute accuracy. The US Government has guaranteed the short term viability of the system, however long term is uncertain. The basic theory relies on the accurate measurement of the time difference of radio signals received from a master and slave transmitters to derive a hyperbolic position line. With two position lines or more, a position fix can be made based on the intersection on lines of position.

a. **Transmitter Chains.** Transmitter chains are always grouped according to their geographical location. A transmitter group consists of the following:

(1) A master station designated as M.

(2) From 2 to 5 secondary stations designated as V (Victor), W (Whiskey), X (X-ray), Y (Yankee), Z (Zulu).

b. **Time Differences.** Loran calculates a Line of Position (LOP) from signals transmitted from a master and a secondary station. The receiver measures the difference in time signal arrival which is in microseconds. For every Time Difference (TD), there is a line between master and secondary station where the TD is constant and therefore where your vessel location may be. A second set of signals enables calculation of position based on intersection of two LOPs.

c. **Group Repetition Intervals.** All chains operate at 100 kHz,; differentiation can only be achieved as each chain is identifiable from a Group Repetition Interval (GRI). Each chain has a unique GRI and the GRI is a variation in timing of pulses.

d. **Accuracy.** Accuracy is the fundamental reason behind any navigation system. Like all systems, Loran-C has errors that must be accounted for.

(1) **Absolute Accuracy.** Absolute accuracy is the ability to uniquely determine position using the receiver. Typical accuracy can be around 0.1 to 0.25 nm. Errors are caused by signal anomalies and conversion to TDs to latitude/longitude.

(2) **Repeatable Accuracy.** This is typically in the range of 50–500 feet. The accuracy in Loran-C is best when TD lines are close and cross at 90°. Accuracy is obviously worst at the limit of the coverage area of about 1200 miles where the signal strength is weak. At these ranges, the angles of intersection are low and accuracy is poor. For a given position the plotted positions can wander around and usually vary up to a 100 meters, and tend to align with the TD lines.

e. **Secondary Station Selection.** Use this secondary station selection criteria:

(1) **Signal Strength.** Always select a secondary station with a strong signal. Weak signals are often distorted by skywaves.

(2) **Crossing Angles.** Selected LOPs should cross as near to perpendicular as possible. Shallow crossing angles increase plotting errors.

(3) **Time Difference Gradients.** The spacing of Loran lines is called the TD gradient, and the closer they are together, the better the gradient. Being hyperbolic, Loran LOPs are not a constant distance apart. Avoid any secondary station that requires operation near the baseline extension.

f. **Fix Errors.** Loran-C is prone to a range of errors:

(1) **Skywave and Groundwave Effect.** Loran signals travel via a groundwave which is the shortest path. Other paths also occur, including several skywave types. Depending on time of day, skywaves may be even stronger than groundwaves, but they always arrive after groundwaves. At chain extremities, the stronger skywaves may be stronger than weak groundwaves giving errors up to 10 nm.

(2) **Lightning Impulses.** Pulses from lightning can distort or corrupt signals.

g. **Installation.** Correct installation is the key to optimum performance.

(1) **Antenna and Coupler Location.** Install away from spars and masts. Clearance is ideally a minimum of 6 feet. On cruising yachts, use a separate whip antenna or an insulated backstay can be used.

(2) **Grounding.** Grounding is as important as the antenna and can be the RF groundplate. The grounding wire should be at least 12 AWG.

(3) **Interference.** Interference is the major cause of fix errors. Loran-C is sensitive to noise in the <90–110 kHz spectrum. Common causes are fluorescent lights, alternators, tachometers, and radars. You can test for noise problems using diagnostics to check Signal to Noise Ratio (SNR).

18.12 Loran Transmission Chains. These are the current stations and GRI number:

5543 – Calcutta - India
5930 – Canadian East Coast
5980 – Russian American
5990 – Canadian West Coast
6042 – Bombay - India
6731 – Lessay - France
6780 – China Sea South - China
7001 – Bo - Norway
7030 – Saudi Arabia South
7270 – Newfoundland East Coast
7430 – China North Sea
7499 – Sylt
7950 – Eastern Russia Chayka
7960 – Gulf of Alaska
7980 – Southeast U.S.
7990 – Mediterranean - Italy
8000 – West Russia
8290 – North Central U.S.
8390 – China Sea East
8830 – Saudi Arabia North
8930 – North West Pacific
8970 – Great Lakes

9007 – Eidi
9610 – South Central U.S.
9930 – East Asia
9940 – Western U.S.
9960 – Northeast U.S.
9990 – North Pacific

18.13 Chart Plotters. Most chart plotters universally now incorporate GPS, or conversely are included within GPS units, and are effectively position fixing devices for many. The chart plotter is essentially a display with processor that decodes the data on the chart cartridges for display on the screen. The information is often layered so that chart areas can be expanded, the lights, buoyage and contours can be called up as required. There are many functions and stored data available. They can include tidal predictions; sun and moon rise and set, Navaids such as lights and buoys (10,000 plus items is typical), waypoints and routes (1000/20 is typical). DSC radio interfacing is also possible showing the location of a vessel in distress.

a. Plotter Displays. Systems now use high-resolution monochrome (gray) LCD displays, or full color active matrix TFT (Thin Film Transistor) displays. Screen display quality or resolution is determined by pixels, i.e. 480 x 350. The greater the number of pixels, the greater the resolution, and consequentially the price also increases. Power consumption is typically 6W (0.5 amps). To get the most from dedicated plotters or computer based software plotting make sure you read and understand the manual and practice. Trying to learn while under way is both dangerous and distracting. There is a trend in larger sailing yachts to have PC based systems with remote sunlight viewable displays. Many displays are flat screen LCD types with SVGA resolution, and external standard ones are waterproof to NEMA 4 standard and are rated up to a brightness of more than 1,600 nit which some 1000% brighter than a laptop.

b. Features. One important feature is sets that use standard chart cartography as some use proprietary software that may not be easily obtainable. There are many features to consider that include Drag and Drop waypoints, Active Route waypoint insertion and deletion, and the ability to store routes, and fast redraw times, typically 0.5 to 3 seconds. Cartography features should include clear chart scale indication, clear direction indication when not in North-Up mode, an indication when over-zoomed, good buoy visibility and identification, clearly visible and identifiable land features and contours, the use of standard labelling conventions on features, the display of drying heights, the ability to edit contours and shaded depth contours. One important factor is the ability to update and also some identification when the chart was last updated, similar to paper charts. It is important that any unit be user friendly, and has intuitive controls, menus, soft keys and dedicated keys. Most sets have variable voltage inputs up to 50 VDC and consume 0.5 to 1.15 amps, along with NMEA inputs and outputs for interfacing GPS. Most power supplies are resilient to electrical system spikes and surges, and on boats with single battery systems load surges during starting can cause problems.

18.14 Cartography Systems. The cartographic chart data systems are based on memory chips or more correctly EPROM (Electronically Programmable Read Only Memory) or CD-Rom. There are many formats that include M-93, S-57, NDI and Navionics.

a. **Raster Charts.** A Raster Chart is identical to the paper chart, and originates from original government master charts.

(1) ARCS (Admiralty Raster Chart Service) charts are supplied on a CD-Rom.

(2) Maptech uses raster scan charts and works with the NOAA to produce official charts. Log on to www.maptech.com.

(3) Seafarer are identical to ARCS and are produced by the Australian Hydrographic Office for Australian waters.

b. **Vector Charts.** The scanning of paper charts to create a raster image produces a Vector Chart. These are then vectorized to store data in layers, which allows easy zooming in on detail. They do not resemble conventional charts. The advantages over raster charts are much faster screen update rates.

(1) C-Map NT uses vector charts. The charts are stored on solid-state memory cards that include C-Cards and PCMCIA cards. Log on to www.c-map.com.

(2) Passport use vector charts based on official charts. Log on to www.nobeltec.com.

(3) Garmin G-Charts & G-Map. These are based on Navionics data for Garmin plotters. G-Map is for the new plotters and is available on CD-ROM also.

(4) Navionics Microcharts & Nav-Charts. These credit card sized cartridges use PCMCIA formats. Nav-Charts are contained on very small memory cartridges.

18.15 Chart Corrections and Updates. Chart corrections and updates are now part of standard services and are offered by Maptech for NOAA charts. They require up to date Notices to Mariners, although you can do that via Internet for free now on www.nms.ukho.gov.co.uk. Cartridges can be updated every couple of years. You must bear this in mind with plotters, and where they are used as a substitute for properly corrected charts.

18.16 Chart Software. Software developments for electronic charting have been rapid with packages offering very powerful navigation tools; the following are leading software packages and salient features. There are varying features and capabilities. Software features include the ability to display multiple chart windows; the ability to rotate charts; passage planning showing hazards; the display of tidal heights and streams and planning; the ability to interface using NMEA to autopilot, radar, GPS and instruments; the ability to overlay radar and display ARPA targets.

a. **dKart.** This software supports ECDIS vector charts from C-Map, CM-93 and the S-57 format. The charts can be rotated Head-up, North-up or Course-up while maintaining text correctly. When used with passage planning there is automatic hazard identification within a nominated distance of the track. There is also the facility to show safe depth contour around the boat, with automatic hazard identification on your track.

b. **seaPRO 2000.** This uses several chart formats that include the European Livechart vector charts. In addition ARCS, Seafarer, Maptech and NOAA/BSB raster charts can also be used. SeaPRO Plus offers fuel consumption calculations, and ARPA radar targets. The Network version will suit large yachts that wish to display the information in other areas or cabins.

c. **Chartview.** This is from Nobeltec and supports ARCS Skipper and Navigator raster charts, NDI, and Maptech formats. A unique feature when using ARCS charts is the ability to automatically place chartlets in the right geographical location on the chart. Also there is seamless chart quilting and rotation. Tidal streams and currents can be accounted for with the built-in tidal prediction tools. This has now been replaced by Nobeltec Admiral and the Visual Navigation suite.

d. **Raytech RNS V6.** This is a Raymarine product and had its basis in the New Zealand developed Kiwitech software. It supports Raster, Vector, 3D and aerial photography cartography from Navionics, Maptech and C-Map NT. It offers additional facilities such as engine monitoring, instrument, fishfinder and radar overlays, and route optimization. It also allows integration with Raymarines' Seatalk system, has touch screen capability, and even Internet access.

Communications Systems

19.0 Global Maritime Distress and Safety System (GMDSS). GMDSS was fully implemented in 1999 for all commercial vessels exceeding 300 GRT. In many areas VHF Channel 16 or 2182 kHz is not monitored and requirements are being dropped. The primary function of GMDSS is to coordinate and facilitate Search and Rescue (SAR) operations, by both shore authorities and vessels, with the shortest possible delay and maximum efficiency. It also provides efficient urgency and safety communications, and broadcast of Maritime Safety Information (MSI) such as navigational and meteorological warnings, forecasts, and other urgent safety information. MSI is transmitted via NAVTEX, International SafetyNet on INMARSAT C, and some NBDP radio telex services.

19.1 GMDSS Operational Details. Worldwide communications coverage is achieved using a combination of INMARSAT and terrestrial systems. All systems have range limitations that have resulted in the designation of four sea areas, which defines communications system requirements.

- **a.** **Area A1.** Within shore-based VHF radio range. Distance is in the range of 20–100 nm. Radio required is VHF operating on Channel 70 for DSC, and Channel 16 radiotelephone. EPIRB required is 406 MHz or L-band unit (1.6 GHz). Survival craft require a 9-GHz radar transponder and portable VHF radio (with Channel 16 and one other frequency).
- **b.** **Area A2.** Within shore-based MF radio range. Distance is in the range of 100–300 nm. Radios required are MF (2187.5 kHz DSC) and 2812 kHz radiotelephone, 2194.5 NBDP, and NAVTEX on 518 kHz. Also needed are the same VHF requirements as Area A1. EPIRB required is 406 MHz or L-band (1.6 GHz). Survival craft requirements are the same as in Area A1.
- **c.** **Area A3.** Within geostationary satellite range (INMARSAT). Distance is in the range of 70°N–70°S. Radios required are MF and VHF as above and satellite (with 1.5-1.6 GHz alerting), or as per Areas A1 and A2 plus HF (all frequencies). Survival craft requirements are the same as in A1.
- **d.** **Area A4.** Other areas (beyond INMARSAT range). Distance north of 70°N and south of 70°S. Radios required are HF, MF, and VHF. EPIRB required is 406 MHz. Survival craft requirements are the same as in Area A1.

19.2 GMDSS Radio Distress Communications Frequencies. The frequencies designated for use under GMDSS are as follows:

- **a.** VHF DSC Channel 70, Channel 16, Channel 06 Intership, Channel 13 Intership MSI.
- **b.** MF DSC 2187.5 kHz, and 2182 kHz.
- **c.** HF4 DSC 4207.5 kHz, and 4125 kHz.

d. HF6 DSC 6312 kHz, and 6215 kHz (CH421).

e. HF8 DSC 8414.5 kHz, and 8291 kHz (CH833).

f. HF12 DSC 12577 kHz, and 12290 kHz (CH1221).

g. HF16 DSC 16804.5 kHz, and 16420 kHz.

19.3 Digital Selective Calling (DSC). DSC is a primary component of GMDSS and is used to transmit distress alerts and appropriate acknowledgments. DSC will improve accuracy, transmission, and reception of distress calls and VHF Channel 70 is the nominated DSC channel.

a. DSC has the advantage that digital signals in radio communications are at least 25% more efficient than voice transmissions, as well as significantly faster. A DSC VHF transmission typically takes around a second, and MF/HF takes approximately 7 seconds, both depending on the DSC call type.

b. DSC requires the use of encoders/decoders, or additional add-on modules to existing equipment. A dedicated DSC watch receiver is required to continuously monitor the specified DSC distress frequency. Affordable VHF DSC radio equipment is a priority for small vessels and Class D controllers are now available. They are from Horizon, Raymarine, Icom, Simrad and others.

c. DSC equipment enables the transmission of digital information based on four priority groupings, Distress, Urgency, Safety, and Routine. The information can be selectively addressed to all stations, to a specific station, or to a group of stations. To perform this selective transmission and reception of messages, every station must possess what is called a Maritime Mobile Selective-call Identity Code (MMSI). Note that Distress "Mayday" messages are automatically dispatched to all stations. A DSC Distress alert message is configured to contain the transmitting vessel identity (the MMSI nine-digit code number), the time, the nature of the distress, and the vessel position where interfaced with a GPS. After transmission of a distress alert, it is repeated a few seconds later to ensure that the transmission is successfully transmitted.

19.4 GMDSS Distress Call (Alert) Sequence. It is important to explain the various elements of GMDSS in an emergency situation.

a. **Distress Alert.** This is usually activated from a vessel to shore. For sailing yachts, this is usually via terrestrial radio, and larger vessels use satellites. Ships in the area may hear an alert, although a shore-based Rescue Coordination Center (RCC) will be responsible for responding to and acknowledging receipt of the alert. Alerts may be activated via an INMARSAT A, B, or C terminal, via COSPAS/SARSAT EPIRB (243/406 MHz), or via an INMARSAT E EPIRB. DSC VHF or MF/HF can also activate alerts.

b. **Distress Relay.** On receipt and acknowledgment of alert, the RCC will relay the alert to vessels in the geographical area concerned, which targets the resources available and does not involve vessels outside the distress vessel area. Vessels in the area of distress can receive appropriate alerts via INMARSAT A, B, or C terminals, DSC VHF or MF/HF radio equipment, or via NAVTEX MSI. On reception of a distress relay the vessels concerned must contact the RCC to offer assistance.

c. **Search and Rescue.** In the SAR phase of the rescue, the previous one-way communications switch over to two-way for effective coordination of both aircraft and vessels.

d. **Rescue Scene Communications.** Local communications are maintained using short-range terrestrial MF or VHF on the specified frequencies. Local communications take place using either satellite or terrestrial radio links.

e. **Distress Vessel Location.** A Search and Rescue Transponder (SART), and/or the 121.5MHz homing frequency of an EPIRB assist in determining the precise location of the vessel in distress.

19.5 GMDSS False Alerts. GMDSS is relatively new, and currently the false alert rate is around 95%. False alerts are not desirable simply because of the load placed on SAR services. False alerts are generally caused by operator errors and incorrect equipment operation. Another cause of false alerts is the improper acknowledgment of distress alerts leading to excessive DSC calls. Training and experience of equipment operation is essential to resolve these problems.

19.6 GMDSS and Sailing Yachts. The installation of GMDSS is not compulsory for pleasure boats, but due to its universal implementation on commercial vessels, most yachts will be forced to install partial GMDSS equipment simply to remain "plugged in" to the system. GMDSS will certainly maximize SAR situations for yachts so in most cases it will enhance offshore safety. GMDSS equipment will accurately identify your own yacht, current position, and type of emergency, and this information will be broadcast automatically. What you get is automatic activation of alarms at coast stations and on other vessels simply by pushing one button. Just as GPS, electronic charting, and the EPIRB have opened up the world to cruisers, so will GMDSS significantly improve sea safety. Few will be able to invest in full INMARSAT terminals. A more advanced training course and operation certificate is also required and is being run in many locations.

19.7 Satellite Communications Systems. Under GMDSS, satellite systems play a major role and prices are becoming more affordable for sailing yachts. INMARSAT was established by the IMO to improve distress and safety of life at sea communications and general maritime communications. INMARSAT is based on satellites placed in geostationary orbit. Under GMDSS all commercial vessels operating in areas outside designated areas of International NAVTEX coverage require a receiver for reception of INMARSAT SafetyNET Maritime Safety Information (MSI). The Standard-C SES is a GMDSS-compliant system that offers compact and lightweight terminals. These systems are designed to support data-only services, not voice. Services are telex, e-mail, Internet access, and computer database access.

19.8 COSPAS/SARSAT System. GMDSS incorporates the COSPAS/SARSAT system as an integral part of the distress communications system. The acronym is based on the former Soviet "Space System for Search of Distress Vessels" and the American "Search and Rescue Satellite Aided Tracking." Under GMDSS if a vessel does not carry a satellite L-band EPIRB in sea areas A1, A2, and A3 (described earlier), a 406 MHz EPIRB is required operating in the COSPAS-SARSAT system. This unit must have hydrostatic release and float-free capability. The system is a worldwide satellite-assisted SAR system for location of distress transmissions emitted by EPIRBs on the 121.5/243 MHz and 406 MHz frequencies. 121.5 kHz is an aircraft homing frequency and 243 MHz is a military distress frequency that enables military aircraft to assist in SAR operations. The Emergency Position Indicating Radio Beacon (EPIRB) is an essential item of safety equipment for any offshore vessel. Earlier EPIRB units relied solely on over-flying aircraft for detection of signals and relay of the position to appropriate SAR authorities; the new systems utilize satellites. The satellite-compatible system relies on satellites inserted in near polar orbits with orbit times of approximately 100 minutes. Accuracy of the system improved from approximately 10 nm for 121.5/243 MHz units to 3 nm for a 406 MHz unit. Note that the 406 MHz units are far more effective at lower latitudes than the 121.5/243 MHz units, and the latter are being phased out. These were not useful in mid ocean and had a high false alarm rate so it was usual to wait for two separate satellite hits before activating SAR. There was no vessel specific identification, and error was around 20km. Aviation frequency is 121.5 kHz, and military 243 MHz. 121.5 kHz is now primarily used on personal locator beacons only.

19.9 Satellite (L-Band) EPIRBs. This system, developed by the European Space Agency, will alert rescue services in distress within 2 minutes, rather than in hours as with current systems. The new system combines position determination along with a distress signal using the INMARSAT geostationary satellites. The system uses special EPIRBs that incorporate GPS receivers and ensure a position fix within 200 meters. The distress signal transmits via one of four Land Earth Stations (LES) and landline links with appropriate rescue coordination centers. Recent testing shows an average of 5-minute delay from activation to reception by rescue services.

19.10 406 EPIRBs. The 406 MHz units have a unique identification code, and information is usually programmed at time of sale, with MMSI or registered serial numbers. Some units also have integral strobes and all incorporate 121.5 MHz for aircraft homing signal purposes. Float-free units are called Category 1; manual bracket units are Category 2. Orbit time is 100 mins using COSPAS satellites so a delay in transmission up to 4 hours near the equator can occur. Accuracy is 2km and uses the Doppler effect so usually requires 2 satellite passes. The system uses a store and forward system so the satellite stores and downloads distress data when in view of a LUT.

19.11 406 MHz EPIRB Registration. If you acquire a yacht with a 406 MHz EPIRB, you must register the EPIRB unit properly and provide all of the appropriate data, including its Unique Identification Number (or MMSI). Registration should be done immediately upon purchase. Failure to do this can cause absolute havoc if you use it, because a vessel may be incorrectly identified or, worse still, not identified at all, which could seriously jeopardize your rescue. Bad information means bad rescue problems for everyone. If you have not registered, contact the organizations listed:

- **a.** **United States of America.** NOAA/NESDIS Tel +1-301-457 5428. Additional information on registration Tel +1-302-763 4680.
- **b.** **United Kingdom.** EPIRB Registry, Marine Safety Agency, Tel +44-1326 211569 (www.msa.co.uk).
- **c.** **Canada.** Canadian EPIRB Registry Director, Search and Rescue, Canadian Coastguard, Tel +1-613-990 3124.
- **d.** **Australia.** Maritime Rescue Co-ordination Center, Australian Maritime Safety Authority. Tel 1 800 406 (406 www.amsa.com.au)
- **e.** **Netherlands** Ministry of Transport Tel +31-50-222111.
- **f.** **New Zealand.** CAA. Tel +64-4-5600400.

19.12 EPIRB Activation Sequence. On activation of an EPIRB the following sequence of events occurs:

- **a.** A satellite detects the distress transmission. With 243/121.5 MHz units a satellite and the EPIRB must be simultaneously in view of the Local User Terminal (LUT).
- **b.** The detected signal is then downloaded to a LUT. (In 406 MHz units the satellite stores the message and downloads to the next LUT in view).
- **c.** The LUT automatically computes the position of the distress transmission. The distress information is then passed to a Mission Control Center (MCC) before going to a Rescue Control Center (RCC) and then to SAR aircraft and vessels.

19.13 EPIRB Operation. Do not operate an EPIRB except in a real emergency, because you could initiate a rescue operation. Do not even operate it for just a short period of time and then switch it off, because authorities may assume your vessel went down quickly before circumstances stopped transmission. With current attitudes changing toward false alarms, it may reflect very badly on boaters as a whole in terms of wasting taxpayers' money. If you activate your EPIRB during an emergency, once rescued, do not leave the EPIRB in the raft or floating as the beacon may continue to transmit for some time.

19.14 Rescue Reaction Times. There is a mistaken belief that rescues are instantaneous after activation of an EPIRB. The reality, however, is a time lag that can average up to 6 hours or more from detection of a signal and physical location, although position is usually confirmed in less than 2 hours. This is dependent on suitable aircraft, weather conditions, and SAR coordinator response times. Every LUT has a "footprint" coverage area, and the closer you are to the edge of that footprint, the longer the delay. Time lags depend on intervals between satellite passes over a given location. There are six polar orbiting satellites and, although random in orbit, their tracks are predictable. If you have to activate, be patient and wait. Remember, you are not a survivor until you're on the deck of a rescue vessel or in the helicopter. Priority one is a survival training course. Have you done one? Have you evaluated and planned a helicopter evacuation procedure?

19.15 Battery Life and Transmit Times. Much concern has been raised over battery transmit life after activation. Always ensure that the battery pack is replaced within the listed expiration date. Nominally a lithium battery has a life of 4 to 5 years depending on the manufacturer. Typical transmit times are 80–100 hours at 5W output. Standards require a minimum of 48 hours.

19.16 EPIRB Maintenance. The only maintenance required is to test the EPIRB using the self-test function every six months in accordance with the manufacturer's instructions. Do not self-test by activating the EPIRB distress function. Do not drop the unit unless it is in the water or damage may occur.

19.17 Personal Locator Beacons (PLBs). PLBs are essentially miniature EPIRBs. They operate on 121.5 MHz, which is the frequency used for homing in by SAR vessels and aircraft. Due to their small size they can be attached onto your wet weather gear, life vest or carried in a pocket or emergency bag. They are not as accurate as other units and will localize your position to around 12 nm, because the transmitters are line of sight only. Some units are configured to activate in water, and most operate for at least a 24-hour period and some work up to 48 hours. The PLB is not a substitute for a 243/121.5 MHz or 406 MHz EPIRB.

19.18 Search and Rescue Transponders (SARTs). Under GMDSS these units are required on all vessels over 300 GRT. These devices are designed for use in search and rescue. An EPIRB will put potential rescue vessels in the area, but the transponder will accurately localize your position to search radars. Units typically have the following characteristics:

a. **Signal Transmission.** The transponder responds automatically and emits a 9200–9500 GHz high-speed frequency sweeping signal which is synchronous with received scanning radar pulse.

b. **Signal Reception.** On reception of the signal, the position is indicated on radar screens as a line of 12 blips giving range and bearing.

c. **Transponder Receiver.** The transponder gives an audible alarm when the radar emission of a search and rescue vessel is detected.

19.19 NAVTEX. NAVTEX is an integral part of GMDSS as well as the Worldwide Navigational Warning Service (WWNWS). It is an automated information system providing meteorological, navigation, and maritime safety information (MSI). Messaging is broadcast in English on a pre-tuned and dedicated frequency of 518 kHz with an additional frequency of 590 kHz now implemented within UK/Europe in local languages. Range is typically around 250 nm. INMARSAT Enhanced Group Calling (EGC) provides long-range information. Each of the 16 Navarea are divided into four groups each with up to 6 transmitters with an allocation of 10-minute transmissions each four-hour period. This is time shared to prevent interference on adjacent areas, and they have limited power outputs. Message reception requires a dedicated receiver. Broadcast times are included within frequency listings.

a. **Message Priorities.** Prioritization is used to define message broadcasts. Vital messages are broadcast immediately, usually at the end of any transmission in progress. Those classified as Important will be broadcast at the first available period when the frequency is not in use. Routine messages are broadcast at the next scheduled transmission time. Those messages classified as Vital and Important will be repeated if still valid at the following scheduled transmission times. Messages incorporate a Subject Indicator code (B2 character), which allows acceptance and rejection of specific information. Navigational and meteorological warnings and SAR information are non-selective so that all stations receive important safety information. B2 codes include Nav Warnings (buoy positions altering, wrecks, floating hazards, oil rig moves, naval exercises, meteorological warnings (gales etc.) ice reports, SAR and anti-piracy info (cannot be rejected), met forecasts shipping and synopsis, pilot service messages, Loran messages, OMEGA messages, GPS messages, other NAVAID messages, Nav warnings additional to A. (Cannot be rejected). A = Nav warnings, B=Gales, D=Distress information, E=Forecasts.

b. **Station Identification.** Navigation information is broadcast from a number of stations located within each navarea. Broadcast times as well as transmitter power outputs are carefully designed to avoid interference between stations. Each station is assigned an identification code (B1 character). This is essential so that specific geographical region stations can tune in.

c. **Operation.** Stations are selected by letter designation such as M-Casablanca, I-Las Palmas, 2 x 24 hr forecasts per day for sea areas; the message format contains the following. The letter Z indicates there are no messages to transit, checks system and is an operational check message.

Format nine characters, header code followed by technical code

ZCZC B1 (Transmitter ID) B2 (Subject ID), B3, B4 (consecutive number)

Time of origin

Series ID and consecutive number

The message text

NNNN (end of message group)

19.20 VHF Radio. VHF is probably the most useful radio system available. It allows easy ship to ship, or ship to shore communications. The disadvantage is that the range is line of sight, typically around 35 miles. All countries have licensing regulations that must be adhered to. Failure to comply may result in prosecution and fines.

a. **Ship Station License.** All VHF installations must possess a station license issued by the appropriate national communications authority, i.e. FCC. On issue of the first license a call sign is issued.

b. **Operator License.** At least one operator, normally the person registering the installation, should possess an operator's license or certificate. Under GMDSS and DSC this has changed. In the UK it is the Short Range Certificate (SRC); similarly the requirements have changed in the US, Canada and Australia. It requires a short one-day course.

19.21 VHF Theory. The spectrum consists of 55 channels in the 156-163 MHz band.

a. **Range.** As VHF is line of sight the higher the two antennas are mounted, the greater the distance. There are theoretical ways to work out the range but for simplicity I will leave them out. Factors such as atmospheric conditions and the installation itself also affect the actual range. Typical range with a coast station can be up to 35–40 nm.

b. **Power Consumption.** Typically units consume 5–6 amps when transmitting. Reception only consumption can add up as the set is on for virtually 24 hours. This is in the range of 1 to 7 amps. In a day, that can add up to 12–19 amp-hours depending on the set. VHF is however one piece of equipment that should be left on regardless of power consumption. The merchant ship that sights you and tries to communicate will do so well before you may be aware of it.

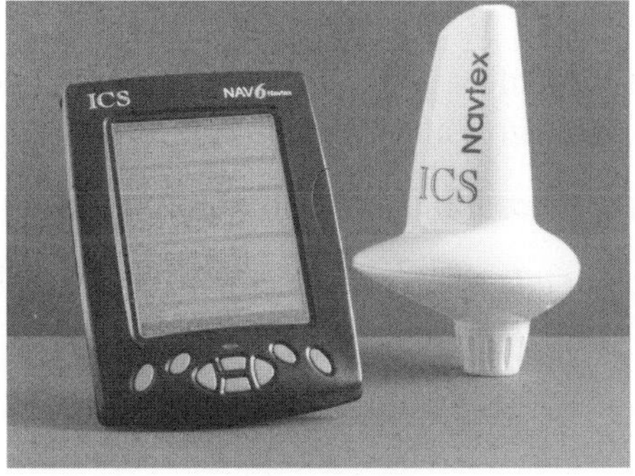

Figure 19-1 Dual Frequency Navtex
(Courtesy ICS Electronics

19.22 VHF Propagation. VHF signals penetrate the ionosphere rather than reflect. There are circumstances in which VHF signals can reflect back from the ionosphere to give "freak" long distance communications such as during very strong solar cycles. This occurred during cycle 19 in 1957/58, cycle 21 in 1980 and cycle 22 in 1990. During these peaks, the monthly sunspot average rose to extremely high values and the ionosphere reflected higher frequencies than normal. VHF can also be reflected from clouds of increased ionization in the E layer of the ionosphere, and during auroras, which are those spectacular light curtains caused by charged particles from the sun.

19.23 VHF Operation. As VHF is widely used by official and commercial operators, it is essential to use your set properly for optimum performance.

a. **Power Setting.** Always use the 1-watt low power setting for local communications, and the 25-watt high power for distance contacts.

b. **Squelch Setting.** Squelch reduces the inherent noise in the radio, but do not reduce the squelch too far.

c. **Simplex and Duplex.** Simplex means that talk is carried out on one frequency. Duplex is where transmit and receive are on two separate frequencies.

d. **Dual Watch.** This facility enables continuous monitoring on Channel 16 and the selected channel.

e. **Talk Technique.** Hold the microphone approximately 2 inches from the mouth and speak at a volume only slightly louder than normal. Be clear and concise and don't waste words. Many newer sets also incorporate noise-canceling microphones.

19.24 Radio Procedure. After selecting the required channel use the following procedures. Procedures are valid for coast stations or other vessels:

a. **Operating Procedure.** Wait until any current call in progress is terminated. Even if you do not hear speech, listen for dial tones or other signals. Do not attempt to cut in or talk over conversations. Sometimes traffic may be busy and patience is required.

   (1) Always identify your vessel and call sign both at the beginning and end of transmission.

   (2) Keep conversations to a minimum, ideally less than three minutes.

   (3) After contact with other vessels, allow at least ten minutes before contacting them again.

   (4) Always observe a three-minute silence period on the hour and half hour. While it is not essential it is good practice.

b. **Coast Station Calls.** Operate your transmitter for at least 7–8 seconds when calling and use the following format:

(1) Call the coast station 3 times.

(2) "This is <vessel name and call sign>" and repeat three (3) times.

(3) Response will be "vessel calling <station name> this is <station name> on Channel <No>." This is usually on VHF 16 or the nominated call channel.

(4) Response "This is <call sign> my vessel name is <name>."

(5) State purpose of business, link call, request for information or advice. "Good evening Sir, I wish to make a transfer charge call". "The number I require is (number)."

(6) On completion of business, "Thank you <station> this is <vessel name> over and out, and listening on Channel 16 or <No>."

19.25 Distress, Safety and Urgency Calls. Channel 16 should only be used for the following.

a. **Mayday.** This distress call should only be used under the direst of circumstances, "grave and imminent danger". Use of the call imposes a general radio silence on Channel 16 until the emergency is over. Use the following procedure, and allow time before repeating:

(1) "MAYDAY, MAYDAY, MAYDAY."

(2) "This is the vessel <NAME>."

(3) "MAYDAY, vessel <NAME>."

(4) "My Position is <latitude and longitude, true bearing and distance from known point>".

(5) State <Nature of Distress> calmly, clearly and concisely.

(6) State type of assistance required.

(7) Provide additional relevant information including number of people on board.

b. **Pan-Pan.** (Pronounced PAHN-PAHN) Use this call to transmit an urgent message regarding the immediate safety of the vessel or crewmember. It takes priority over all traffic except Mayday calls. The call is used primarily in cases of injury or serious illness, or man overboard:

(1) "<All Ships>."

(2) "PAN PAN, PAN PAN, PAN PAN."

(3) "This is the vessel <NAME>."

(4) Await response and transfer to working channel.

c. **Security.** (Pronounced SAY-CURE-E-TAY). For navigational hazards, gale warnings, etc as follows:

(1) "SAY-CURE-E-TAY, SAY-CURE-E-TAY, SAY-CURE-E-TAY."

(2) "This is the vessel/station <NAME>."

(3) Pass the safety message.

d. **Medical Services.** Use of this call is to advise of an urgent medical emergency. It takes priority over all traffic except Mayday calls.

(1) "PAN PAN, PAN PAN, PAN PAN."

(2) "RADIOMEDICAL" or "MEDICO."

(3) "This is the vessel <NAME, CALL SIGN, NATIONALITY>."

(4) "My position is <latitude and longitude.> Diverting to <location>."

(5) Give patient details, name, age, sex, and medical history. Give present symptoms, advice required, and medication on board.

e. **Phonetic Alphabet**

A.	ALFA	N.	NOVEMBER
B.	BRAVO	O.	OSCAR
C.	CHARLIE	P.	PAPA
D.	DELTA	Q.	QUEBEC
E.	ECHO	R.	ROMEO
F.	FOXTROT	S.	SIERRA
G.	GOLF	T.	TANGO
H.	HOTEL	U.	UNIFORM
I.	INDIA	V.	VICTOR
J.	JULIETT	W.	WHISKEY
K.	KILO	X.	X-RAY
L.	LIMA	Y.	YANKEE
M.	MIKE	Z.	ZULU

Phonetic Numbers.

1.	WUN	6.	SIX
2.	TOO	7.	SEVEN
3.	THUH-REE	8.	AIT
4.	FO-WER	9.	NINER
5.	FI-YIV	0.	ZERO

Table 19-1 United States and Canada VHF Channels

Channel	Channel Designation
01	Port Operations – Ship to Ship
02	Port Operations – Ship to Ship
03	Port Operations – Ship to Ship
04	Port Operations – Ship to Ship
05	Port Operations – Ship to Ship
06	**INTERSHIP SAFETY - SAR Communications – Ship to Ship**
07	Commercial Ship to Ship
08	Ship to Ship Commercial
09	US Calling Channel (Ship to Ship)
10	Commercial Ship to Ship
11	Harbor – Ship to Ship
12	Port Operations, Traffic advisory, USCG Coast Stations
13	**Bridge and Locks, Ship to Ship (1 watt only) Intracoastal Waterway (ICW)** Commercial vessels. No call signs, abbreviated operating procedures only. Maintain dual watch 13 and 16
14	Port Operations - Bridge and Lock Tenders
16	**DISTRESS, SAFETY and CALLING**
17	Maritime Control
18	Commercial Ship to Ship and Harbor
19	Commercial Ship to Ship and Harbor
20	Port Operations (Duplex)
22	**USCG and Marine Information Broadcasts**
24	Public Correspondence – Marine Operator (Duplex)
25	Public Correspondence – Marine Operator (Duplex)
26	Public Correspondence – Marine Operator (First priority) (Duplex)
27	Public Correspondence – Marine Operator (First priority) (Duplex)
28	Public Correspondence – Marine Operator (First priority) (Duplex)
60-62	Port Operations, Public (Duplex)
63	Port Operations, Ship to ship
64	Port Operations, Public (Duplex)
65	Port Operations Ship to Ship
66	Port Operations Ship to ship
67	Navigational (1 watt only)
68	Ship to Ship and Harbor
69	Ship to Ship and Harbor
70	**DIGITAL SELECTIVE CALLING ONLY (DSC)**
71	Ship to Ship and Harbor
72	Ship to Ship (Non-commercial)
73	Port Operations
74	Port Operations
77	Ship to Ship
78, 79, 80	Ship to Ship and Harbor
81-83	USCG Auxiliary
84	Harbor – Ship to Ship - Public Correspondence (Duplex)
85	Public Correspondence (Duplex)
86	Public Correspondence (Duplex)
87	Public Correspondence (Duplex)
88	Public Correspondence - Ship to Ship
WX1-7	NOAA Weather broadcasts – Receive Only

Table 19-2 International VHF Channels

Channels	Channel Designation
1	Public Correspondence, Port Operations, Ship to Ship, Movement
2	Public Correspondence, Port Operations, Ship to Ship, Movement
3	Public Correspondence, Port Operations, Ship to Ship, Movement
4	Public Correspondence, Port Operations, Ship to Ship, Movement
5	Public Correspondence, Port Operations, Ship to Ship, Movement
6	**SAR,** Ship to ship, Movement, Public, Port Operations
7	Public Correspondence, Port Operations, Ship to Ship, Movement
8	Public Correspondence, Port Operations, Ship to Ship, Movement
9	Public Correspondence, Port Operations, Ship to Ship, Movement
10	**SAR,** Ship to Ship, Movement, Public, Port Operations
11	Public Correspondence, Port Operations, Ship to Ship, Movement
12	Public Correspondence, Port Operations, Ship to ship, Movement
13	**Navigation Safety Communications,** Ship to ship
14	Public Correspondence, Port Operations, Ship Movement
15	Public Correspondence, Port Operations, Ship Movement
16	**DISTRESS, SAFETY and CALLING**
18	Public Correspondence, Port Operations, Ship Movement
19	Public Correspondence, Port Operations, Ship Movement
10	Public Correspondence, Port Operations, Ship Movement
21	Public Correspondence, Port Operations, Ship Movement
22	Public Correspondence, Port Operations, Ship Movement
23	Public Correspondence, Port Operations, Ship Movement
24	Public Correspondence, Ship Movement
25	Public Correspondence
26	Ship Movement, Ship to Ship
27	Ship Movement, Ship to Ship
28	Ship Movement, Ship to Ship
60 to 66	Public Correspondence
67	**SAR,** Ship to Ship
68	Ship Movement, Ship to Ship
69	Ship Movement, Ship to Ship
70	**DIGITAL SELECTIVE CALLING ONLY (DSC)**
71	Port Operations, Ship Movement
74	Port Operations, Ship Movement
72	Ship to Ship
77	Ship to Ship
73	**SAR,** Ship to Ship, Port Operations
78	Public Correspondence, Port Operations
81	Public Correspondence, Port Operations
79	Ship Movement, Public Correspondence, Port Operations
80	Ship Movement, Public Correspondence, Port Operations
82	Public Correspondence
83	Public Correspondence
84	Public Correspondence
85 to 88	Public Correspondence

19.26 VHF Frequencies. Rapid growth in cellular phone use has significantly reduced link call activity resulting in the closure of many coast stations. In the US the Maritel Company has opened a private network, and in the future there will be substantial US coastal VHF coverage with automated link call capabilities. In Europe Channel 06 is for intership business, Channel 77 for intership chat only. Where the boat is navigating within Vessel Traffic System (VTS) zones make sure you have the correct frequencies for contacting control stations.

a. US GREAT LAKES – CANADA ST LAWRENCE.
The Great Lakes and approaches operate on various frequencies. Call on Channel 16. Channels 24, 26, 27, 28 and 85 are the most often used. Traffic, harbor, port and bridge control usually 11, 12, 13 and Channel 14 for locks. Weather is broadcast continuously on WX1, WX2, WX3. **Buffalo** (Weather 22 0255, 1455); **Rochester** 25, 26; **Ripley** 17, 84, 86; **South Amherst/Lorain** (Weather 17 0002, 1102, 1702, 2302); Erie 25; **Cleveland** 28, 86, 87; **Toledo** 17, 25, 84, 87; **Detroit** 26, 28 (Weather 22 0135, 1335); **Port Huron** 25; **Grand Haven** (Weather 22 0235, 1435); **Harbor Beach** 17, 86, 87; **Bath City** 28; **Spruce** 17, 84, 87; **Frankfort** 28; **Milwaukee** (Weather 22 0255, 1455); **Michigan City** 25; **Chicago** 26, 27; **Port Washington** 17, 85, 87; **Sturgeon Bay** 28; **Hessel** 17, 84, 86; **Sault Sainte Marie** 26 (Weather 22 0005, 1205); **Grand Marais** 28; **Marquette** 28; **Copper Harbor** 86, 87; **Duluth** 28. **St Lawrence**: most stations 16 and primary channels 24, 26 or 27.

b. GULF of MEXICO
WLO on Ch 25, 28, 84, 87 with traffic lists on the hour. **St Petersburg** (Weather 22 1300, 2300); **Marathon** 24; **Naples** 25; **Cape Coral** 26; **Venice** 28; **Palmetto** 25, 27; **Tampa Bay** 86; **Clearwater** 24, 26; **Crystal River** 28; **Cedar Key** 26; **Panama City** 26; **Mobile** (Weather 22 1020, 12220, 1620, 2220); **Pensacola** 26; **Pascagoula** 27; **Gulfport** 28; **New Orleans** 24, 26, 27, 87 (Weather 22 1035, 1235, 1635, 2235); **Venice** 24, 27, 28, 86; **Leeville** 25, 85; **Houma** 28, 86; **Morgan City** 24, 26; **Lake Charles** 28, 84; **Port Arthur** 26, 27; **Galveston** 25, 86, 87 (Weather 22 1050, 1250, 1650, 2250); **Port Lavaca** 26, 85; **Corpus Christi** 26, 28 (Weather 22 1040, 1240, 1640, 2240); **South Padre Island** 26.

c. ATLANTIC COAST
Southwest Harbor 28 (Weather 22 1135, 2335); **Camden** 26, 27, 84; **Portland** 24, 28 (Nav 22 1105, 2305); **New Hampshire** 28; **Gloucester** 25; **Boston** 26, 27 (Weather 22 1035, 2235); **Hyannis** 28, 84; **Nantucket** 27, 85, 86; **Woods Hole** (Weather 22 1005, 2205); **New Bedford** 24, 26, 87; **Providence** 27, 28; **Bridgeport** 27; **Riverhead** 28; **Long Island Sound** (Weather 22 1120, 2320); **Bay Shore** 85; **Moriches** (Weather 22 0020, 1220); **New York** 25, 26, 84 (Weather 22 1050, 2250); **Sandy Hook** 24 (Weather 22 1020, 2220); **Bayville** 27; **Atlantic City** 26; **Cape May** (Weather 22 1103, 2303); **Philadelphia** 26; **Wilmington** 28; **Dover** 84; **Delaware Bay** 27; **Salisbury** 86; **Cambridge** 28; **Baltimore** 24 (Nav 22 0130, 1205); **Point Lookout** 26; **Norfolk** 25, 26, 27, 84; **Hampton** 25, 26, 27, 84; **Georgetown** 24; **Charleston** 26 (Weather 22 1200, 2200); **Savannah** 27, 28; **Brunswick** 24; **Jacksonville** 26; **Daytona Beach** 28;

Cocoa 26; **Vero Beach** 27; **West Palm Beach** 28, 85; **Boca Raton** 84; **Fort Lauderdale** 26, 84; **Miami** 24, 25 (Weather 22 1230, 2230); **Miami Beach** 85; **Homestead** 27, 28; **Key West** 24, 84 (Weather 22 1200, 2200).

d. **PACIFIC COAST (CANADA –CG)**
Vancouver 26; **Van Inlet, Barry Inlet, Rose Inlet, Holberg, Port Hardy, Alert Bay, Eliza Dome, Cape Lazo, Watts Pt, Lulu Is, Mt Parke, Port Alberni** Ch 26; **Dundas, My Hayes, Klemtu, Cumshewa, Naden Harbour, Calvert, Nootka, Mt Helmcken, Mt Newton, Bowen Is, Texada, Discovery Mt** Ch 84. Weather continuous broadcast on WX1, WX2, WX3, 21B.

e. **PACIFIC COAST (US)**
Bellingham 28, 85; **Camano Island** 24; **Seattle** 25, 26 (Weather 22 CG 0630, 1830); **Tacoma** 28; **Cosmopolis** 28; **Astoria** 24, 26 (Nav 22 CG 053, 1733); **Portland** (Nav 22 CG 1745); **Newport** 28; **Coos Bay** 25; **Brookings** 27; **Humboldt Bay** (Weather 22 CG 1615, 2315); **Casper** 28; **Point Reyes** 25; **San Francisco** 26, 84, 87; **Santa Cruz** 27; **Monterey** 28 (Weather 22 CG 1615, 2345); **Long Beach** (Weather 22 CG 0203, 1803); **Santa Barbara** 22, 86; **Avalon/San Pedro** 24, 26; **San Diego** 28, 86 (Nav 22 CG 0103, 1703).

19.27 Caribbean VHF Channels

a. **MEXICO. Chetumal** 26, 26; **Cozumel** 26, 27; **Cancun** 26, 27; **Veracruz** 26, 27. Channel 68 is channel for local cruiser nets.

b. **BERMUDA.** 27, 28. Coastal forecasts on Channels 10, 12, 16, 27, 38 at 1235 and 2035.

c. **BAHAMAS. Nassau** 16, 27; **8 Mile Rock** 27; **Exuma** 22 CG; **Marsh Harbor** 16. Forecasts every odd hour Ch 27. Cruisers Net operates on Channel 68 at 0815 with weather forecasts etc.

d. **CAYMAN ISLANDS.** Radio Cayman 1205, 89.9, 105.3 at 0320, 1130, 1220, 1230, 1330, 1910, 2320.

e. **JAMAICA. Kingston** 16, 26, 27. Forecasts for SW, NW, and Eastern Caribbean, and Jamaica coastal waters forecast at 0130, 1430, and 1900 on Channel 13.

f. **PUERTO RICO (USCG).** 16 **Santurce** 16, 26. NOAA forecasts broadcast continuously on VHF WX2 and VHF 22 at 1210 and 2210.

g. **VIRGIN ISLANDS (US). St Thomas** 16, 24, 25, 28, 84, 85, 87, 88. Forecast West North Atlantic, Caribbean and Gulf of Mexico on Ch 28 at 0000 and 1200. Channels 16, 24, 25, 28 (Traffic Lists), 84, 85, 87 and 88.

h. **VIRGIN ISLANDS (UK). Tortola** 16, 27. Weather on ZBVI Radio 780 at 0805, and every H+30 0730-1630, 1830-2130 LT.

i. **WINDWARD ISLANDS. Martinique (Fort-de-France), Guadeloupe** 16, 11. Warnings on receipt odd H+33 and VHF 26 and 27 every odd H+30. Weather messages VHF 26 and 27 at 0330 and 1430.

j. **BARBADOS.** 16, 26. Forecasts at 0050, 1250, 1650, 2050. Warnings on receipt and every 4 hrs for Caribbean, Antilles, Atlantic waters on Channel 26.

k. **GRENADA. St George's** 16, 06, 11, 12, 13, 22A. Forecast on request.

l. **TRINIDAD and TOBAGO.** 16, 24, 25, 26, 27. Forecast at 1340 and 2040.

19.28 Europe, UK, Mediterranean VHF Channels

a. **UNITED KINGDOM.** VHF MSI on Channels 10, 23, 84 or 86. (Inshore Forecast Times in Brackets). **Shetland** 0710 - 1910 (0110, 0410, 1010, 1310, 1610, 2210) **Aberdeen & Forth** 0730 - 1930 (0130, 0430, 1030, 1330, 1630, 2230) **Humber & Yarmouth** 0750 - 1950 (0150, 0450, 1050, 1350, 1650, 2250) **Thames & Dover** 0710 - 1910 (0110, 0410, 1010, 1310, 1610, 2210) **Solent & Portland** 0730 - 1930 (0130, 0430, 1030, 1330, 1630, 2230) **Brixham & Falmouth** 0710 - 1910 (0110, 0410, 1010, 1310, 1610, 2210) **Swansea & Milford Haven & Holyhead** 0750 - 1950 (0150, 0450, 1050, 1350, 1650, 2250) **Liverpool** 0730 - 1930 (0130, 0430, 1030, 1330, 1630, 2230) **Belfast** 0710 - 1910 (0110, 0410, 1010, 1310, 1610, 2210) **Clyde** 0810 - 2010 (0210, 0510, 1110, 1410, 1710, 2310) **Stornaway** 0710 - 1910 (0110, 0410, 1010, 1310, 1610, 2210).

b. **NETHERLANDS.** VHF 13 is for intership communications. 24 hour watch on **Schiermonnikoog** 5; **Brandaris** 5; **Den Helder** 12; **IJmuiden** 88; **Scheveningen** 21; **Hoek van Holland** 1, 3; **Ouddorp** 74; **Vlissingen** 14, 64. Weather forecasts for Dutch Coastal Waters and IJsselmeer at 08.05, 13.05 and 23.05 local time. (Local Time is UTC + 2 hrs March to October and UTC +1 October to March). Gale Warnings and Safety Messages are on Channel 23 and 83, after prior announcements on Ch16. Scheduled broadcast times are at 03.33, 07.33, 11.33, 15.33, 19.33 and 23.33 UTC.

c. **IRELAND.** Forecast on **Valentia** at 0103 and every 3 hours to 2203 on Channel 24. **Malin Head** on Channel 23 for **Fastnet, Shannon** and Irish coastal waters. **Valentia** MF on ITU Ch 278 and 280, **Malin Head** on ITU Ch 244 and Ch 255.

d. **CHANNEL ISLANDS (Jersey).** Weather at 0645, 0745, 1245, 1845, 2245 on Channels 25, 82.

e. **FRANCE.** Weather Bulletins all in French. Storm, Gale and Nav Warnings (times in brackets) in English and French. **Gris-Nez 79** (H+03); **Jobourg** 80 (H+03); **Corsen** 79 (H+03); **Etel** 80 (H+03); **Soulac** 16, 15, 67, 68, 73; **Agde, LaGarde, Corsica** 16, 11, 67, 68, 73; **Monaco** 16, 20, 22, 23, 86. Weather 0903, 1403, 1915.

f. **SPAIN.** Weather channel and time in brackets. **Bilbao** 26; **Santander** 24 (Ch11 @ 0245, 0645, 1045, 1445, 1845, 2245; **Cabo Peñas** 26; **Coruña** 26 (Ch26 @ 0803, 0833, 2003, 2033); **Finisterre** 01, 02 (Ch11 @ 0233 every 4 hrs); **Cádiz** 26; **Tarifa** 82 (Ch10, 74 @ 0900, 2100); **Málaga** 26; **Cabo Gata** 27; **Almeria** (Ch10, 74 @ every H+15); **Cartagena** 04; **Alicante** 0; **Valencia** (Ch10 @ every even Hr +15); **Algeciras** (Ch15, 74 @ 0315, 0515, 0715, 1115, 1515, 1915, 2315; **Ibiza** 03; **Palma** 07; **Menorca** 87; **Las Palmas**. Weather Forecast on 04, 05, 26, 28 at 0903, 1203 and 1803. **Arrecife** 25; **Fuerteventura** 22, 64; **Tenerife** 27; **La Palma** 22.

g. **PORTUGAL: Arga** 25, 28, 83; **Arestal** 24, 26, 85; **Monsanto** (Ch11 @ 0250, 0650, 1050, 1450, 1850, 2250); **Montejunto** 23, 27, 87; **Lisboa** 23, 25, 26, 27, 28; **Atalaia** 24, 26, 85; **Picos** 23, 27, 85; **Estoi** 24, 28, 86; **Sagres** (Ch11 @ 0835. 2035); **Acores**. Forecast on 16, 23, 26, 27, 28 at 0935 and 2135. **Madeira** 25, 26, 27, 28.

h. **ITALY: (Sardinia) Monte Serpeddi** 04; **Margine Rosso** 62; **Porto Cervo** 26; **(West Coast) Monte Bignone** 07; **Castellaccio** 25; **Zoagli** 27; **Gorgona** 26; **Monte Argentario** 01; **Monte Cavo** 25; **(West Coast/Sicily): Posillipo** 01; **Capri** 27; **Sera del Tuono** 25; **Forte Spuria** 88; **Cefal** 61; **Ustica** 84; **Erice** 81; **Pantelleria** 88; **Mazarra del Vallo** 25; **Gela** 26; **Siracusa** 85; **Campo Lato Altoi** 86; **Lampedusa Ponente** 25; **Crecale** 87; **Capa Armi** 62; **Ponta Stilo** 84; **Capo Colonna** 88; **Monte Parano** 26; **Abate Argento** 05; **Bari** 27; **Monte Calvario** 01; **(Adriatic). Silivi** 65; **Monte Secco** 87; **Forte Garibaldi** 25; **Ravenna** 27; **Monte Cero** 26; **Piancavallo** 01; **Conconello** 83. (Weather Bulletins on Channel 68 @ 0135, 0735, 1335, 1935; Nav and Gale Warnings on receipt H+03 and H+33, continuous in Northern Adriatic).

i. **CROATIA. Senj, Pula, Zadar** 10, 16; **Rijeka** 04, 16, 20, (24) 0535, 1435, 1935; **Split 16**, 21 (0545, 1245, 1945); **Dubrovnik** 07, 63, (04) (0625, 1320, 2120). Channels 67, 69, 73 continuous weather forecasts for Northern and Central Adriatic Sea updated 3 times/day in English.

j. **GREECE. Kerkyra** (02) 03, 64; **Kefallinia** 26, (27), 28; **Koryfu** 87; **Petalidi** 23, (83), 84; **Kythira** (85), 86; **Poros** 27, 28, 88; **Gerania** 02, 64; **Perama (Piraeus)** 25, 26, 86, 87; **Parnis** (25), 61, 62; **Lichada** 01 **Pilio** 03, (60); **Sfendami** (23), 24; **Tsoukalas** 26, 27; **Thasos** 25, 85; **Limnos** (82), 83; **Mytilini** (01), 02; **Chios** 85; **Andros** 24; **Syros** 03, (04); **Patmos** 84; **Milos** 82; **Thira** 26, 87; **Kythira** 85, 86; **Astypalea** 23; **Rodos** 01, (63); **Karpathos** 03; **Sitia** (85), 86; **Faistos** 26,27; **Moystakos** 04; **Knossos** (83), 84. Hellas Channels in brackets at 0600, 1000, 1600, 2200. Hellas Channel 86 has Wx Bulletin from Perama.

k. **TURKEY. Akcakoca** 01, 23; **Keltepe** 02, 24, 82; **Sarkoy** 05, 27; **Camlica** 03, 07, 25, 28; **Mahyadagi** 04, 26; **Kayalidag** 01, 23; **Akdag** 02, 24, 28; **Izmir** 16, 04, 24; **Antalya** 25, 27; **Dilektepe** 03, 07, 25; **Palamut** 04, 05, 26; **Yumrutepe** 01, 23; **Anamur** 03, 25; **Cobandede** 02, 26; **Markiz** 02, 24.

l. **CYPRUS. Olympos** 16, 26, 24, 25, 26; **Kionia, Pissouri, Lara** 25, 26, 27.

m. **MALTA.** 01, 02, 03, 04, 16, 28. Traffic Lists on 04.

19.29 **South Africa VHF Channels**

Capetown. Forecasts at 1333, 0948, 1948 on 01, 04, 23, 25, 26, 27, 84, 85 and 8. **Kosi Bay** 01; **Sodwana Bay** 03; **Cape S Lucia** 25; **Richards Bay** 28, 24; **Durban** 01, 26; **Mazeppa Bay** 28; **East London** 26; **Port Elizabeth** 87; **Kynsna** 23; **Albertinia** 86, 03; **Capetown** 01, 25; **Saldanha** 27; **Doringbaai** 87; **Port Nolloth** 01; **Alexander Bay** 04; **Walvis Bay** 16, 23, 26, 27.

19.30 **New Zealand VHF Channels**

Channel 16 and Working Channels 25, 67, 68, 69 or 71 Safety Information and Weather at 0533, 0733, 1033, 13333, 1733, 2133. **Cape Reinga** 68; **Kaitaia** 71; **Whangarei** 67; **Great Barrier Island** 25, 67, 68, 71; **Plenty** 68; **Runaway** 71; **Tolaga** 67; **Napier** 68; **Wairarapa** 67; **Wellington** 71; **Picton** 68; **Kaikoura** 67; **Akaroa** 68; **Waitaki** 67; **Chalmers** 71; **Bluff** 68; **Stewart Island** 71; **Puysegur** 67; **Fiordland** 71; **Fox** 67; **Greymouth** 68; **Westport** 71; **Farewell** 68; **D'Urville** 67; **Wanganui** 69; **Cape Egmont** 71; **Taranaki** 71; **Auckland** 71.

19.31 **Australia VHF Channels**

Channel 67 monitored in Darwin, Townsville, Mackay, Tin Can Bay, Mooloolaba, Whyte Island, Southport, Newcastle, Lake Macquarie, Sydney, Nowra, Melbourne, Perth, Rottnest Island.

1. **QUEENSLAND.** Bureau of Meteorology VHF Marine Forecasts **Townsville**. (Ch 72); **Home Hill** (80) at 2215, 0215, 0715 (0815, 1215, 1715 EST); **Rockhampton** (Ch21) at 1920, 2105, 2320, 0205, 0705 (0520, 0705, 0920, 1205, 1705 EST); **Rockhampton** (Ch 82) at 1940, 2050, 0150, 0650 (0540, 0650, 1150, 1650 EST); **Cairns** (Ch 81) at 1945, 2145, 2345, 0145, 0345, 0545, 0745 (0545, 0745, 0945, 1145, 1345, 1545, 1745 EST); **Mackay, (Brampton Is)** (Ch 21) 2015, 0215, 0645 (0615, 1215, 1645 EST); **Mackay (Knight Is)** (Ch 80) 2015, 0215, 0645 (0615, 1215, 1645 EST).

   Seaphone and Weather (0633 1633 EST) **Weipa** 03; **Torres Strait (Mia Is)** 26; **Thursday Is.** 66; **Torres Strait (Darnley Is)** 60; **Lochhart River** 28, 26; **Cooktown** 61; **Cairns** 27, 24 **Townsville** 26 23; **Ayr/Home Hill** 60; **Whitsunday Is.** 25, 28, 83, 86; **Shute Harbour** 66; **Mackay** 65 (0733 1803 EST); **Port Clinton** 01/04; **Yeppoon** 61; **Gladstone** 27, 24 **Fraser Island** 62; **Sunshine Coast** 25, 28; **Brisbane Central** 02; **Gold Coast** 23, 26.

2. **NEW SOUTH WALES** Seaphone and Weather (0648 1818 EST) **Coffs Harbour** 27; **Camden Haven** 62; **Port Stephens** 25, 28; **Lake Macquarie** 0; **Hawkesbury River** 02/05; **Sydney** 23, 26, 63; **Sydney South** 84; **Nowra** 27; **Eden** 86.

3. **VICTORIA** Seaphone and Weather (0803 1733 EST) **Lakes Entrance** 27; **Port Welshpool** 60; **Melbourne** 23, 26.

4. **TASMANIA** Seaphone and Weather (0803 1733 EST) **North Tasmania** 28; **Hobart** 07; **Bruny Island** 24, 27; **St Marys** 26.

5. **SOUTH AUSTRALIA** Seaphone and Weather (0748 1718 CST) **Adelaide** 23, 26; **Kangaroo Island** 61; **Port Lincoln** 24, 27.

6. **WESTERN AUSTRALIA.** Bureau of Meteorology VHF Marine Forecasts **Carnarvon** (Ch 73) 2145, 2345, 0415, 0815 0545, 0745, 1215, 1615 WST); **Esperance** (Ch72) 2235, 0435, 0835 (0635, 1235, 1635 WST); **Geraldton** (Ch 73) 2215, 0015, 0415, 0915 (0615, 0815, 1215, 1715 WST); **Broome** (Ch 72) 2240, 0910 (0640, 1710 WST).

   Seaphone and Weather (0633 1703 WST) **Rottnest Island** 60; **Jurien Bay** 62; **Geraldton** 28; **Carnarvon** 24; **Dampier** 26; **Port Headland** 27; **Broome** 28.

7. **NORTHERN TERRITORY** Seaphone and Weather (0803 1833 CST) **Darwin** 23, 26; **Gove** 28.

19.32 VHF Aerials. The majority of sailing yachts use whip aerials. The aerial length is directly related to the aerial gain, and the higher the gain, the narrower the transmission beam. A high gain antenna has a greater range, but during vessel rolling and pitching the lower gain antenna is more reliable with a greater coverage pattern. Half-wave whip aerials are typified by the stainless steel rod construction. The radiation pattern has a large vertical component, which suits boats under heel conditions. These antennas can also come in the form of a whip with lengths varying between 1 to 3 meters. The fiberglass whip effectively increases the height, and therefore the range of the radiating element. The gain is typically 3 dB. Helical aerials have a gain slightly less at 2.5 dB, but do have a characteristically wider signal beam-width. The higher the gain the more directional the emitted signal becomes. Sailing yachts usually have 6 dB aerials as they are more directional.

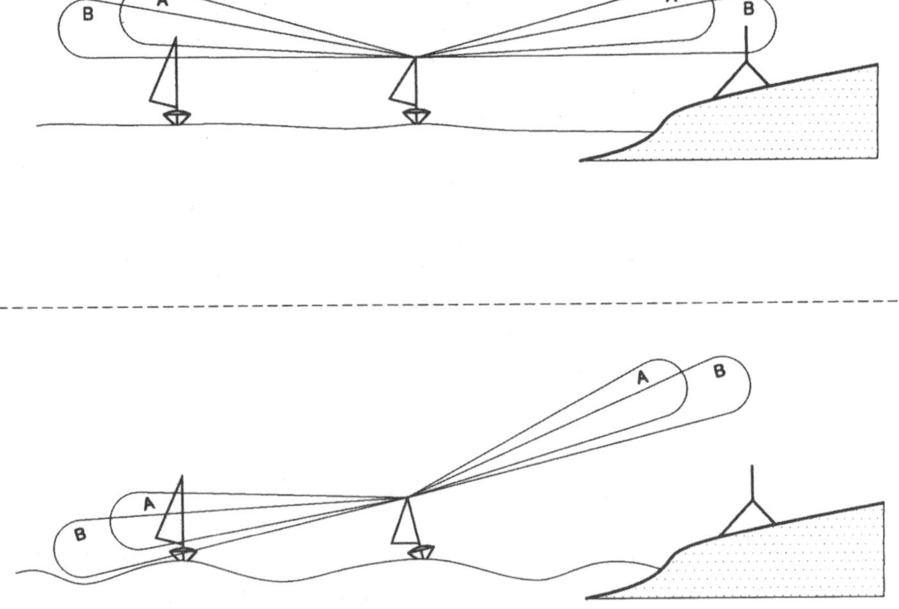

Figure 19-2 VHF Antenna Radiation and Aerials

19.33 Aerial Cables and Connections. Cables and bad connections are the principal causes of degraded performance. Avoid using thin RG58U coaxial cable where possible as the attenuation is increased and large signal loss can occur. The amount of signal that is transmitted depends on low losses within the cable, and the connections. For cabling aerials always use RG213/U or RG8/U 50 ohms for minimum attenuation. Ensure that the cable has no sharp bends that may affect the attenuation of the cable. The typical cable attenuation of both types for a 100-foot run is as follows:

a. **RG58/U.** This is a nominal 7.1 dB, a signal loss of aproximately 75–80%.

b. **RG8/U and RG8/X (RG213/U).** This is a nominal 2.6 dB, a signal loss of approximately 45%.

19.34 VHF Installation Testing. Many VHF installations operate poorly, with often undiagnosed problems. Many boaters install their own cables, connectors, and aerials, but in the majority of cases the installation is never tested. If the maximum range is to be realized, the installation requires proper testing. With the increasing reliance on new technology, in particular, with DSC VHF units, reliability is of crucial importance. In an earlier chapter I highlighted the importance of installing the correct coaxial cable to reduce losses. The attenuation inherent within the cable is only part of the loss equation, and the following should be observed.

Voltage Standing Wave Ratio (VSWR). When a signal is transmitted via the cable and aerial a portion of that signal energy will be reflected back to the transmitter. The effect is that coverage is reduced due to the reduced power output. Measure the VSWR with a meter. Up until recently you had to hire a technician to bring along an expensive meter (I am fortunate to possess a Bird meter), but you can now use a Shakespeare meter, the ART-1. This allows easy fault diagnosis and timely repairs, and is highly recommended. A number of problems can reduce the VSWR. Regular testing of reflected power and detection of excessive values will alert you to potential installation problems. It may even save your life.

(1) **Damaged or Cut Ground Shields.** This is common where the cable has been jointed, or improperly terminated at the connector. Make sure the shield is properly prepared and installed.

(2) **Dielectric Faults.** This common problem occurs when cables are run tightly around corners, through bulkheads, and through cable glands. Make sure the cables are bent with a relatively large radius. The tighter the bend, the more dielectric narrowing will occur with increased reflected power.

(3) **Pinched Cable.** This common problem occurs where a cable has not been properly passed through a bulkhead with the gland or connector. This impinges on the cable and reduces the dielectric diameter. Radio waves pass along the outside of the central core and along the inner side of the braiding, so any deformation will alter the inductance and reduce power output.

(4) **Connector Faults.** The most common problem is that of connectors not being installed or assembled correctly. Ensure that connectors are properly tightened, that pins are properly inserted, and that the pin-to-cable solder joint is sound and not a dry joint. Ensure that shield seals are properly made. Many connectors appear good at the time of assembly, but deteriorate very quickly when exposed to rain, salt spray, and corrosion. Check the status with a multimeter between the core and screen for short circuits.

(5) **Antenna Faults.** If an antenna is out of specification or suffered storm damage, or if a new antenna has been damaged in transit, the functional efficiency will decrease and losses increase. Inspect the antenna and connectors regularly. I always wrap the aerial connection with self-amalgamating tape to reduce ingress of moisture and salt air.

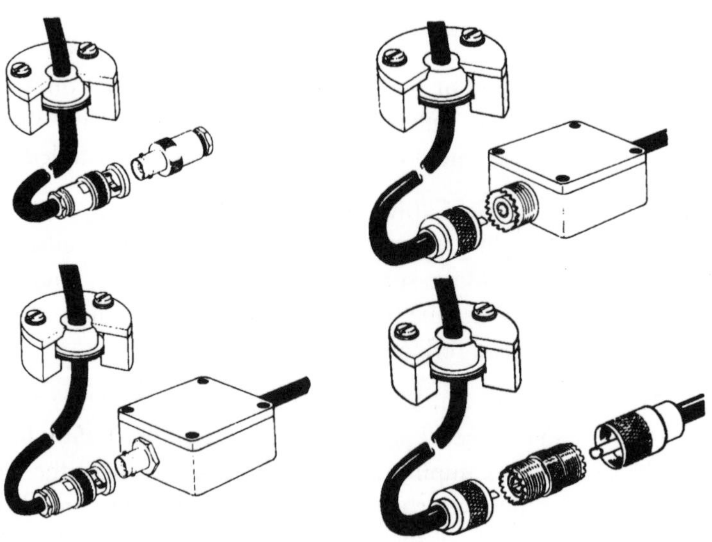

Figure 19-3 VHF Mast Connections

19.35 SSB/HF Radio. While HF is part of GMDSS many boaters keep using sets without the capability, in particular for e-mail and weatherfax. Long-range radio communications depend on radio frequencies in the High Frequency (HF) spectrum of 2 to 24 MHz.

a. **Signal Propagation.** Skywaves travel up until they reach the ionosphere and are bent and reflected back over a wide area. The ionosphere exists at a height of 30 to 300 miles (50 km to 500 km) above earth, and is formed by the ionization of air atoms by incoming UV ionizing radiation from the sun. The ionosphere is a weakly ionized plasma that is constantly changing and the changes alter the propagation characteristics of the radio waves. This is typified by the differences in night and day time transmission characteristics. Good HF communications depend on the utilization of these changing conditions with use of optimum frequencies. The ionosphere structure is divided into layers D, E, F1 and F2 in order of increasing height. There is a software package called HF-Prop, which is useful at predicting optimum transmission times worldwide.

(1) **F Layer.** The main reflecting layer is called the F layer and is approximately 200 mile (320 km) high. The layer is permanently ionized but during daylight hours energy from the sun causes the intervening layers E and D to form, and at night these reflect the highest radio frequencies in HF bands.

(2) **E and D Layers.** The signals reflected from these layers have lower ranges. Frequencies of 3 MHz or less are absorbed by the D layer, eliminate skywave propagation, and therefore 2 MHz is not favored.

(3) **Ground Wave.** Ground wave signals travel along the earth's surface but are absorbed or masked by other radio emissions.

(4) **Skip Zone.** The skip zone is the area between the transmission zone and the zone where the signal returns to earth, with generally negligible signal.

b. **Propagation Changes.** The ionosphere will affect each frequency differently. Extreme Ultra-Violet EUV radiation is responsible for forming and maintaining the ionosphere and depends on solar sunspot activity. When sunspot activity is lowest, the solar cycle EUV radiation is also weak and the density of charged particles in the F region is lowest. In this state only lower frequency HF signals can be reflected. At sun spot cycle peaks the EUV and ionosphere density are high and higher frequencies can be reflected. The season, time of day, and the latitude also affect HF radio communications. Solar flares produce high levels of electromagnetic radiation, and the X-ray component increases the D layer ionization. As HF communications use the F layer above they must transit the D layer twice during any signal skips. During a major solar flare the increased ionization results in a higher density of neutral particles and absorption of signal in the D layer. This is called sudden ionosphere disturbance (SID). It is characterized by increased attenuation of HF signals at lower frequencies.The disturbance is also called short wave fadeout, SSWF for sudden and GSWF for gradual. These events are synchronized with solar flare patterns. They are characterized by rapid onsets of just several minutes and declines of up to an hour or greater.

19.36 Space Weather Effects and HF Radio. Space weather has become very important in the satellite communications age. The underlying factor ruling space weather, at least in this end of the galaxy is our sun. The sun is by nature prone to dramatic and violent changes, with events such as solar flares, and the resultant blast streams of radiation and energized particles that stream towards earth. Space weather is caused by changes in the speed or density of the solar wind, and this is the continuous flow of charged particles that flow from the sun past the earth. The flow tends to distort the earth's magnetic field, compressing it in the direction of the sun and stretching it out in the opposite direction. The solar wind fluctuations cause variations in the strength and direction of the magnetic field near the earth's surface. Sudden variations are called geomagnetic disturbances. The electrical layers of the ionosphere are disrupted.

19.37 Solar Cycles. Space weather depends on an 11-year solar cycle. Cycles vary in both intensity and length, and the solar activity is characterized by the appearance of sunspots on the sun. Sunspots are regions of stronger magnetic field, and the solar maximum is the time when maximum spot numbers are visible. Sunspot numbers are quoted for average numbers over a 12-month period, and are the traditional measure of solar cycle status. Peak sunspot activity is calculated a little like the highest rainfall and is for recorded worst cases. Five of the last six solar cycles have been of high magnitudes. Cycle 19 in 1957 peak had a sunspot number of 201, the largest on record. Cycle 21 in 1979 had a peak sunspot number of 165 and was the second largest. Cycle 22 in 1989 was the third largest. The present Cycle 23, due to end in early 2007, peaked in 2000 and any major solar disruption should gradually diminish.

Frequency Preferences. The best ocean frequencies are on 4 MHz with ranges of up to 300 miles in day conditions, and thousands of miles at night without static at 2 Mhz. The principal characteristics are:

(1) **Sunset.** At sunset, the lower layer ionization decreases, and the D layer will disappear.

(2) **Dusk.** At dusk, the range increases on 2 MHz over thousands of miles, almost instantaneously. The interference levels are dramatically reduced.

(3) **Night.** The reflecting layer of the ionosphere rises at night, increasing the ranges for 4–6 MHz so lower frequencies are best at night. High frequencies are not good at night.

(4) **Day.** Low frequencies are weak during daytime. High frequencies are used in the daytime.

Table 19-3 SSB Optimum Transmission Times

Frequency (MHz)	Sunrise 0600	Noon 1200	Sunset 1800	Midnight 2400
22000	Average 100-2000 nm	Good 2000 nm plus	Good 2000 nm plus	Average 100-2000 nm
12000	Good 2000 nm plus	Good 2000 nm plus	Good 2000 nm plus	Good 2000 nm plus
8000	Good 2000 nm plus	Average 100-2000 nm	Average 100-2000 nm	Good 2000 nm plus
6000	Good 2000 nm plus	Average 100-2000 nm	Average 100-2000 nm	Good 2000 nm plus
4000	Average 100-2000 nm	Bad 50 nm	Bad 50 nm	Good 2000 nm plus
2000	Good 2000 nm plus	Bad 50 nm	Bad 50 nm	Good 2000 nm plus

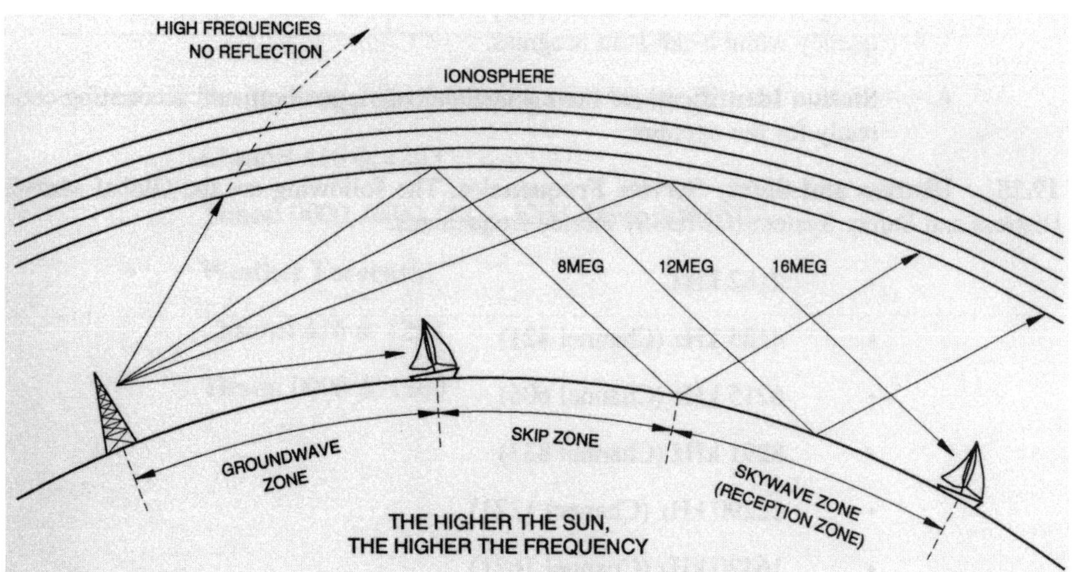

Figure 19-4 HF Radio Wave Behavior

19.38 Operation Requirements. There are certain legal requirements and operational procedures to observe.

a. **Ship Station Licensing.** Every vessel must have a license issued by the relevant communications authority. Transmitters must also be of a type approved by the appropriate authority. The issued call sign and vessel name must be used with all transmissions.

b. **Operator Licensing.** An operator's certificate is required and a test is given that covers knowledge of distress and safety procedures, and related marine communications matters.

19.39 HF Radio Frequencies and Bands. Always consult a current list of radio signals. The UK Admiralty List of Radio Signals (ALRS) is by far the most accurate, so invest in the relevant volume for any world location.

a. **Listen to Station.** If you can hear traffic clearly on the band you will probably have relatively good communications on that band.

b. **Monitor Bands.** Monitor the various bands and channels and determine the best peak period for communications. If the signal strength is good but the channel is busy, use a second channel if available, or wait. Do not tune equipment while a call is in progress.

c. **Station Identification.** Have name, call sign, position, and accounting code ready for the operator if required.

19.40 United States SSB Weather Frequencies. USCG CAMSPAC (Master Station Pacific) Honolulu (NMO) and Point Reyes (NMC) have very good weather transmissions.

Table 19-4 US Coast Guard Channels

ITU Channel Number	Receive Frequency	Transmit Frequency
	2182.0	
424	4426.0	4134.0
601	6501.0	6200.0
816	8764.0	8240.0
1205	13089.0	12242.0
1625	19314.0	16432.0

19.41 United States, Canada, Caribbean HF Frequencies. The U.S. Coast Guard broadcasts National Weather Service offshore forecasts and storm warnings on 2670 kHz after an initial announcement on 2182 kHz. Visit http://www.navcen.uscg.gov/marcomms

UNITED STATES

(1) Mobile (WLO) www.wloradio.com
Tx/Rx (kHz) 4077/4369, 4104/4396, 6218/6519, 8264/8788, 8280/8806, 12263/13110, 12305/13152, 16378/17260, 16480/17362, 22108/22804
ITU Channels 405, 414, 607, 824, 830. 1212, 1226, 1607, 1641, 2237
Weather and Traffic List Broadcast Times UTC Gulf of Mexico – 0400, 1300, 1600, 2200. SW N Atlantic – 0500, 1300, 1700, 2300. Caribbean Sea – 0600, 1300, 1800, 0000. East Pacific – 0300, 1400, 2000, Alaska Offshore – 0800, 1500
Radio Email Simplex Rx/Tx 6416.0, 8473.0, 12886.5, 13051.5, 116997.5, 22688.0

(2) Seattle (KLB) (Pacific and Alaska) www.wloradio.com
Tx/Rx4113/4405, 8207/8731/, 12254/13101, 16429/17311 kHz
ITU Channels 417, 805, 1209, 1624

(3) Chesapeake (NMN)
Frequencies 4125, 6215, 8291, 12290 kHz.
Weather Forecasts at 0330, 0515, 0930, 1115, 1530, 1715, 2130, 2315

(4) New Orleans (NMG)
Frequencies 4125, 6215, 8291, 12290 kHz.
Weather Forecasts at 0330, 0515, 0930, 1115, 1530, 1715, 2130, 2315

(5) Point Reyes (NMC) (MMSI 003669905)
Frequencies 4426, 8764, 13089, 17314, 4125, 6215, 8291, 12290
Weather Forecasts at 0430, 1030, 1630, 2230

(6) Kodiak (NOJ)
Frequencies 6501 kHz.
Weather Forecasts at 0203, 1645

(7) Honolulu (NMO) (MMSI 003669990)
Frequencies 6501, 8764, 13089 4125, 6215, 8291, and 12290 kHz.
Weather Forecasts at 0005, 0600, 1200, 1800

(8) Guam (NRV)
Frequencies 6501, 13089
Weather Forecasts at 0330, 0930, 1530, 2130

(9) Miami (NMA)
Frequencies 4125, 6215, 8291 and 12290 kHz.
Weather Forecasts at 0350, 1550.

(10) Boston (NMF)
Frequencies 4125, 6215, 8291 and 12290 kHz.
Weather Forecasts at 1035, 2235.

US NAVTEX. Coverage is relatively continuous on the East, West and Gulf coasts of the United States, and also around Kodiak Alaska, Guam and Puerto Rico. There is no coverage in the Great Lakes region. Much of the Great Lakes are covered by the Canadian Coast Guard.

NMF - Boston (Cape Cod) (F) 0445, 0845, 1245, 1645, 2045, 0045

NMN - Savannah (E) 0040, 0440, 0840, 1240, 1640, 2040

NMA - Miami (A) 0000, 0400, 0800, 1200, 1600, 2000

NMG - New Orleans (G) 0300, 0700, 1100, 1500, 1900, 2300

NMN - Chesapeake (N) 0130, 0530, 0930, 1330, 1730, 2130

NMR - Isabella San Juan (R) 0200, 0600, 1000, 1400, 1800, 2200

NMC - Pt Reyes CA (C) 0000, 0400, 0800, 1200, 1600, 2000

NMQ - Cambria CA (Q) 0045, 0445, 0845, 1245, 1645, 2045,

NMW - Astoria (W) 0130, 0530, 0930, 1330, 1730, 2130

NOJ - Kodiak (J) (X) 0300 (0340), 0700, 1100, 1500, 1900, 2300

NMO - Honolulu (O) 0040, 0440, 0840, 1240, 1640, 2040

NRV - Guam (V) 0100, 0500, 0900, 1300, 1700, 2100

CANADA

(1) **Prince Rupert (VAJ)**
Frequency 2054. Weather/Nav. 0105, 0705, 1305, 1905
Navtex (D) 0030, 0430, 08030, 1630, 2030

(2) **Tofino (VAE)**
Frequencies 2054, 4125. Weather/Nav. 0050, 0500, 0650, 1250, 1930, 1850, 2330.
Navtex (H) 0110, 0910, 1310, 1910, 2110.

(3) **St Lawrence (St Johns, Halifax, Sydney Placentia, Port Aix Basques, Rivière-au-Renard, S.Anthony)**
Frequencies 1514, 2538, 2582. Traffic lists on 2749, 2582.

(4) **Fundy (VAR) (MMSI 003160015)**
Frequencies 2182, 2749.
Weather/Na 0140, 1040, 1248, 1625, 1930, 1948, 2020.
Navtex (U) 0320, 0720, 1120, 1520, 1920, 2320.

(5) **Sydney (VCO)**
Frequencies 2182, 2749.
Weather/Nav 0033, 0733, 1433, 1503, 2133.
Navtex (Q) 0255, 0655, 1055, 1455, 1855, 2255.

(6) **Rivière-au-Renard (VCG)**
Frequencies 2182, 2598, 2749.
Weather/Nav 0437, 0847, 0937, 1407, 1937.
Navtex (C) 0020, 0420, 0820, 1220, 1620, 2020.

CARIBBEAN

(1) **Bermuda (Bermuda Harbor) (MMSI 003100001)**
Frequencies 2182, 2582 (ITU 410, 603, 819, 1220, 1618).
Weather Coastal forecast 1235, 2035 on www.weather.bm.
Navtex 0010, 0410, 0810, 1210, 1610, 2010.

(2) **Bahamas (Nassau)**
Frequencies 2182, 2522, 2588, 2522/2126.
Weather Forecasts every odd hour on 2522, storm and hurricane warnings are issued on receipt. Radio Bahamas 1540/1240/810, and 107.9-MHz broadcast detailed shipping weather reports M-F at 1205 hrs. Daily weather messages/synopsis 0815, 1315 and 1845 hrs.

(3) **Jamaica (Kingston)**
Frequencies 2182, 2587, 2590, 3535 ITU 405, 416, 605, 812, 1224.
Weather Coast Guard on 2738 kHz at 1330, 1830 hrs for SW, NW, and Eastern Caribbean, and Jamaica coastal waters forecast. Radio Jamaica on Montego Bay 550/104.5. Weather messages M-F 0015, 0340, 1104, 1235, 1909, and 2004 hrs. Jamaica B.C. on 560/620.700/93.3 MHz, fishing and weather forecast M-F 2248 hrs.

(4) **Puerto Rico (USCG) San Juan**
Frequencies 2182, 2670 (Santurce) 2182, 2530.
Weather Forecast at 0030, and 1430.
Navtex (R) 0200, 0600, 1000, 1400, 1800, 2200.

(5) **US Virgin Islands (St Thomas)**
Frequencies 2182, 2506/2009 ITU 401, 604, 605, 804, 809, 1201, 1202, 1602, 1603, 2223.
Weather Forecast West North Atlantic, Caribbean and Gulf of Mexico on 2506 at 0000 and 1200. Also at 1400, 1600, 1800 and 2000 forecasts for Virgin Islands Eastern Caribbean. Virgin Islands Radio on VHF 28, 85 at 0600, 1400, 2200. Detailed Caribbean weather reports. WIVI FM 99.5 MHz 0730, 0830, 1530 and 1630.

(6) **Curaçao (Netherlands Antilles)**
Frequencies 2182, 8725.1.
Weather Forecast at 1305.
Navtex (H) 0110, 0510, 0910, 1310, 1910, 2110.

(7) Barbados
Frequencies. 2182, 2582, 2723, 2805. ITU. 407, 816 (Traffic Lists), 825, 1213, and 1640.
Weather Forecasts at 0050, 1250, 1650, 2050. Warnings on receipt and every 4 hours for Caribbean, Antilles, and adjacent Atlantic waters. Caribbean Ham Weather Net (8P60M). Broadcasts out of Barbados on 21.400 MHz daily at 1300 hr. Receives positions 1300-1330. Translates RFI WFs 1330-1400.

(8) Martinique (Windward Islands)
Frequencies 2182, 2545.
Weather Warnings odd H+33. Weather messages at 2545 at 1333.

(9) Grenada (Windward Islands)
Frequencies 2182, 1040, 3365, 5010, 1508, 7850.
Weather Forecast at 2100-0215 (1040), 2230-0215 (3365), 2100-2230 (5010), 2100-0215 (15085). GBC Radio on 535 and 15105. Hurricane warnings on receipt and every H+30 after news 0200, 1030, 1130, 1630, 2030, 2230.

(10) Trinidad and Tobago
Frequencies 2182, 2735, 2049, 3165.
Weather 1250 and 1850.

(11) Caribbean SSB Weather Nets. Synoptic forecasts and analysis including hurricane information and tracks for all of Caribbean. Times are all UTC. Frequencies 4003 kHz at 1215 to 1230, 8104 kHz at 1230 to 1300, in the hurricane season also 8107 kHz at 2215 to 2245.

19.42 English Channel and Atlantic Frequencies. The following are selected frequencies for navigational warnings, weather forecast and working frequencies.

(1) NETHERLANDS (CG Radio) (MMSI 002442000)
Frequencies 2182, MF DSC 2187.5, 3673.
Weather North Sea forecasts at 09.40 and 21.40 hrs UTC on 3673 kHz. Gale warnings are made on receipt. Scheduled broadcast times at 03.33, 07.33, 11.33, 15.33, 19.33 and 23.33 UTC.
Navtex (P) 0230, 0630, 1030, 1430, 1830, 2230.

(2) BELGIUM (Oostende Radio) (MMSI 0020050480)
Frequencies 2182, 2761.
Weather Forecast at 0820 and 1920.
Navtex (M) 0200, 0600, 1000, 1400, 1800, 2200 (for Dover Straits).

(3) UNITED KINGDOM
MF MSI broadcasts, Shipping Forecasts etc **Shetland** (2226kHz) 0710 - 1910 (0110, 0410, 1010, 1310, 1610, 2210) **Aberdeen & Forth** (2226kHz) 0730 - 1930 (0130, 0430, 1030, 1330, 1630, 2230) **Humber & Yarmouth** (2226kHz) 0750 - 1950 (0150, 0450, 1050, 1350, 1650, 2250) **Brixham & Falmouth** (2226kHz) 0710 - 1910 (0110, 0410, 1010, 1310, 1610, 2210) **Clyde** (1883kHz) 0810 - 2010 (0210, 0510, 1110, 1410, 1710, 2310) **Stornaway** (1743kHz) 0710 - 1910 (0110, 0410, 1010, 1310, 1610, 2210.

NAVTEX
Cullercoats (G) Gale Warnings 0100, 0500 0900, 1300, 1700, 2100 24 Hr Synopsis 0900, 2100.
Niton (E) Gale Warnings 0040, 0440, 0840, 1240, 1640, 2040 24 Hr Synopsis 0840, 2040.
Portpatrick (O) Gale Warnings 0220, 0620, 1020, 1420, 1820, 2220 24 Hr Synopsis 0620, 1820.

(4) CHANNEL ISLANDS
Jersey Radio. Frequencies 2182, 1926.
Weather. 0645, 0745, 1245, 1845, 2245.

(5) IRELAND
Valentia Radio. Frequencies 2182, 1952.
Weather. 0233, 0303, 0633, 0903, 1033, 1433, 1503, 1833, 2103.
Navtex (W). 0340, 0740, 1140, 1540, 1940, 23403.

(6) FRANCE
Boulogne-sur-Mer. Frequencies 2182, 1970, 1692, 1694.
Weather. 1970 - H + 03 and H + 33 and 1692 @ 0703, 1833.
Gris-Nez (Cross) (MMSI 002275100); 1650, 2182, 2677.
Etel (Cross) (MMSI 002275000); **Soulac (Cross)** 2182, 2677.
Brest Radio. 1635, 1671, 1876, 2691, 1862.
Weather 1635 - H + 03 and H + 33 and 1671, 1876, 2691, 1862 @ 0733, 1803 and 1671, 1876 @ 0600.
St Nazaire Radio. 2182, 1671, 1876, 1922, 2691, 2740.
Weather. 0333, 0733, 0803, 1133, 1533, 1833, 1933, 2133.

(7) SPAIN (Port Operations on Ch 18, 19, 20, 21, 22, 79, 80).
Coruna Radio. 2182, 1698.
Weather 0803, 0833, 1233, 1933 (Bay of Biscay), 2003.
Navtex (D). 0030, 0430, 0830, 1230, 1630, 2030.
Finisterre Radio. 2182, 1964.
Weather 0803, 0833, 1203, 1903 (Bay of Biscay), 2033.
Tarifa Radio. 2182, 1904.
Weather 0803, 0833, 1233, 1933, 2003.
Navtex (G). 0100, 0500, 0900, 1300, 1900, 2100.
Navtex (X). 0350, 0750, 1150, 1550, 1950, 2350 (Valencia).
Islas Canarias (Las Palmas Radio) 2182, 1689, 2045, 2048, 2114, 2191. ITU 406, 604. Gale warnings and forecast on 1689, 2820, 4372, 6510 at 0903, 1203 and 1803.
Navtex (I). 0120, 0520, 0920, 1320, 1920, 2120.

(8) **PORTUGAL**

Lisboa. 2182, ITU 802, 813, 1203, 1207, 1615, 1632, Traffic lists on 13083 at even H +05.
Apulia Radio. 2182, 2657. Weather 0735, 1535, 2335.
Sagres Radio. 2182, 2657. Weather 0835, 2035.
Navtex (R) 0250, 0650, 1050, 1450, 1850, 2250.
Acores (Faial) Horta Radio. 2182, 1663.5, 1683.5, 2657, 2742, 2748, 4434.9/4140.5. Weather Warnings and forecast 0935 and 2135.
Navtex (F) 0050, 0450, 0850, 1250, 1650, 2050.
Madeira Radio. Frequencies 2182, 2843, 2657. Weather 0905, 2105.

19.43 Mediterranean Radio Frequencies. The following frequencies are for the principal Mediterranean areas.

(1) **SPAIN**

Tarifa 2182, 1904/2129.
Malaga 2182, 1656/2081.
Cabo de Gata Radio 2182, 1967. Wx.0803, 0833, 1233, 1933, 2033.
Palma Majorca (Palma Radio) 2182, 1955.
Weather. 0803, 0833, 1203, 1903, 2033.

(2) **FRANCE**

Nice Radio 1350. Weather. 0725, 1850.
Marseille Radio 675. Weather. 0725, 1850.
Monaco (UTC + 1) 2182, 4363, 8728, 13146 (ITU Med - Ch 403, 804, 1224, 1607, 2225; Atlantic 403, 830, 1226, 1628, 2225).
Weather. On receipt and at H+03. Forecast at 0903, 1403, 1915. On 8728 @ 0715 and 1830, 13146 on request, Atlantic Bulletin on 8806, 13152, 19232 and 22846 @ 0930. Coastal continuous on 161.750M. Nav info at 0803 and 2103.

(3) **ITALY**

Cagliari
Frequencies 2182, 1922 (Traffic list on 2680, 2683).
Weather 0125, 0725, 1325, 1925.
Navtex (T). 0310, 0710, 1110, 1510, 1910, 2310.
Navtex (V). 0330, 0730, 1130, 1530, 1930, 2330 (Augusta)

Porto Torres (*2719*) **Genova** (1667, 2642, *2722*) **Livorno** (1925, *2591*) **Civitavecchia** (*1888*, 2710, 3747) **Napoli** (1675, *2632*, 3735) **Palermo** (*1852*) **Mazara** (1883, *2211, 2600*) **Lampedusa** (*1876*) **Augusta** (1643, *2628*) **Crotone** (1915, *2663*) **S.Benedetto** (*1855*).

Messina Radio 2182, 2789.
Weather 0135, 0233, 0633, 0735, 1133, 1335, 1533, 1933, 1935.

Roma 4292, 8520, 13011, 19160.8.
Weather 0348, 0948, 1518, 2118; Fleet Wx 0830, 2030.
Navtex (R). 0250, 0650, 1050, 1450, 1850, 2250

Bari Radio (Adriatic). 2182, 2579. Wx. 0125, 0725, 1325, 1925.

Ancona Radio (Adriatic). 2182, 2656. Wx. 0148, 0748, 1348, 1948.

Trieste Radio. 2182, 2624. Wx. 0848, 1218, 1648, 2048.
Navtex (U) 0320, 0720, 1120, 1520, 1920, 2320.

Venezia Radio 2182, 2698. Wx. 0135, 0403, 0735, 0903, 1303, 1335, 1935, 2103.

(4) CROATIA (Adriatic)
Rijeka Frequencies. 2182, 1641, 1656.
Dubrovnik Radio. Frequencies 160.95M. Wx. 0625, 1320, 2120.
Split. Frequencies 160.95M. Wx. 0545, 1245, 1945.
Navtex (O) 0240, 0640, 1040, 1440, 2240.

(5) **GREECE** (Hellas Radio MMSI 002371000)
Kerkyra. 2182, 2830 (Traffic 2607, 2792, 3613).
Weather 0703, 0903, 1533, 2133.
Navtex (K). 0140, 0540, 0940, 1340, 1940, 2140.
Limnos 2182, 2730.
Weather 0033, 0633, 1033, 1633.
Navtex (L). 0150, 0550, 0950, 1350, 1950, 2150.
Rodos 2182, 2624.
Weather 0703, 0903, 1533, 2133.
Iraklion Kritis 2182, 1615.5/2150.5, 2799, 1942, 3640.
Weather 0703, 0903, 1533, 2133.
Navtex (H). 0110, 0510, 0910, 1310, 1910, 2110.
Athinai 2182, 1695, 1967, 2590, (8743).
Weather 0703, 0903, (1215), 1533, (2015), 2133.

(6) TURKEY
Izmir 1850, 2182, 2760.
Weather 0333, 0733, 1133, 1533, 1933, 2333.
Navtex (I). 0120, 0520, 0920, 1320, 1920, 2120.
Antalya. 2182, 2187.5, 2670.
Navtex (F). 0050, 0450, 0850, 1250, 1650, 2050.

(7) CYPRUS (MMSI 002091000)
Frequencies. 2182, 2187.5, 2670, 2700, 3690. ITU 406, 414, 426, 603, 807, 818, 820, 829, 1201, 1208, 1230, 1603.
Weather. 2700 kHz @ 0733, 1533.
Navtex (M). 0200, 0600, 1000, 1400, 1800, 2200.

(8) MALTA
Frequencies. 2182, 2625. ITU 410, 603, 832, 1216, 1233.
Weather. 0103, 0603, 1003, 1603, 2103.
Navtex 0220, 0620, 1020, 1420, 1820, 2220.

19.44 Australia and New Zealand Radio Frequencies.

(1) Australian Radio Frequencies and Weather Forecasts.

24 hour Distress, Safety, Urgency watches on 4126, 6215, 8291 kHz and navigation warnings on 8176 kHz twice daily. Special announcements on VMC/VMW 5 minutes to every hour (25 minutes after the hour CST).

Charleville VMC (Australian Weather East)
(0700-1800) 4426, 8176, 12365, 16546 (1800-0700) 2201, 6507, 8176, 12365. Coastal waters forecasts and warnings for QLD, NSW, VIC, TAS and SA. Every hour commencing 0000 EST (0030 CST). High Seas Forecasts and Warnings for Northern, NE and SE Areas. Every hour commencing 0000 EST (0030 CST).

Wiluna VMW (Australian Weather West)
(0700-1800) 4149, 8113, 12362, 16528 (1800-0700) 2056, 6230, 8113, 12362. Coastal waters forecasts and warnings for QLD Gulf, NT, WA and SA. Every hour commencing 0000 WST (0030 CST).). High Seas Forecasts and Warnings for Northern, Western and SE Areas. Every hour commencing 0000 WST (0030 CST).

Penta Comstat (VZX) (http://web.mac.com/pentacomstat/)

Penta Comstat provides SailMail Association email over HF radio. Voice communications ceased on 30 September 2006 after 30 years of operation. The continuous SailMail service is on 12 frequencies - 2824, 4162, 5085.8, 6357, 8442, 10476.2, 12680, 13513.8, 14436.2, 16908, 18594, 22649 for Eastern Australia and the South Pacific. http://www.sailmail.com/

(2) New Zealand (Taupo Maritime Radio) MMSI 005120010

Distress and Safety 4207.5, 6312.0, 8414.5, 12557.0, 16804.5.
Frequencies 2182, 2207, 4125, 4146, 6215, 6224, 8297, 12290, 12356, 16420, 16531.
Weather (NZST) Coastal Warnings and Bulletins 0133, 0533, 1333, 1733 on 2182, 4125, 6215. Coastal Reports 0803, 1203 2003 on 2182, 4125, 6215. Oceanic Warnings at 0303 and 1503 on 6215, 12290, 6224, 12356. At 0333, 133 on 8291, 16420, 8297, 16531. At 1503, 1533, 2103, 2133, 2103, 2133 on 6215, 12290, 6224, 12356. At 0903, 2103 on 6215, 12290, 6224, 12356. At 0933, 2133 on 8291, 16420, 8297, 16531.

19.45 South Africa Radio Frequencies.

SOUTH AFRICA (Capetown) DSC: MMSI 006010001
Frequencies 2182, 1964, 4435, 2191, 19338, 22711.
ITU 405, 421, 427, 801, 805, 821, 1209, 1221, 1608, 1621, 1633, 2204, 2206, 2221.
Weather at 1333, 0948, 1948.
Navtex (C) 0020, 0420, 0820, 1220, 1620 and 2020 UTC.

19.46 HF Radio Tuner Units. The function of the tuner unit is to match the antenna length to the frequency being used:

a. **Manual.** There are still manual tuner units around although they have been largely phased out by fully synthesized systems with automatic tuner units. These require matching the antennas by adjusting tune and load controls using a built-in tune meter.

b. **Fully Synthesized Units.** The new synthesized radio sets with automatic tuner units enable boaters to communicate easily. Units consist of a full range of ITU EPROM controlled frequencies. The tuner unit essentially consists of inductors and capacitors that are automatically switched in series or parallel with the antenna to achieve the correct tuned length.

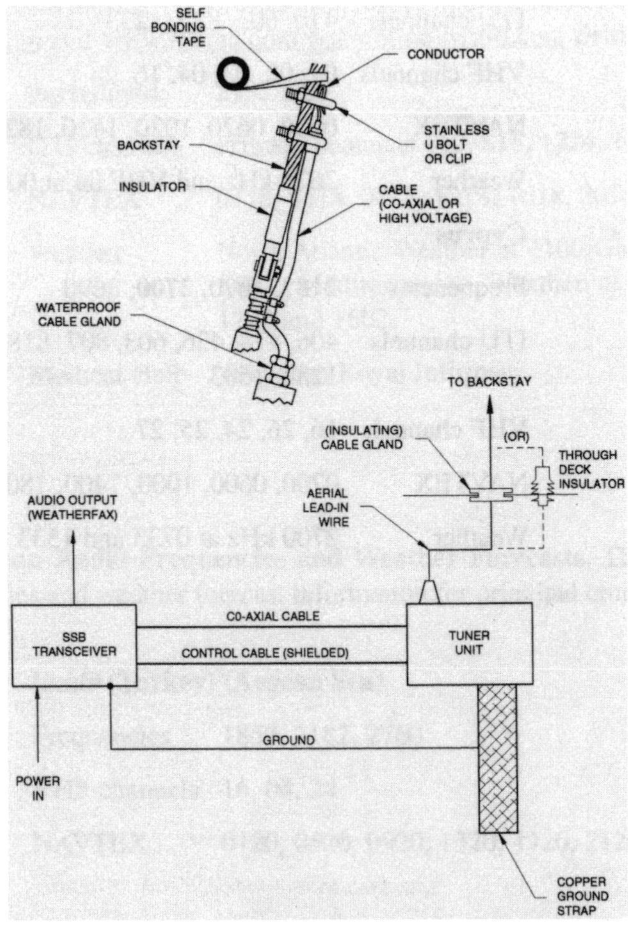

Figure 19-5 Tuner Unit and Aerial Connections

19.47 HF Radio Aerials. The aerials are crucial to proper performance of the HF radio. The whip generally operates over a wider frequency range than wire line aerials seen on sailing boats.

a. **Loaded Whip.** These aerials have loading coils, and are generally very long.

b. **Unloaded Whip.** These whips have a similar performance to long wire backstay aerials and the ATU provides the required aerial length. As the voltage and currents can be significant at the base, it is essential that high quality insulators and well insulated feed line cables are used to minimize losses. A very low resistance ground system is required.

c. **Backstay.** The insulated backstay is the most common on sailing boats. They do find some use on trawler motor yachts as well, often in a triatic stay arrangement and they are most efficient in the 2–8 meg range. Losses can occur here as well, as signal radiates into the mast and rigging. It should be at least 11 meters long for an effective aerial. The insulators should be free of chips and have long leakage paths.

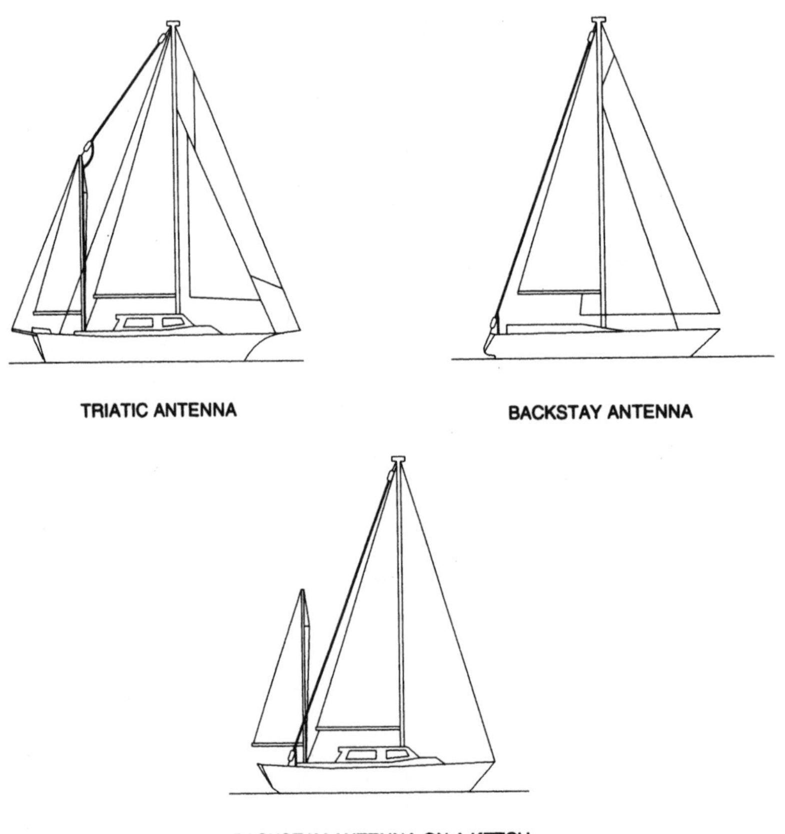

Figure 19-6 SSB Aerial Arrangements

d. **Aerial Feed Line.** The feed line to the aerial is very important as resistance degrades the transmission signals.

(1) **Feed line Cables.** Thin conductors and bad joints result in conductor heating and losses. Ideally the cable should not run close to metal decks or hull.

(2) **Insulation Quality.** Insulation losses also occur through conductors and deck feed insulators. Use cables with good insulation values, such as a silicon insulated high voltage cable.

(3) **Deck Transits.** Poorly insulated leads close to metal decks and hull can cause arcing or induction losses. External cables can also leak when the insulation cracks due to UV rays. The best system in steel vessels is the use of through deck insulators. These offer long leakage paths and less signal loss, but they must be kept clean.

(4) **Backstay Connections.** The feed line to aerial connection must be made properly.

19.48 HF Radio Grounds. HF radio problems of transmission and reception are often caused by inadequate grounding systems. Remember that the ground plane is an integral part of the aerial system. If it is poor, you may not be able to tune properly to required frequencies.

a. **Ground Shoes.** Ground shoes are the most effective method of providing an RF ground on fiberglass and wooden boats. They provide half of the required aerial length and are an integral part of the radiating system.

b. **Internal Copper Mesh.** Glass and wooden vessels may avoid the installation of ground plates by glassing in a large sheet of copper mesh (Newmar).

c. **Copper Straps.** The interconnecting copper strap from tuner unit to ground plane is essential. It must be strap, not cable and the surface area is the critical factor. To be effective, a low resistance is required and is the cause of many performance drops and interference. The ground strap should be 2" wide at least. The copper strap should be run clear of bilge areas.

19.49 HF Radio Maintenance. There are regular maintenance tasks that will ensure good radio performance.

a. **Aerial Connections.** The lead wire aerial connections should be regularly checked for deterioration. If exposed, the wire may degrade, and introduce resistance into the circuit. Always tape up the connection with self-amalgamating tape.

b. **Insulators.** Always clean the insulators to remove salt deposits that encrust and cause surface leakages. This should include the upper insulator on wire antennas. A damp rag is the best tool.

c. **Ground Connection.** Check the RF ground connections. Clean and tighten the bolts and connection surfaces. After this, apply a light smear of petroleum jelly to prevent deterioration in the bilge area. Always check and keep this area clean and dry, as reaction between the copper strap and metalwork can cause corrosion problems.

19.50 HF Radio Troubleshooting. Basic troubleshooting is as follows:

Table 19-5 - HF Radio Troubleshooting

Symptom	Probable Fault
No reception	Wrong channel selected
	Propagation problems
	Aerial lead wire broken
	Aerial connection corroded
	Tuner unit fault
Poor reception	Propagation problems
	Aerial connection corroded
	Insulators encrusted with signal leakage
	Aerial grounding out
No transmission	Tuner unit fault
	Aerial connection corroded
	Insulators encrusted with signal leakage
	Aerial grounding out
	Aerial lead wire broken
	Ground connection corroded
	Low battery voltage
	Transceiver fault
Poor transmission	Propagation problems
	Aerial connection corroded
	Insulators encrusted with signal leakage
	Aerial grounding out
	Tuner unit fault
	Ground connection corroded

19.51 Standard Time Frequencies. The NIST broadcasts time and frequency from WWV and WWVH in the US. This is also known as the time tick and is used in celestial navigation. It also broadcasts an hourly voice broadcast of current high seas storm warnings for the Atlantic, Pacific and Gulf of Mexico which are provided by the NWS. Discrimination between the 2 stations is easy, WWV has a male voice and WWVH a female voice.

a. **WWV (Fort Collins).** Times 8th, 9th and 10th minute past the hour on 2.5, 5, 10, 15 and 20 MHz. Atlantic high seas warnings 8 and 9th minute and Pacific 10 mins past the hour.

b. **WWVH (Kauai, Hawaii).** Times 48–51 minutes past the hour on 2.5, 5, 10 and 15 MHz. Pacific high seas warnings 48–51 mins.

c. **VNG (Llandilo, Australia).** Times are announced continuously on 2.5, 5.0, 8.638, 12.984 MHz. 2200-1000 UTC on 16 MHz. Voice broadcasts on 2.5, 5 and 16 MHz on 15th, 30th, 45th and 60th minute.

19.52 Amateur (Ham) Radio. Ham radio is the realm of a worldwide group of radio enthusiasts. Ham operators have been involved in many life saving efforts with sailors, but regrettably ham operators and the system have been badly abused. Ham operators were crucial in the recent tsunami and many other disasters. Ham radios are a major communication source in the cruising world. In the US, about 70% of cruisers sail with ham, while for the UK and Australia it is probably around 10%. There are a number of important factors to consider.

a. **Operator Licensing.** It is the operator not the station that is licensed. There are a number of levels that give either partial or full access to frequencies. Levels require examination in Morse, radio theory, rules and regulations with respect to operations. Fear of technical matters and theory as well as the Morse test frightens off many would be amateurs. A general class license will be required for access to Maritime Mobile Nets in the 15-, 20- and 40-meter bands.

b. **Penalties.** You must be licensed for the country of operation. Be aware that in some third world countries where communications are controlled, jail and vessel loss can occur if it is used in port and without authorization. In many cases you will not be acknowledged on ham bands unless you are licensed and have a call sign.

c. **SSB vs. Ham.** This is a perpetual argument with both systems having a place. Carry both or a combined Ham/SSB unit.

(1) **SSB.** Radio sets are generally easier to operate for non-technical people, and with automatic tuning, it is simple to punch in a channel number and talk. Additionally radios have automatic emergency channel selection, and radios are type approved for marine communications. Only a restricted license or permit is required. You may operate a SSB radio on amateur frequencies if you have a ham licence. One of the disadvantages of SSB on ham frequencies is that synthesizers are programmed in 0.10kHz steps. Ham communications may be at frequencies outside of that so that SSB sets can be marginally off frequency. Most SSB sets operate on upper side band (USB) while most frequencies below 40 meters are lower side band (LSB).

(2) **Ham.** The ham operator must have a license appropriate to the frequency band being worked. Access to GMDSS emergency frequencies is illegal except in emergencies. It is illegal to operate non type-approved radios such as ham radios on marine frequencies. Ham allows the use of casual conversation which marine SSB does not. Ham allows full access to information packed nets, and a world-wide communications network. Ham does not readily allow access to telephone networks, although some stations and net controllers may offer phone patches.

19.53 Ham Nets. A good receiver allows listening to ham nets and valuable information as marine SSB sets cannot access them. The following maritime mobile net times could vary an hour either way depending on summer time changes in respective countries, 14.314 is monitored virtually 24 hours, and is the de facto maritime mobile international calling frequency.

Table 19-6 - Atlantic/Caribbean/ Mediterranean Nets

Zulu	Frequency	Call Sign	Net Name and Area
	14.325		Hurricane Net
0100	3.935		Gulf Coast Hurricane Net
0200	14.334		Brasil Net
0230	14.313	K6QTR	Seafarers Net
0530	14.303		Sweden Net
0645	12.353		Greece Net
0530	7.088		Eastern Med Net
0700	7.085		Mediterranean MM net
0700	14.313		German MM Net
0700	14.303		International Net
0800	14.303		Caribbean - UK Net and 1200, 1800
0900	14.313	G3TJY	Mediterranean Net
0900	7.080		Canary Island Net (Atlantic)
1000	14.303		German Net
1030	3.815		Caribbean WX Net
1030	14.265		Barbados Cruising Net
1100	7.241		Caribbean Net St Croix
1100	14.283		Caribus
1110	3.930		Puerto Rico/Virgin Islands Weather
1125	7.086		Caribbean Weather Net Details
1130	3.815		Antilles Emergency Weather Net
1130	14.320		South Africa MM Net (South Atlantic)
1200	14.118		Le Reseau du Capitaine Net Canada
1220	3.696		Bahamas Weather Net
1245	14.121		Mississauga Maritime Net
1245	7.268		Waterway Net WRCC (US East Coast/Caribbean)
1300	21.400		Transatlantic Net (operates in crossing season)
1300	7.085		Central American Breakfast Net
1330	8.152		Cruiseheimers Net – E Coast to Caribbean
1400	8.188		Northwest Caribbean Cruisers Net
1400	7.292		Florida Coast Net
1600	14.313		US Coast Guard Net
1600	14.300		Maritime Mobile Service Net
1630	14.313	G3TJY	German MM Net (Worldwide Winlink/Echolink)
1900	7.240		Bejuka Net. (Central America)
1800	14.303		MM Net (Atlantic weather forecast)
1800	14.303		UK MM Net
1945	12.359	N6GYR	Herb Hilgenbergs Southbound II Net
2000	14.297		Italian MM Net (Atlantic, Brasil, Africa)
2000	14.303		Sweden Net
2040	7.190		Admirals Net (US West Coast)
2200	14.300		UK Mobile M Net
2200	21.404		Central American Net
2330	3.815		Barbados/Trinidad Net

Table 19-7 - Pacific/Asia/Indian Ocean Nets

Zulu	Frequency	Call Sign	Net Name and Area
0025	14.323		MM Net SE Asia (Japan to Seychelles)
0100	21.407	W6BYS	MM Net (Pacific/Indian Ocean)
0200	7.290	KH6B	Hawaii Inter-island Net (MF)
0200	14.305		California Hawaii Net
1200	21.402		Jerrys Happy Hour
0220	14.315	VK9JA	John's Weather Net (Norfolk Is and Pacific)
0230	14.313		Pacific Seafarers Net
0300	14.313	VE7CEM	DDD (Doers, Dunners and Dreamers) Net
0400	**14.340**		**MARITIME EMERGENCY NET**
0400	12.353		Russell Radio NZ Bora Bora to Australia (also 1630 hrs)
0400	14.318		Arnolds Net (Weather Pacific) From Cook Islands
0400	12.356		Onerahi Yacht Club Weather from New Zealand
0400	14.313		Pacific Maritime Net
0500	14.316		Tonys Net
0500	8.101		Radio Peri-Peri East Africa (also 12.353 at 1500)
0500	21.200	VK3PA	Aus/NZ/Africa Net (Indian and Pacific Ocean)
0530	14.314	WH6ANH	Pacific MM Net (Covers all Pacific via relay stations)
0630	14.316		South Africa MM Net (Moves to 7.045)
0630	14.330	255MU	Durban Net (Indian Ocean)
0630	14.180	VR6TC	Pitcairn Net
0700	14.220		Pacific Net
0700	14.310		Mariana - Guam Net
0715	3.820	ZL1BKD	Bay of Islands Net (Sth Pacific/Australia)
0800	14.315	P29JM	Pacific Inter-island Net (Micronesia/Hawaii)
0800	7.238		PST Baja California MM Net (PDT Time)
1000	14.320	HG3BA	Dixie's Net MM (Philippines, Weather NW Pacific)
1000	14.330		Pacific Gunkholers Net
1000	14.315		Robbys Net (Australia) also at 2300 hrs
1115	14.320		Roys Net (West Aust) NW Indian Ocean
1130	14.316	WB8JDR	South Africa MM net
1200	14.320	WB8JDR	SE Asian Net
1430	3.963	WA6VZH	Sonrisa Net (Baja California)
1430	7.294		Chubasco net
1545	14.340		Marquesas Net
1600	11.825		International SW Tahiti
1630	21.350		Pitcairn Net
1730	14.115		DDD Pacific Net (Canadians)
1800	**14.340**		**MARITIME EMERGENCY NET**
1900	14.329		MM Hawaii Net
1900	14.115	VR6TC	Pitcairn Net
1900	14.329	KH6OE	Skippers Net
1900	14.340	K6VDV	California Hawaii Net
1900	14.305	N6GYR	Confusion Net
1900	14.340	KA7HYA	Manana MM Net (West Coast to Hawaii)
1930	14.115	VE7CEM	Jerry's Net
1800	14.282	WA2CPX	South Pacific Sailing Net
1900	21.390		MM's Pacific Net
1900	7.285		Shamaru Net (Hawaii)
1900	14.329	ZL1BKD	Bay of Islands Net
1900	7.288		Friendly Net (Hawaii)
1900	3.990		Northwest MM Net (NW Pacific)
1900	14.340		Mariana Net
2030	7.087		Comedy Net Southwest Pacific Australia
2100	14.315	ZL1ATE	Tony's Net South Pacific
2130	14.318		Daytime Pacific Net
2200	21.412	KH6CO	Pacific Maritime Net (2300 in winter)
2400	14.320	VS6BE	SEA MM Net (Rowdy's Net, SW Pacific/SE Asia)

19.54 E-Mail Services. For many "snail mail" is a thing of the past. For most boaters an INMARSAT terminal is not a viable economic alternative, although GMDSS inspired changes make communications improvements essential. If, you have a quality SSB radio on board, that valuable piece of equipment is your means to get connected to the world.

a. **HF E-Mail System Components.** The basic components are:

(1) **SSB Radio.** Not all SSB radios are configured for e-mail and may require modification to operate, with the addition of an audio output jack. This should provide a line level output signal of 100mV RMS. Radios such as the ICOM M710 are e-mail ready. Radios must be able to transmit full power signal without damage, however older sets including ICOM M700, SEA 235, SGC SG2000 cannot do this, so they must be operated at reduced output power levels. A good power supply is essential to maintain constant transmission and battery voltages must be up and power supply connections sound. Aerials and ATU grounds must also be good to ensure optimum transmission and reception.

(2) **HF/SSB Modem.** Modems are generally part of the service providers' systems, although those using other non-service company systems such as packet radio enthusiasts use what is called a terminal node controller (TNC); the most common modems are those from Kantronics such as the KamPlus and the Kam98. A modem has a power input, data port, and radio port, along with operating software. The recommended SailMail modem is the SCS PTC-II, as it is compact and has lower power consumption and it also has faster speeds. The audio cable to the SSB consists of 4 wires: transmit audio (TxD); receive audio (RxD), push-to-talk (PTT), and the audio signal ground. The audio cable must be shielded with the shield being connected at both ends. Pre-wired cables are available from Kantronics. Clip-on Ferrites must also be fitted at both ends to reduce RF interference, and also coax line isolators (ungrounded T-4 model), and these are available from www.radioworks.com.

(3) **Notebook/Laptop Computer.** Many boaters are incorporating this as an essential part of the equipment inventory so the addition of an e-mail function further enhances the investment.

(4) **Software.** Software is required and AirMail is a proven Windows based message package. You prepare messages using the text editor and attach word processing files with point-and-click simplicity, as well as automating the radio link. You can download AirMail for use with SailMail from the SailMail website.

b. **Transmission System Modes and Configurations.** Both the principal service providers and alternative systems utilize different methods for handling e-mail traffic. Although similar equipment is used, the systems cannot communicate with each other.

(1) **Clover.** These modems are used by PinOak and are made by HAL Communications in the US. They use a four-tone signal and are used in the PinOak PODLink-e service. Currently Globe establishes a link in SITOR (marine telex) and then switches over to Clover mode. PinOak does not use SITOR but establishes links either in Clover or PacTOR 2.

(2) **PacTOR 2.** These modems are made by SCS in Germany. They use a two-tone signal and are far more effective and reliable with data transfer in noisy environments. Effectively, they are a hybrid Packet/Amtor modem. They are becoming the preferred modem type for use in most marine HF e-mail systems. The new PinOak PODLink-f service utilizes PacTOR modems. PacTOR is replacing Amtor communications due to improved capabilities and is supported by many Aplink stations.

19.55 E-Mail Service Providers. There are several ways to connect to e-mail via SSB/HF, SailMail and WinLink being the best. They offer comprehensive services with GMDSS that require the installation of satellite systems, such as weather and navigational warnings.

a. **Sailmail.** Check out www.sailmail.com for complete details. SailMail system works best with the SCS PTC-IIpro or SCS PTC-IIex. The PTC-IIpro has a radio control output which allows remote tuning of same SSB radios.
WRD719, Palo Alto, CA 5881.4, 7971.4, 10343, 13971, 18624 kHz.
WQAB964, San Diego, CA 2759, 5740, 7380, 10206, 13874, 18390, 23060
WHV861, San Luis Obispo, CA 2800.4, 5861.4, 8020.4, 10320, 10982, 13946, 18296
WHV382, Friday Harbor, WA 2794.4, 5830, 7995, 10315, 13940, 8277
KUZ533, Honolulu, HI 2710.4, 5836, 7957.4, 10325, 13930, 18264
KZN508, Rockhill, South Carolina 2656.4, 5876.4, 7961.4, 7981.4, 10331, 13998, 18618, 18630 kHz.
WPTG385 Corpus Christi, TX:2720.8, 5859.4, 7941.4, 10361.4, 13906.4, 13926.4, 18376.4, 22881.4
WPUC469 South Daytona, FL 2807.8, 5897.4, 8009.4, 10366.4, 13921.4, 18381.4, 22961.4
XJN714, Lunenburg, NS, Canada 4805, 7822, 10523, 13937, 1443.2, 18234, 21866
VZX, Firefly, NSW, Australia 2824, 4162, 5085.8, 6357, 8442, 10476.2, 12680, 13513.8, 14436.2, 16908, 18594, 22649.
RC01, South Africa 12689, 13992, 14588, 18630, 22212, 27777
V8V2222 Brunei Bay, Brunei 5212, 6305, 8399, 10323, 13426, 14987, 16786, 18893, 20373, 22352

OSY Brugge, Belgium 63309.5, 8422, 12580.5, 16684.5
HPPM1 Chiriqui, Panama 2650, 4075, 5735, 8185, 10450, 13880, 18240, 18440, 18460, 23050
CEV773 Los Lagos, Chile 2828.5, 5266.5, 10620, 10623, 13861.5, 13875
SSM678Abu Tig Marina, El Gouna Egypt 2824.5, 4162.5, 6239.5, 8325.5, 12394.5, 16598.5, 18866.5, 22645.5

b. **WinLink 2000.** This is a radio digital message transfer system for e-mail transfer with attachments, position reporting, graphic and text-based weather bulletins and emergency communications and is available to amateur radio operators by linking radio to the Internet. http://winlink.org/

c. **AMTOR (Amateur Teletype Over Radio).** This is probably the cheapest option that I have seen in wide use. The system uses what is termed Amtor Packet Link (Aplink). These Aplink stations are ham stations configured for automatic reception, storage, and transmission of Amtor messages. Messages are transferred between stations until the designated destination station is reached. Addressing mail requires the recipient MBO (Electronic Mail Box) details. What I found most attractive with on-board systems using this system is the ability to talk with other vessels on a chat net. Log on to www.airmail2000.com and www.shortwave.co.uk for useful information and www.win-net.org for Ham e-mail shore stations.

d. **PinOak Digital.** Stations located worldwide include Galapagos, Falkland Islands, Capetown, Cape Verde Islands, Grand Banks, West Greenland, Eastern Mediterranean, Sri Lanka, Hawaii, Tahiti, Wellington, South China Sea, Perth, and others. Users are charged a subscription fee, which allows a specific amount of data transfer, and then a charge per kilobit transferred.

e. **Globe Wireless.** They are primarily providers with stations in San Francisco, New Orleans, Hawaii, Bahrain, Sweden, Newfoundland, Australia, and New Zealand. Service offered is called GlobeEmail, along with GPS position reporting tied with USCG AMVER system. Users are charged a subscription fee, which allows a specific amount of data transfer, and then a charge per kilobit transferred. Message reception is similar with automatic notification. Log on to www.globewireless.com.

f. **Alternative E-Mail Systems.** The main systems offer a seagoing system, but there are other useful options.

(1) **Pocketmail.** I have some friends who regularly message me from their boat in Europe using this system. PocketMail is a mobile e-mail service that allows access from anywhere and can be used with a GSM cell phone. Use the PocketMail composer, dial a toll free number, then hold to the phone receiver. Also it can be used with Palm PDA's. Log on to www.pocketmail.com.

(2) **SeaMail.** Xaxero provides unrestricted service to the Pacific. Free software at http://www.seamail.org/ It operates through Kumeu Radio ZMH302 Auckland New Zealand to the Pacific. In USB kHz 6380.3, 8485.8, 12706.3, 16952.3. Use a PACTOR and PACTOR II. Log on to www.xaxero.com for details and software downloads.

(3) **CruiseEmail.** They have 9 operating stations in the U.S. and Caribbean, with extended service onto the North and South Pacific. They also offer Internet access, mail forwarding, free propagation bulletins, current weather condition, updates and offshore surface analysis. http://www.cruiseemail.com/

g. **Wireless Hot Spots.** This system still requires a notebook computer with an appropriate modem card installed or wireless equipped and enabled. Wireless hotspots are expanding. I have used many times an unsecured wireless network and sent mail. Switch on this function and refresh your network list. In Europe there are chains developing in waterways and marinas. Check these sites out for useful location information. http://www.wififreespot.com/ and http://www.wi-fihotspotlist.com/ also in the UK http://www.teleadapt.com/shop/wireless.php and check out http://www.hotspot-locations.com/ and http://metrofreefi.com/

h. **Internet Cyber Cafes and E-Mail Centers.** Many marinas now offer access to e-mail and Internet. Log on to www.ipass.com for information on how to access your ISP back home when traveling by calling a local telephone number. In most cases, you can find an Internet café virtually everywhere, just log on to Yahoo or Hotmail and create an account. I have switched to Yahoo as they have 6 meg accounts as compared to Hotmail and 2 megs. I am at MrCruisine@yahoo.com. Say hello and give me some feedback.

19.56 Satellite Services. There are a variety of services available and these are described with their salient features. The system selected will depend on many issues that include the services required, the coverage area, the initial installation costs, antenna sizes, and the call costs. Log on to www.heavens-above.com for information on satellite orbits and tracking.

a. **INMARSAT.** This is based on INternational MARitime SATellite Organization. The system comprises four satellites in geostationary orbit (35,600km high), and the satellites remain in the same position relative to the earth. There are 4 Ocean Regions (Atlantic East, Atlantic West, Pacific and Indian). Currently new generation satellites are entering service. The new I-4 satellites have an additional 228 narrow spot beams, which will form the backbone of future broadband services. This will include the Broadband Global Area Network (BGAN), which will deliver Internet and intranet content, video-on-demand, videoconferencing, fax, e-mail, phone and LAN access at speeds of up to 492kbps.

(1) **Inmarsat A.** Provides two-way direct-dial phone connection and fax, telex and data services at 9.6kbps up to 64kbps. Services are being replaced by new digital services and will cease to be supported on December 31, 2007.

(2) **Inmarsat B.** Supports global voice, telex, fax and data at speeds from 9.6kbps to 64kbps, as well as GMDSS-compliant distress and safety functions.

(3) **Inmarsat M.** provides global telephone and 2.4kbps data on a medium-sized antenna, includes voice, data transfer and fax.

(4) **Inmarsat C and Mini-C.** Has two-way, store-and-forward packet data communication via a lightweight, low-cost terminal. This is the only system that can allow ships to meet most GMDSS communications requirements. Antennas are small (25cm) and gyro-stabilized. They use SIM card pre-paid call billing options. They are not global coverage and utilize "spot beam" technology and the signal is beamed to specific areas, footprint coverage is typically 200 nm offshore.

(5) **Inmarsat Mini-M.** The smallest telephone it offers direct-dial voice, fax and data communication to 2.4kbps and uses spot beam.

(6) **Inmarsat D+.** This is a 2-way data communication service using small terminal. They are fitted with GPS-assisted satellite roaming. Used for data transfer, remote monitoring, tracking.

b. **Globalstar.** It consists of 48 low earth orbit (LEO) (1500km high) satellites. This uses a handset similar to cell phone with larger antenna or a remote one. Services include high quality voice, short messaging services (SMS) and roaming. Also dial-up fax and data services are available. Coverage is limited to coastal areas to around 200nm offshore, which will suit most sailing people. They now have an enhanced service covering Florida, the Gulf of Mexico, and the Caribbean, including the Bahamas. This is not a GMDSS system. www.globalstar.com.

c. **Iridium.** It consists of 66 satellites in low earth orbit (LEO) and offers global voice, data, fax, and paging services. The system is now operated by Boeing for the US Department of Defense. This is not a GMDSS system. It offers voice services to 85% of the world landmass. With computer can provide Internet access and e-mail, and also fax and enhanced messaging service and messaging services. Log on to www.iridium.com.

d. **ORBCOMM.** The system consists of 30 LEO satellites. The service is digital data two-way communications only and offers paging, e-mail, etc. It is not a GMDSS system and details are at www.orbcomm.net.

e. **Thuraya.** This system has two and a planned third satellite in geostationary orbit offering some 13,500 telephone channels, on 250 spot beams. Service offers voice, fax, data, short messaging. www.thuraya.com/

f. **Emsat.** Provides voice and fax communication, low speed data transmission and positioning. It was originally developed for EU fishing boats. Uses a single geostationary orbit satellite Artemis. Coverage is limited to the Mediterranean and Northern Europe and operated by Eutelsat. www.eutelsat.com.

g. **VSAT.** (Very Small Aperture Terminals) They use Ku-band geostationary satellites such as Eutelsat Atlantic Bird 1 and W3A and e-mail service, Voice and data simultaneous connections, Voice over IP (VoIP) and Internet access. Transmission bit rate from 8 Kb/s to 128 Kb/s and reception bit rate of up to 2 Mb/s (comparable with ADSL). The antenna radome is 1 meter across and is large for a yacht. www.eutelsat.com.

h. **Motient.** It is a satellite-based, digital mobile communications system which offers voice, fax and data services. They use 2 MSAT satellites and coverage is in North and Central America, Northern South America, the Caribbean, Hawaii and in coastal waters.

19.57 Satellite System Installation. Systems consist of the radome, which encloses a stabilized antenna dish, a pedestal control unit (PCU) and the RF unit. Follow installation instructions in the user's manual precisely as warranty may be voided if installation is incorrect.

a. **Radome Installation.** The radome must be located as far as practicable from any HF and VHF antenna, and preferably a minimum of 5 meters from all other communications and navigation receiver antennas. Do not mount the unit in any location subject to vibration, and as some are mounted on stern posts make sure it's well supported. Safe compass distance is a minimum of 1 meter. Systems also require radiation precautions and 2 meters above to avoid excessive microwave radiation. The radome should be outside the beamwidth of radar antennas, typically 10 degrees on each side of the central plane. The radome should be properly aligned parallel with the boat's axis. As beamwidth is around 10°, a clear line of sight is required from 5° elevation and above. Obstructions will create blind spots, and disrupt communications. Obstructions less than 15cm are acceptable within 3m of the antenna, but note that marginal signal strengths are vulnerable to them. The azimuth and elevation angles must be considered at all times. Normal cable installation rules apply, and must be observed to prevent mechanical damage. The antenna unit uses double-screened 50 ohm coaxial cable. This is usually RG223/U and RG214, with maximum lengths of 13m and 25m respectively to achieve 10dB/0.6 ohm maximum losses and attenuation. All cables must be shielded, and the shield grounded. Peripheral equipment must also be grounded. Co-axial connectors must be put on correctly, and this is a frequent cause of problems.

b. **Operation.** In normal operation, the dish auto-tracks the satellite and must be aligned correctly. The nominated satellite is based on the boat's position, and then selection of a relevant satellite within area coverage. The vessel heading is required to give correct azimuth heading and the gyro or fluxgate input provides this. The azimuth angle is the angle from North and horizontal satellite direction. The elevation angle is the satellite height above the horizon with respect to the vessel. At power-up, the system must locate a satellite and synchronize with it. This is either by automatic or manual initiated hemispheric scan for selected or ocean region satellite. The dish does a search pattern until the satellite signal is located in the relevant ocean region. Systems also carry out a self-test at initialization. Systems default to last settings on gyro, azimuth and elevation, and if the vessel position is lost, this data is required. Under GMDSS the default LES and Distress Alarm address must be configured. During operation, displays on handsets show signal quality, and signal strength Signal/Noise Ratio (S/N). Bit error rates (BER) decrease with increased signal quality.

19.58 Weatherfax Receivers. Weather facsimile gives access to many stations that transmit weather charts and the charts are a lot easier to interpret than foreign language voice forecasts. Transmitted data is varied and includes ocean current positions, sea temperatures charts, and current weather charts every six hours, forecasts up to five days in advance, sea state and swell forecasts and ionosphere propagation forecasts.

a. **Weather Facsimile Components.** A facsimile transmission consists of a number of distinct components:

(1) **Continuous Carrier.** This single tone is emitted before the start of any broadcast. It allows the receiver to be tuned to maximum signal strength prior to data reception.

(2) **Start Tone.** Also called the Index of Co-operation (IOC) select tone, this enables receivers to recognize the start of a transmission and to select the appropriate IOC drum speed.

(3) **Phasing Tone.** This tone is used to synchronize the edge of the transmitted image.

(4) **Scale Tone.** Some systems enable the tone variations within the broadcast to be selected or varied.

(5) **Body of Transmission.** This characteristic rhythmic "crunching" tone is the facsimile data being transmitted for decoding into an image.

(6) **Stop Tone.** The stop tone is similar to a start tone and indicates the end of the transmission.

(7) **Close Carrier.** This tone follows the conclusion of the transmission.

b. **Decoders.** To obtain weatherfax data it is necessary to obtain signals via a SSB or short-wave radio, and decode it so that it can be displayed on a laptop computer, or printer as required. The basic function of a decoder is to convert transmitted audio signals into data. The audio signal is taken from the audio jack if fitted or a terminal on the rear of the SSB set.

c. **Printers.** An ink-jet printer can be utilized from the laptop computer. As a plain paper printer, it is significantly cheaper to operate than thermal paper roll types. One of the factors to consider is both the ease of printing and the size and quality required. Make sure you carry enough spare paper and ink cartridges for your trip, as often these items are hard to procure.

d. **Discrete Systems.** On larger vessels an integrated decoder and printer system is often used. The most common are Furuno and the smaller ICS Fax-2. A paper roll lasts a considerable period. The unit also has a number of useful features. An additional aerial can be added for full Navtex reception and the reception of RTTY and FEC signals is possible. Like most weatherfax units, programming of specific reception times is possible which takes all the worry out of looking up and catching broadcast times.

e. **Power Consumption.** The power consumption rate is relatively low, although it should take into account SSB consumption as well if a unit is left on permanently to capture programmed transmissions. If power consumption is an issue, you will have to power up before the required broadcast and after receiving shutdown again. The combination of decoder and SSB over 24 hours can be at least 25–30 amp hours, which is considerable. Typical power rates are as follows:

   (1) **Standby Listening Mode.** The ICS Fax-2 unit has a drain of only 2.5 watts. The SEA SSB unit power consumption is 2 amps, while the 322 model is only 1.0 amp.

   (2) **Print Mode.** The power drain increases to approximately 4 amps when printing. The SSB drain remains the same unless the audio is turned up and it can be around another 0.5A.

19.59 Computer Based Weather Systems. The computer offers a range of weather information options:

a. **MScanMeteo and MScanFAX.** This is an economical and simple package with reception limited to weatherfax, RTTY, CW and Navtex, which suits most users. A useful function is the reception of SYNOP, which is the current weather reports, transmitted over RTTY, and is the raw data used by forecasters. It plots observational data as it is transmitted in real time. Data can include wind speed and direction, temperature, barometric pressure and cloud cover. When all information is displayed, the package automatically generates isobars and isotherms on screen. These signals are decoded into text reports and a plot is automatically made to give the very latest weather situations.

b. **ICS Fax 6.** This uses advanced digital signal processing to track weak signals with high resolution on-screen graphics of charts. It is designed for use with any quality HF radio receiver and incorporates facility for the remote control of several models. Compatible with most versions of the Windows operating system, this software also utilizes the sound card. The built-in automatic reception scheduler allows fully automated operation across a wide range of radio frequencies and message types over a 24-hour period.

c. **Xaxero Weather Fax 2000.** A simple package, this is an easy-to-use hardware and software system for PCs. It offers multitasking system with weather fax frequency database, Navtex, Morse Code and telex capability.

d. **HF Fax.** This leading package uses a demodulator and has auto tuning. Signal tracking is possible. Once an image is received, you can zoom in, scroll or save to disc, or print out if required to name a few functions. The programs are also capable of receiving and printing RTTY, FEC and CW (Morse Code) modes, as well as Navtex transmissions. System has mouse control capability, and can automatically control frequency of Lowe and Icom receivers.

e. **SeaTTY.** Version v1.60 of this program receives weather reports, navigational warnings and weather charts transmitted in RTTY, NAVTEX and HF-FAX (WEFAX) modes on longwave and shortwave bands. No additional hardware is required and a receiver and computer with a sound card is a prerequisite.

f. **Computer Fax Problems.** HF reception criteria apply to receiving quality signals. The greatest problem with computer fax reception is noise generated from the computer and picked up through the audio demodulator line.

g **Troubleshooting.** Make sure that the frequency is accurately tuned. If instability and drifting occur the signal will also be inconsistent. Often weatherfax reception problems can indicate that aerial and earth connections in the SSB system are defective. Check these out first. The frequencies being used may be affected by adverse propagation conditions, which effect all HF transmissions. It is advisable to tune to another frequency and try again, or wait until conditions improve. If this is a regular problem, obtain propagation forecasts. Check all sources of noise, that will include fluoro lights, motors etc.

Table 19-8 Weather Facsimile Frequencies

Station	Frequencies
Kodiak (US) NOJ	2054, 4298, 8459, 12412.5
Point Reyes (US) NMC	4346, 8682, 12786, 17151.2, 22527
Honolulu (Hawaii) KVM70	9982.5, 11090, 16135
Boston (US) NMF	4235, 6340.5, 9110, 12750
New Orleans (US) NMG	4317.9, 8503.9, 12789.9, 17146.4
Halifax (Canada) CFH	122.5, 4271, 6496.4, 10536, 13510
Northwood (UK) GYA	2618, 4610, 8040, 11086.5
Northwood (Gulf) GYA	3289.5, 6834, 14436, 18261
Charleville (Aust) VMC	2628, 5100, 11030, 13920, 20469
Wiluna (Australia) VMW	5755, 7535, 10555, 15615, 18060
Wellington (N. Zealand)	3247.4, 5807, 9459, 13550.5, 16340.1
Cape Naval (South Africa)	4014, 7508, 13538, 18238
Nairobi (Kenya)	9044.9, 17447.5
Offenbach (Germany)	3855, 7880, 13882.5
Skamlebaek (Denmark)	5850, 9360, 13855, 17510
Rome (Italy)	4777.5, 8146.6, 13597.4
Athens (Greece) SVJ4	4481, 8105
Moscow (Russia)	3830, 5008, 6987, 7695, 10980, 12961, 11617
Tashkent (Uzbekistan)	3690, 4365, 5890, 7570, 9340, 14982.5
JJC (Japan/Singapore)	4316, 8467.5, 12745.5, 16971, 17069.6, 22542
Tokyo (Japan) JMH	3622.5. 7305, 13597
Beijing (China) BAF	5526.9, 8121.9, 10116.9, 14366.9, 16025.9, 18236.9
Shanghai (China) BDF	3241, 5100, 7420, 11420, 18940
Seoul (Korea) HLL2	5385, 5857.5, 7433.5, 9165, 13570
Taipei (Taiwan)	4616, 5250, 8140, 13900, 18560
Bangkok (Thailand)	7396.8, 17520
New Delhi (India)	7404.9, 14842
Rio de Janeiro (Brasil)	12665, 16978
Valparaiso (Chile) CBV	4228, 8677, 19146.4

NOTE: In upper side band mode, adjust frequency 1900Hz (1.9kHz) lower, if in lower side band mode adjust 1900Hz (1.9kHz) higher.

19.60 Cellular Telephones. The rapid development of the mobile cellular telephone has made personal communications in coastal waters much easier. Many of us are fortunate to be able to utilize GSM technology. I can use my phone in, and offshore of, more than 30 countries. This technology has not been without a price. The rapid drop in placement of link calls has meant the closure of many coast stations and repeaters. I have had some interesting contacts, on one occasion I had a call from a yacht regarding a charging problem, and after enquiring about his location, he said he was some 2000 miles to the north, and 10nm offshore. On another occasion a skipper in a major offshore yacht race called me during the race and described his problem and we nursed him to the finish line. More recently there have been text message SOS calls from Indonesia to UK setting off SAR via Australia, and people in life rafts calling for help. The cellular must be put into perspective and it must be emphasized that the cell phone is not a substitute for VHF or HF marine communications systems.

a. **Distress Calls and Cellular Phones.** There are many reasons why cell phones are not good for distress unless as a last resort. A vessel in distress cannot communicate with other potential rescue vessels in the area. This has the effect of delaying rescues considerably, and uses greater resources and increases the risks to all involved. If you are in distress, you simply may not get through to an appropriate authority, and as documented, put on hold, or you may be at the outside of the cell range and drop out repeatedly. Vessels in distress who cannot provide exact position information cannot be located using VHF direction-finding (DF) equipment. Vessels in distress cannot activate priority distress alerting using cell phones. Rescue scene communications can be severely disrupted because normal cell phone communications can only occur between two parties. Most rescue vessels and SAR aircraft do not have cellular phones. These communications problems and resulting message passing have the potential to cause disruptions or delays to the extent that a safe rescue opportunity is lost with catastrophic results. It is worth noting that a number of system operators have introduced services for vessel cellular phone users.

b. **Coverage.** If you use your cellular phone on coastal trips, you will have a few problems with dropouts. The problems will occur at the outer range of the transmission cell, and this is more pronounced at sea. Range is typically 8 miles, which is a lot less than VHF. If you really need to use a cellular phone regularly, install an external aerial for maximum range. If you wish to install a set on board it is worth considering buying a higher 8-watt set such as from Motorola, and installing it to the vessel power supply and adding an external whip aerial to maximize range and power. It is also worth considering a dual or tri-band phone to maximize operational range.

c. **SIM Card Change Outs.** Global roaming rates are not cheap. If you are in one place for any length of time, and make lots of local calls consider buying a pay-as-you go SIM card and number. This offers real long-term savings in most.

d. **SMS Text Messaging.** Short text messaging is a useful low cost way of staying in touch. Make sure you can both send and receive text messages if you intend using this method. I have had problems in some European areas. Also in some cases the memory will fill quickly so messages must be deleted to ensure you don't lose new ones.

Instrument Systems

20.0 Instrument Systems. Discrete stand-alone instruments are becoming relatively rare. Integrated instrument systems are in the forefront due to the rapid advances in microprocessor computing power, miniaturization, and appropriate software developments. Low cost fluxgate compasses, networking and quality multifunction displays have all contributed to systems that can show 75 or more separate parameters, as well radar and charts and more, all on the same display. The basic architecture of instrument systems varies between manufacturers.

a. **Discrete Instrument Systems.** These systems have a transducer serving each dedicated instrument head. The head processes and displays the information. Data is exchanged between each instrument on a dedicated network using NMEA or manufacturer communications protocols for computing related data.

b. **Networked Systems.** These systems interface single or multifunction instrumentation displays, transducers, radar, GPS, Navtex, Chart Plotter, autopilot, fishfinder to a network. All data is transferred over the network. Transducers are connected to local pods and these are then connected to the network. The Raymarine H6 typifies the totally integrated system. These systems have an intuitive menu structure, split screen capabilities and use a simple keypad with trackball and point and click control.

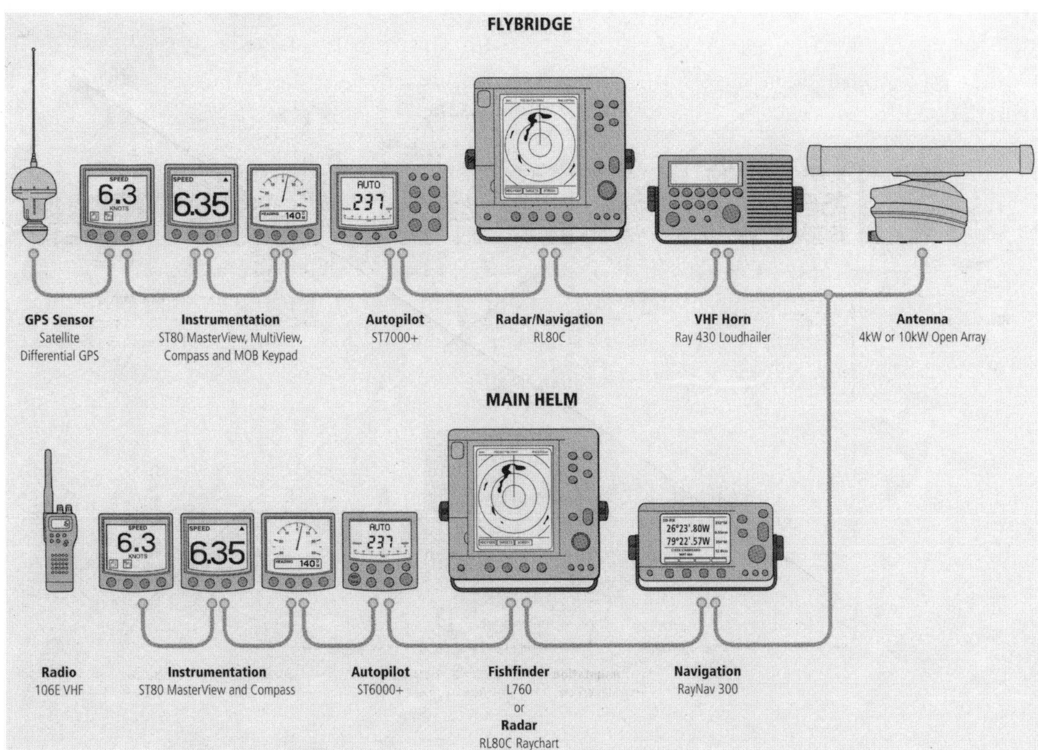

Figure 20-1 Integrated Instrument Systems

c. **Central Server Systems.** These systems have a Data Processing Unit (DPU) or server to which all transducers and some external data are connected. Instrument displays are connected on a daisy chain. The daisy-chain cable can convey data in NMEA sentences (or a manufacturer's protocol) and supply power to each instrument head.

d. **Active Transducer Systems.** Each transducer has a microprocessor in it where all raw data is processed. The transducers are all connected by a single cable network and all data is available through user-definable instrument displays. These multifunction displays can be configured with simple key strokes.

e. **Wireless Systems.** Each transducer transmits data back to the processor. This is the TackTick instrument system. Masthead units are powered by a solar cell.

20.1 Interfacing. Interfacing is the process of interconnecting various electronic equipment and systems so that digitally encoded information can be transferred between them and used for processing tasks or display. Manufacturers have to consider the type of physical equipment involved such as connectors and cable, the voltages, impedances, current values and signal timing. At a more technical level, there is the data structure and transfer rate, and the protocol, which determines the information to communicate, the time to communicate, the frequency and error correction. The data messages must also have compatible structures and content. The US National Marine Electronics Association (NMEA) devised the first general digital standard in 1980 (NMEA 0180). This was developed for position fixing systems to autopilot communication to transfer cross-track error. NMEA 0182 and NMEA0183 followed. The current comprehensive standard is NBMEA2000.

a. **NMEA 2000.** The NMEA-2000 interface standard has been developed in conjunction with International Electrotechnical Commission (IEC). It is a low cost, bi-directional serial data protocol permitting multiple talkers and listeners to share data. It allows GPS, radar, chart displays, sounders, autopilots, engine monitoring and entertainment systems to exchange digital information over a single channel. NMEA-2000 is based on the Controller Area Network Protocol (CAN) originally developed for the auto industry.

b. **NMEA 0183.** This standard was designed to enable transfer of a variety of information between position fixing systems, radar, compass, plotters and autopilots as well as any other systems either sending or requiring data. NMEA is what is called a single talker, multiple listener architecture. Compliance with the standards is a voluntary one, and there are cases where the implementation of the standards has been technically flawed and communication poor or impossible. The NMEA has standard message sentences. They may be divided into input and transmit sentences, where many are simply transmitted as inputs to processors, while other information is transmitted to appropriate systems or display. Message sentences have the following formats, eg. HDM = Compass heading, magnetic, WPL = Waypoint Location, XTE = Cross Track Error. There are as many as there are parameters, and listing them all does not serve any practical purpose. One important recommendation was the use of opto-isolation on circuits. The opto-isolator is commonly used in many high noise environments, and an LED and phototransistor are used to provide total electrical or galvanic isolation. This prevents transfer of noise into equipment circuits.

c. **Communications Protocols.** The trend is for implementation of in-house communications protocols. The main reason is that fast broadband data transfer is required to enable transfer of video and graphics images such as radar, plotter and fishfinder screens. NMEA will remain an important data transfer protocol for external communication between peripherals. The major protocols in use are :

(1) **Raymarine SeaTalk & High Speed Bus (HSB & HSB2).** These protocols are used for total systems compatibility between all equipment. ArcNet is used as the backbone for the Pathfinder HSB network. The system allows addition of any equipment, radar, chart plotters, GPS, logs, sounders, fishfinders and J1939 compatible engines and more. NMEA 0183 as NMEA 2000 requires a separate interface. The newer SeaTalk 2 is based on the CANBus protocol.

(2) **Furuno NavNet.** This uses an Ethernet 10BaseT (twisted pair) system, which is common in many shore data systems. Systems have a star topology, with each device having a separate set of wires radiating out from the hub. When a fault arises it is contained to that one device or cable. Ethernets have high data rates, and cables must be UTP (unshielded twisted pair) standard to ensure data integrity. Make sure cables are routed well clear of fluorescent lights, transformers etc., to avoid interference.

(3) **Controller Area Network (CAN).** This is a fast serial bus designed as an efficient and reliable link between sensors and actuators. CAN uses a twisted pair cable for communications at speeds up to 1Mbit/s with up to 40 devices connected. Originally, Bosch developed the electronics standard for automobiles. The system requires an interface for NMEA communications. Features include any node access to the bus when the bus is quiet, and use of 100% of band width without loss of data and automatic error detection, signaling and retries.

(4) **Simrad.** Simrad instruments use an in-house developed protocol called SimNet. It allows high speed data transfer and is NMEA 2000 compatible. Simrad also now have a wireless controller that allows remote control of VHF communications, autopilot and instrument systems from a handset.

(5) **B & G NETWORK.** This protocol is also used for total systems communications. NMEA interfaces are provided. B & G is now owned by Simrad.

(6) **DO LOGIC.** This protocol is used for inter-instrument communications, but all instrument heads have an NMEA 0183 output to allow easy connection to other systems.

d. **Interface Installation and Problems.** Virtually all problems with interfacing occur at installation, and most systems are simple plug-and-plug making errors hard to make. The majority of faults are related to the following:

(1) **Connections.** Unless an equipment manufacturer supplies the interface cable and connector, make sure that the correct pins are used on the output port connector. These vary between equipment and manufacturers. Check with the supplier, or get them to make up the cable and connector. All connections should observe the correct polarity with respect to ground references. Incorrect connections mean no signals. If the system is fiber optic, ensure the connection is properly inserted, rotate to lock them but do not force them on.

(2) **Grounding.** Ensure screens and reference grounds are properly terminated and connected. In many cases data corruption occurs, or it simply does not work.

(3) **Set-up.** At commissioning, ensure that the appropriate interface output ports are selected with the correct NMEA or output format selected. In many cases problems are directly attributable to this, and many manuals do not clearly explain the process. In most cases carefully go through the set-up procedures.

(4) **Cables.** All cables should be shielded, twisted pair unless stated otherwise. Using other cables may lead to data corruption due to induced "noise" from adjacent electrical cables and radio transmissions. Flat cables are generally untwisted and round ones are. Use only Cat 3, 4 or 5 with data networks. This is usually 100 ohms impedance and 22 to 26 AWG.

e. **Interfacing Cable Designations.** There are a number of variations in designating interface cable connections. The standard NMEA terminology is signal (positive) and return (negative). NMEA output port variations can be very confusing and obviously there is no real "standard" notation. Equipment NMEA ports are configured in what is termed a "balanced pair," with both wires carrying the signal. The signal level is the difference in voltage between the pair, and is also known as a differential data signal. The connection of wires is simple, the transmitting device has the transmit connected to the receive port of the other. The receive port is similarly connected to the transmit port of the other. No connections should be made to boat ground or the DC negative.

(1) **Data Signal Output:** Data O/P; Tx; Tx hot; A Line; Positive data; Signal O/P; NMEA O/P; NMEA Sig Out; O/P Sig; Data Out; Tx - ve; Tx Data O/P.

(2) **Data Return Output:** Gnd; Tx Cold; Ground; Signal Rtn; Return Out; O/P Return; NMEA Rtn; Data Rtn; I/P Gnd; Ref; Negative.

(3) **Data Signal Input:** Signal I/P; NMEA Sig In; I/P Sig; NMEA I/P; Rx Data I/P.

(4) **Data Return Input:** Signal Return In; Signal Rtn; I/P Rtn; NMEA Rtn; Gnd; Negative; Reference; Ref.

20.2 Selection Criteria. When selecting a system, consider the following factors:

Display Types. Ergonomic design is important but the major decision is whether you want digital, which is the most common, or analog. You will confront a confusing array of digital displays. The aviation and motor vehicle industries have invested heavily in researching easier assimilation of data as primary safety factor. They still maintain analog presentations. Many of the display types are combination LCD and analog and most now are made to FCC and CE EMC standards:

(1) **Multi-Function Network Displays.** These screens vary in resolution from color 640 x 480 (VGA) 800 x 600 (SVGA) to marine monitor TFT LCD 1024 x 768 60 kHz XGA screens. Typical power consumption is 20–30 watts at full brightness.

(2) **Digital Liquid Crystal Display (LCD).** Many displays use a 7-segment display with characteristic chunky numerals. Some displays are difficult to read at wide angles or in bright sunlight and you should consider this, although technology is improving with higher contrasts and wider viewing angles. All units generally have a multi level backlit illumination system. Many multifunction displays use a dot-matrix LCD display. Displays average around 65mA power consumption without illumination on.

(3) **Analog Display.** The analog display is still seen on some instruments, and it is practical on ergonomic grounds as it can make overall instrumentation displays easier to monitor; a changed needle position is easier to see than an altered digit. I personally have a preference for analog displays, particularly on depth displays going into coral reefs with the sun behind, they are easier to see. It is good to see many manufacturers such as Raymarime, Simrad, Silva and VDO making analog repeaters part of the range.

(4) **LCD Analog Display.** Some manufacturers are incorporating an analog display using the LCD.

20.3 Electronic Compasses. The principal compass is the fluxgate, and this is already being surpassed with electronic units.

a. **Fluxgate Compasses.** A fluxgate sensor detects the earth's magnetic field electronically, sampling hundreds of times per second. The sensing part of the compass consists of coils mounted at right angles in a horizontal plane. Each coil is fed a precisely controlled current that is subsequently modified by the earth's magnetic field. The processor compares the signals within each coil, automatically correcting for variation. The resulting analog output is subsequently converted to digital signals for processing.

b. **Electronic Compasses.** These are entirely solid state, and are made by KVH and Ritchie. The purely electronic sensing overcomes the problems of analog to digital conversion by the output and processing of a digital signal. KVH have developed a compass called GyroTrac. This combines a digital magnetic compass and a three-axis gyro sensor. These meet more demanding requirements of satellite communications and TV systems, MARPA radars and autopilots. When interfaced with GPS True North is also available. Displays consist of microprocessor-controlled analog rotating cards (450mA), or analog rotating needle (180mA) or digital supertwist LCD (90mA). Power consumption is with lights on. Unlike gyros with long settling times, these warm up immediately.

c. **Sensor Location.** The sensor must be mounted in the area of least magnetic disturbance, so that no interference is induced into it resulting in errors and degrading accuracy. It must also be positioned close to the center of vessel motion, as errors are caused by vessel heeling and pitching. Remember steel vessels pose problems and it must be at least 5 feet above the deck. Accuracy is dependent on proper location clear of interference. Accuracy is typically plus or minus 1 degree although some self compensate to 0.5 degrees, but the display accuracy is still 1 degree.

d. **Damping.** Typically these can be from 5 to 10 levels for some models. The rougher the sea state, the more damping required. A low damping level will result in erratic or rapidly altering headings.

e. **Power Consumption.** Current drains are very low and are typically up to around 300-400mA.

f. **Compensation.** Many have automatic deviation compensation and some will require steering in a circle at commissioning. The compensation takes place with respect to current magnetic deviation. The deviation may vary if you have electrics running, but with electronic compasses, re-compensation is simple and quick.

20.4 Speed/Logs. The log has the obvious function of indicating speed through the water and distance traveled. Not so long ago, the first merchant vessels I served on in the mid-seventies had a towed Walkers log. As soon as we were clear of port and full away on passage, the turbine was streamed and the mechanical counter was mounted on the poop rail. Logs are now part of the integrated instrument system and normally interfaced to other instruments.

a. **Paddle-Wheel Logs.** The common paddle wheel has magnets imbedded in the wheel blades, and a detector giving a pulse that can be counted and processed. Earlier units had a glass reed switch that was prone to impact induced mechanical failure; new units have a Hall Effect device. The signal pulses are normally seen as a voltage change, such as 0 and 5 volts, to give a stepped characteristic that can be counted. The result is directly proportional to the speed and distance traveled. The transducer may count either the pulses per second or the pulse length.

b. **Ultra Sonic Logs.** B & G and Echopilot have sonic speed sensors, which are significantly more accurate with near linear outputs. The transducer consists of two 2 MHz piezo electric crystals. These transmit short pulse acoustic signals simultaneously and reflect the signals off water particles approximately 6" away, which is clear of the turbulent boundary layer. The water particles pass through the forward then the aft beam, and the transmission time of the acoustic sound signal between the two crystals is then measured. The time delay is used to determine precise speed based on the known distance between the two transducers.

c. **Doppler Logs.** Unlike other logs which give speed through the water, these logs report actual speed over the ground by transmitting acoustic pulses which reflect off the bottom.

d. **Electromagnetic Logs.** These systems measure changes to the magnetic field in the water, which alters with boat speed.With this system, there is no impeller to foul up.

e. **Dual Transducer Systems.** Catamarans often require a transducer in both hulls, and some monohulls use a dual system to compensate for heeling. Gravity switches are commonly used in racing monohulls to turn on the appropriate transducer for port or starboard tacks, but in multihulls where heel angles are less, a switch that activates when the mast rotates may be used.

f. **Trailing Logs.** It has been a long time since I encountered someone with a trailing log. Unlike earlier versions such as the reliable unit from Stowe, these trailing logs do not have a rotating line, but have a sensor at the end of a 10-meter cable that sends a signal to the freestanding control box. There are a few basics to remember when using these logs. Prior to streaming the log, make sure that the line is hooked onto the log. (This is a common error!) When streaming, pay out the line quickly and at a constant speed *before* launching the turbine. Do not pay out the turbine first and allow the line to follow; its rotation will cause tangling. When recovering the challenge is to retrieve line and turbine without tangles. Ideally, you should slow the vessel to reduce drag on the line, and initiate a small turn to put some slack into the line. As soon as this is done, disconnect the line from the log and pay it under or over the stern pulpit. This will take out the turns put in the line by the turbine before the turbine is recovered. Dry out the rope before stowing.

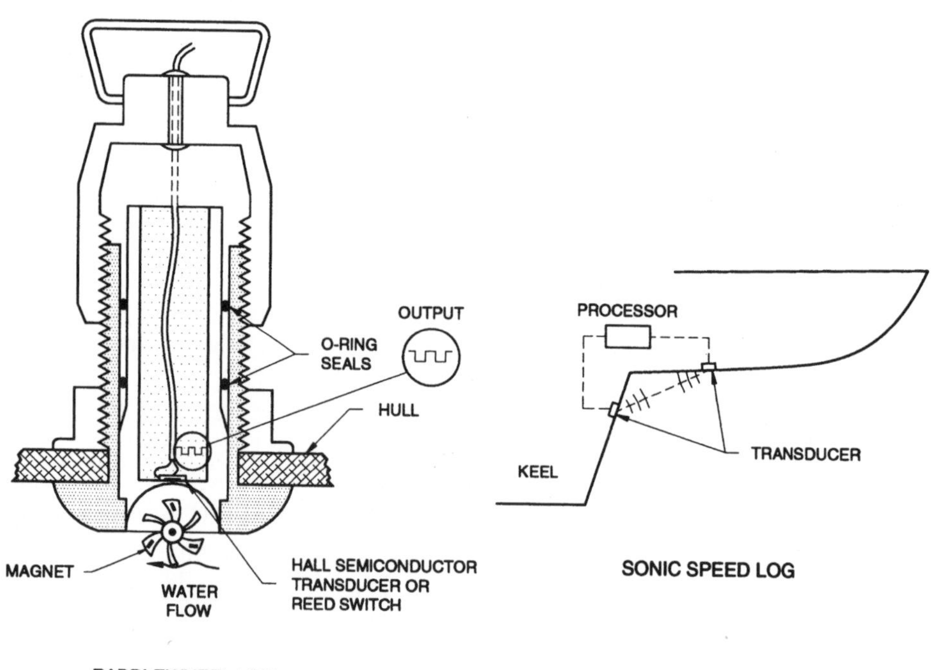

Figure 20-2 Log Transducers

g. **Log Installation.** Correct installation is essential if the log is to be accurate and reliable. Observe the following:

(1) **Location.** The log transducer is normally mounted in the forward third of the hull and must be in an area of minimal turbulence, called the boundary layer.

(2) **Cabling.** Do not run depthsounder and log cables together as interference may result.

h. **Log Calibration.** Calibrating a log normally requires the use of a measured mile. These are always clearly marked on charts. Many new logs are self-calibrating or have an optional manual calibration. The calibration run should be carried out at slack water in calm, wind-free conditions to minimize inaccuracies. Before making a run, check that the vessel is on the correct magnetic course, and this means making appropriate corrections for variation and compass deviation. Make the runs under power at a constant throttle setting. Ensure that your transits are accurately observed at the start and finish of each run. Speed under sail may be different as heeling errors and leeway come into play. The formula for determining log error is as follows:

(1) $$\frac{\text{Runs } 1 + 2 \text{ (ground measurement)}}{\text{Runs } 2 + 2 \text{ through water}} = \text{Correction K}$$

(2) The resulting figure will show either under or over reading, which is used either to calibrate log or correct readings.

i. **Log Transducer Maintenance.** Logs in general need little maintenance, though paddlewheels require more than most. Perform the following checks:

(1) Regularly remove the paddlewheel to see that it is rotating smoothly and freely. Apply some light oil to the spindle.

(2) Check to see if the O-ring seals are in good condition to prevent leakage into the bilge.

j. **Log Transducer Troubleshooting.** To test whether the transducer or instrument head is at fault:

(1) Disconnect the log input cables to the instrument head or processor.

(2) Using a small piece of wire, rapidly short out the terminals and observe whether a reading is indicated. If there is, the transducer is faulty. If there is no reading, the instrument head is probably at fault.

20.5 Wind Instruments. The typical wind system comprises an integral windspeed and direction masthead unit, an instrument head, and usually a combination analog and digital display unit. Masthead wind transducers are now very lightweight and made of carbon fiber or aluminum. They have low friction sealed bearings and very low startup wind speeds.

a. **Wind Speed.** The anemometer is essentially a rotating pulse counter similar to the log. The pulses are counted and processed to give speed.

b. **Wind Direction.** This part of the masthead unit consists of a simple windvane. A number of methods can be used to measure the angle and transmit the signals to the instrument head or processor. Some units use an electromagnetic sensing system. Others use an optical sensing system to identify coded markings that relate to the windvane position and direction.

(1) **Apparent Wind Direction.** The measured wind direction is apparent wind. The display indicates the close-hauled angles and gybe points.

(2) **True Wind Direction.** True wind data is a result of the instrument processing vessel course and speed and apparent wind direction and speed.

c. **Mast Rotation Transmitters.** Boats with rotating mast have mast rotation inputs to accurate apparent wind angle data. TackTick have a wireless transmitter which will make installation considerably easier.

d. **Masthead Unit Installation.** The masthead unit is always mounted at the end of a boom in front of the mast to reduce turbulence. The position is not perfect—the masthead unit is subject to updrafts and turbulence from the sail—but it is still the best alternative.

(1) **Fastening.** It is important that the unit be properly fastened down, especially as masthead units are often installed in a simple bracket assembly and are removable. Check fore-and-aft alignment to reduce inaccuracies in angle readings. Besides birds and lightning, the main cause of masthead damage is vibration.

(2) **Electrical Connections.** Make sure the cable connector is securely fastened. It is good practice to put a few wraps of self-amalgamating tape around it to prevent water entry. If you must put on petroleum jelly or silicone grease, do not smother the socket as many do. It simply gets pushed into the masthead unit and contributes to poor electrical contact. It is better to keep electrical connections dry with tape, as suggested. You can put grease on the screw threads to minimize seizing.

e. **Mast Base Installation.** Make cable connections in a water-resistant instrument connection box and be sure all connections are tight. Every 6 months:

(1) Check securing bolts and frame, and tighten as required.

(2) Check cable connector for moisture and water, as well as for signs of corrosion on the pins. Smear a small amount of petroleum jelly or silicone grease around the threads when replacing it, and rewrap with self-amalgamating tape. Examine cable insulation for signs of chafing at any mast access point.

(3) Check that the anemometer rotates freely without binding or making any noises, which may indicate bearing seizure or failure. Check the cups for splitting or damage, which frequently is caused by birds.

(4) Apply a few drops of the manufacturer's light oil into the lubrication hole and rotate to ensure that it is penetrates the bearing.

(5) Check the connections in the connection box at the mast base. They should be tight and show no corrosion.

f. **Velocity Made Good (VMG).** A sailboat's VMG to a mark or waypoint is an important piece of data with respect to steering and sail trim. VMG is derived from calculation of true wind, course, and speed and is usually combined with one of the wind instruments. Monitoring VMG enables the helmsman to sail the optimum course so that maximum speed is made toward the destination. The following are used to achieve optimum VMG, which is indicated with a higher reading:

(1) **Sail Trim.** Adjusting sail trim will increase or decrease speed and VMG.

(2) **Course Adjustment.** Changing course off the wind or into it will also change the VMG reading.

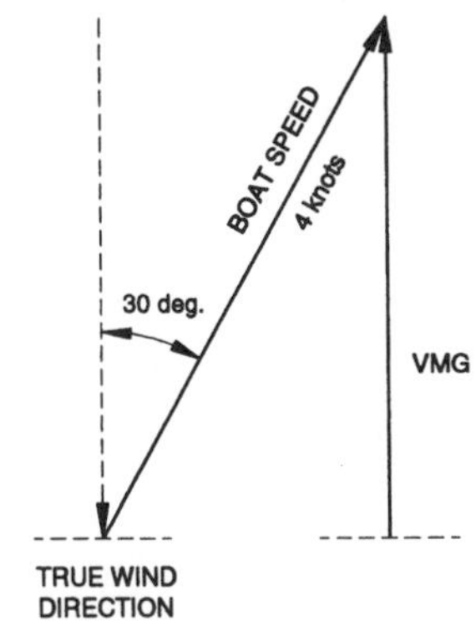

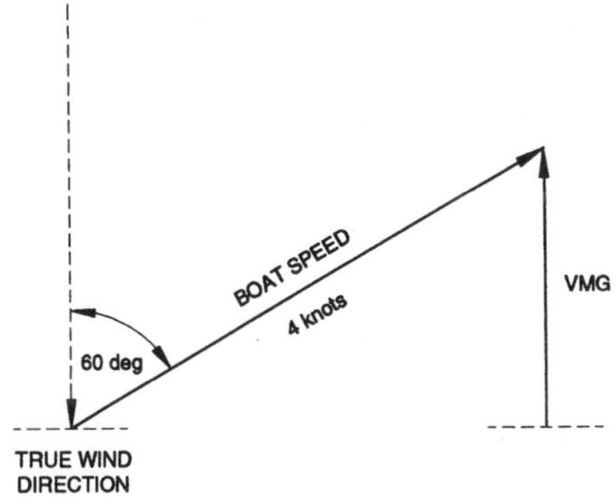

Figure 20-3 VMG Vectors

20.6 Depthsounders, Echosounders and Fishfinders. The echo or depthsounder is an important and indispensable piece of electronics. I have spent considerable time working with underwater acoustics systems in the offshore oil industry, and a couple of years working on a submarine sonar program, and I can say that this technology can be very complex. Equipment performance depends on the output power of the transmitter, the efficiency of the transducer, the sensitivity of the receiver along with the processing software that filters out the spurious noise. Many fishfinders have user selectable noise filters to enhance noise rejection processing. The price of equipment reflects all of these elements, with the most expensive systems having the highest performance specifications on all factors. The word SONAR is derived from SOund, NAvigation and Ranging and has its origins in World War II anti-submarine warfare. The depthsounder normally projects the acoustic signal directly downwards at a set beam angle so that a cone of coverage is made with respect to the bottom or contours being passed over. Most depthsounders operate at a frequency of 200 kHz; lower transmission frequencies of around 50 kHz give greater depth capability, although the EchoPilot operates at 150kHz and B & G at 183kHz.

a. **Digital.** The most common depth instrument is a vertical unit with a digital or analog display incorporating depth alarms, anchor watch alarm facilities, etc. The information displayed is generally several seconds old due to signal processing times.

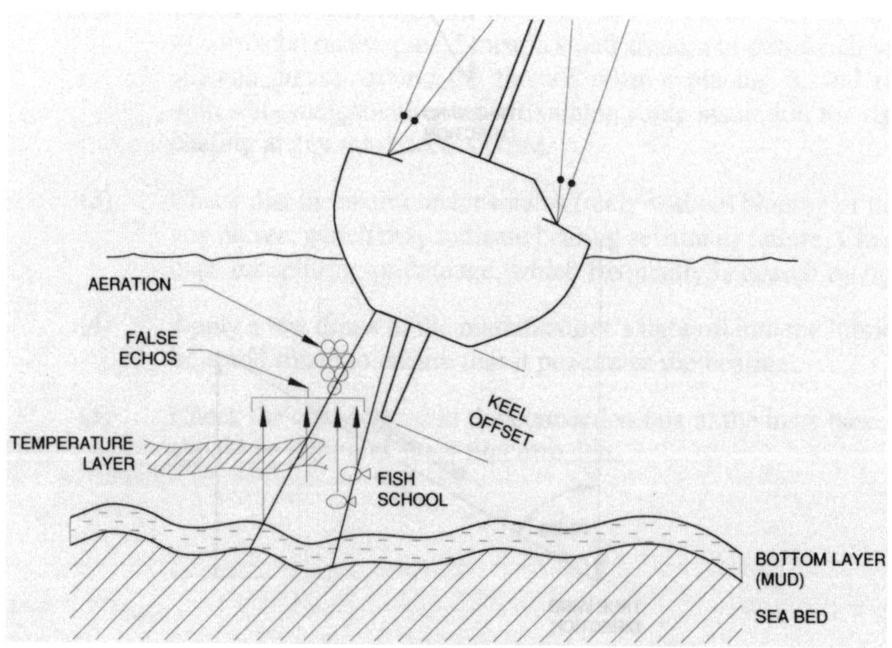

Figure 20-4 Depth Sonic Cone

b. **Accuracy.** Acoustic signals suffer from propagation delays and attenuation as water and various bottom formations cause absorption, scattering, refraction and reflection. Biological matter such as algae and plankton as well as suspended particulate matter such as silt, dissolved minerals and salts can cause this. The water density and salinity levels as well as water temperatures all affect signal propagation. While cold layers of water called thermoclines can affect signal, this is more relevant to deep water. Bottom formations consisting of sand and mud, or large quantities of weed beds will absorb or scatter signal; hard bottoms that comprise shale, sand and rock will reflect signal with strong returns. The power output of a unit is also important with respect to range and resolution, the higher the power the greater the depth range and signal return.

20.7 Sonar Theory. The basic principle is that an electrical signal is converted to an acoustic signal via a piezoelectric element (crystal) and is transmitted towards the sea bottom. Transducers are typically constructed of a crystal composed of various elements that include lead, zirconate, barium, titanate and conductive coatings. Some fishfinders have transmission power ratings up to 1000W. When the transducer transmits the acoustic signal it expands to form a cone shaped characteristic. When the acoustic signal strikes a surface such as the sea bottom or a fish it is reflected back. The shape and diameter of a transducer determines the cone angle. The acoustic signal strength is at maximum along the center axis of the cone, and decreases away from it.

20.8 Cone Angles. The acoustic signal strength is at maximum along the center axis of the cone, and decreases away from it. The cone angle is based on the power at the center to a point where the power decreases to -3db, with the total angle being measured from -3db point on each side. Most manufacturers offer models with a variety of cone angles. Wide cone angles have less depth capability with wider coverage, and small cone angles give greater depth penetration with reduced area coverage. High Frequency transducers (190 kHz) are available in either wide or narrow cone angles. Low frequency transducers that are found in fishfinders have cone angles in the range 30–45 degrees. The further away from the centerline of the cone, the less strong return echoes are. This can be improved by increasing the sensitivity control.

20.9 Transducer Cavitation. Cavitation affects transducer performance, and is caused by water turbulence passing over a transducer head. At slow speeds the laminar flow is smooth without any interference. At speed air bubbles are created over the transducer face affecting acoustic signal transmission and reception. The effect is to interfere with transmitted acoustic signals that reflect back off the bubbles, which effectively causes noise and masks signals. Turbulence is caused by hull form or obstructions, water flow over the transducer, turbulence from propulsion. Transducers must be mounted in areas of little turbulence or clear of hull flow areas, which is not always easy.

20.10 Frequencies and Power Output. Transmission frequency affects both the depth range and cone angle. The speed of sound in water is a constant of 4800 ft per second, and the time between the transmission and reception of the returned signal is measured to give a range or depth figure. Shallow waters less than 300 feet give the best results with high frequency transducers of 200kHz and wide cone angles up to 20 degrees. In depths greater than 300 feet low frequency transducers of 50kHz with small cone angles of 8 degrees are the best option. Power outputs are quoted in watts, some quote peak-to-peak. The use of watts RMS is the more accurate, typically within the range 100–600 watts.

20.11 Sounder Installation. Be very careful not to bump the transducer and possibly damage the crystal element. Most installations are through hull mounted on a fairing block to ensure that beam is facing directly down on an even keel, and to reduce any water flow turbulence. Locate in an area of minimal turbulence. Water bubbles from turbulence are a common cause of problems. In some cases they are mounted inside the hull within an oil bath or epoxy fastened to the hull on GRP boats. There is a sacrifice in maximum depths, which can reach 60–70% reduction in range and therefore should be avoided where possible. Always ensure that cables are installed clear of heavy current carrying cables or radio aerial cables. Never install next to log cables as is generally done, as the interference problem can be significant.

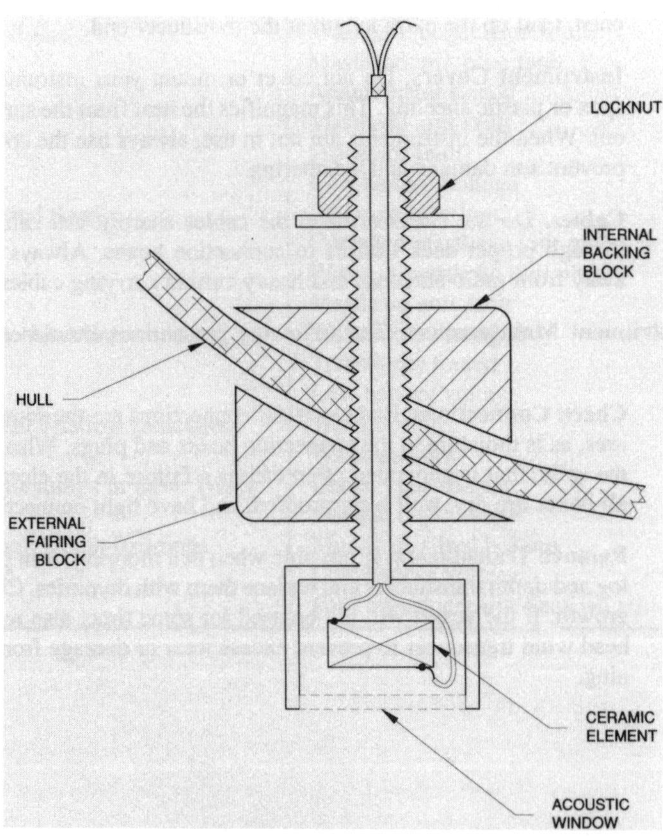

Figure 20-5 Depth Transducer

20.12 Sounder Maintenance and Troubleshooting. The transducer is the only item that can be maintained, and if not will dramatically reduce performance. Inspect the transducer for damage, marine growth, antifouling paints, and clean off the surfaces using soapy water. Do not use heavy abrasives or chisels to clean the faces. Do not bump it or apply any impact to the surface. Avoid applying antifouling to the transducer surface, as it includes small voids and air bubbles, which will reduce sensitivity. If necessary, smear on a very thin layer with your finger. Troubleshooting often entails reading the manual and determining whether settings are correct and operating procedures are also correct. Go into settings or options menu and ensure settings are on auto or defaulting to factory settings. If possible when at a mooring or in port, remove the log and depth transducers and replace them with dummies. Clear them of any growth. If the vessel will not be used for some time, remove the masthead wind transducer to prevent excess wear or damage from birds or lightning.

a. **Cables and Connections.** Check all connectors and connector pins for damage, and make sure that they are straight and not bent. If straightening the pins there is a risk of breakage as they are brittle. Connectors not properly inserted or tightened up are prone to saltwater ingress and corrosion. Check all cables for damage, cuts or fatigue. The transom mounted transducer cables are prone to damage and on some smaller fishing boats the transducer hull cables may be damaged.

b. **Power Supply.** Connection problems are the major cause, either at the supply panel, or at the battery. Check the power at the plug using a multimeter set, and check should be made with engine on or off. If the engine voltmeter shows normal charge voltages, and battery checks out, then it is in the intermediate connections.

c. **Interference.** If the fishfinder has interference, turn off all other equipment and then turn engine off. Progressively start up engine and then other equipment to determine the source, and the power supply may require suppression. Check that two fishfinders are not being run at the same time, and two vessels in very close proximity may also cause mutual interference if using similar acoustic frequencies. If the interference is present with all systems off, the fishfinder automatic noise rejection facility may be malfunctioning.

20.13 Forward Looking Sonar. The most identifiable is the very impressive EchoPilot. These systems have benefited greatly from developments in processing power and speed. The units consist of a powerful processing unit, which enables real time processing of data. This is different to the normal depthsounder, which has an inherent delay of typically up to 16 seconds. The transducer head scans from vertical to horizontal with a beam width of 15 degrees and can see up to 150–200 meters forward. Maximum range depends on water depth and seabed contours, with a bottom that is shoaling being easier to see than one that is level or deepening. Like all sonar and depth transducers the head must be clear of turbulence. These units operate at 200kHz, and it must be noted that two units operating close together will cause corruption of data. Also it is important to note other depthsounders and fishfinder frequencies. If all 3 are at 200kHz problems will arise and only one can be reliably operated and others switched off.

20.14 Instrument Installation. The following should be observed when installing an instrument system:

a. **CPU Location.** Always install the CPU or data box in a clean, dry area that permits easy access to transducer cables. Mount the CPU unit well away from fluxgate compasses, SatNav, Loran, Decca, and GPS receivers, and VHF, SSB, and AM/FM radios. The CPU must be one-meter minimum from a magnetic compass.

b. **Transducer Cables.** Transducer cables should not be lengthened or shortened. Coil up the extra length at the transducer end.

c. **Instrument Covers.** Do not cover or mount your instruments behind perspex or plastic sheeting. This magnifies the heat from the sun and burns them out. When the instruments are not in use, always use the covers provided to prevent sun damage and weathering.

d. **Cables.** Do not stress or bend the cables sharply. All cables must be run through proper deck transits to connection boxes. Always run cables well away from radio antennas and heavy current carrying cables.

Table 20-1 Instrument Troubleshooting

Symptom	Probable Fault
No display	Loss of power
	Cable connection fault
	Instrument fault
Partial display	Processor fault
	LCD fault
	Transducer fault
Erratic readings	Connection degradation
	Interference from radios, electrical, etc.
	Low battery voltage
No or low boat speed	Transducer fault
	Transducer not installed
	Transducer not connected
	Fouled transducer
	Transducer misaligned
	Paddlewheel seizing
High boat speed	Electrical interference
No wind speed	Mast base connection fault
	Masthead unit plug fault
	Anemometer seized
	Masthead unit fault
	Processor fault
	Low battery voltage
Erratic wind angle	Loose connections
	Corroded masthead unit plug
	Water in masthead unit plug
	Masthead unit fault
No depth indication	Transducer damaged
	Transducer fouled
	Low battery voltage
Intermittent shallow indication	Weed or fish
	Water aeration
Shallow readings in deep water	Check your charts!
	Outside depth range
Inconsistent depth readings	Muddy or silted bottom
	Low battery voltage
	Poor transducer interface (in hull only)

Interference

21.0 Interference. Interference is the major enemy of electronics systems, corrupting position fixes and causing general performance problem. It is often the cause of electronics damage. Interference and noise superimpose a disturbance or voltage transient onto power or signal lines, and this corrupts or degrades the processed data. The following describes problems and some solutions:

a. **Voltage Transients.** The voltage transient is the most damaging and comes from many sources. The best known is the corruption of GPS and Loran data where the power is taken off an engine starting battery. If a significant load is applied, there is a momentary voltage drop (brown out condition), followed by an increase. This under-voltage disturbance can exceed 100 volts in some cases, damaging power supplies, wiping memories or corrupting data. The same applies to two battery systems where the house bank supplies items such as electric toilets and large current equipment. A starting battery voltage can have a 3–4 volt dip on starting. Transients are also caused by the variation or interruption of current in the equipment power conductor.

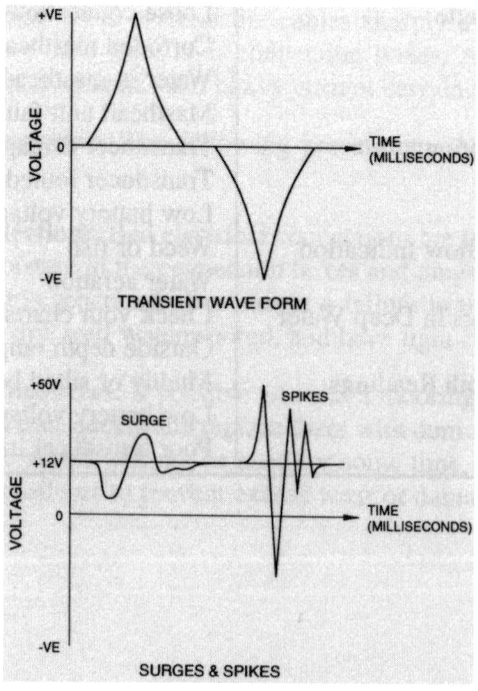

Figure 21-1 Transient Waveforms

b. **Induced Interference.** Electrical fields are radiated from cables and equipment and this is induced into other closely located cables or equipment. The most common causes of this are cables running parallel or within the same cable bundle, and is also called mutual coupling. Always run power supply cables and data cables separately and make cable crossovers at 90°. In particular run power cables to sensitive equipment separate to main power cables to reduce inductive and capacitive coupling to signal conductors.

21.1 Noise Sources. There are a number of noise sources on sailing yachts. Noises can be classified as Radio Frequency Interference (RFI) or Electromagnetic Interference (EMI). Noise also occurs in differing frequency ranges, and similarly equipment may only be prone to problems within a particular frequency range. Multiple noise sources can cause a gradual degradation of electronics components and when the cumulative effects reach a certain point the devices fail.

a. **Arcing Noise.** These are repetitive spikes caused by commutators and sparking of brushes. Charging systems and loose connections commonly cause this. The most frequent cause is loose or poor engine return paths for alternators, when the negative path arcs across points of poor electrical contact. Also ignition systems from distributors and spark plugs being impressed on a DC system, often through radiation to adjacent cables, will cause the problem.

b. **Induced Coupling Interference.** Wiring that is installed in parallel with others can suffer from inductive coupling interference. This occurs like a transformer with a single turn primary and secondary coil, with the magnetic effects causing the induction. Low ground impedances and unbalanced circuits are the most problematic, and serial data, multi-cable control and coaxial cables being the most susceptible ones.

c. **Capacitive Coupling Interference.** This is most common in high frequency circuits and in high impedance to ground circuits such as balanced pair systems.

d. **Ripple Noise.** Ripple is created in any rectifier bridges (diode, SCR, etc.) such as alternators (diode bridges), chargers, fluorescent lights and inverters. It is usually a high-pitched whine. Good equipment has suppressed electronics. Ripple badly degrades communications audio quality.

e. **Static Charges.** These have a number of sources:

(1) **External Charges.** This type of interference can arise due to static build-ups in rigging. On reaching a certain voltage level, the static discharges to the ground, causing interference. Another common cause is when dry, offshore winds occur, and a static charge builds up on fiberglass decks. The problem is prevalent on larger fiberglass vessels with large deck areas. A lightning protection system can help ground out these charges.

(2) **Engine and Shaft Charges.** This type of interference can arise due to static build-up both induced and due to moving parts in the engine. The static charge discharges to ground and causes interference. Shaft interference can arise due to static build-ups on propeller shafts. The static will discharge to ground when it reaches a high voltage level, and cause interference. Typical cures are grounding of the shaft with a brush system.

f. **Surge (Electromagnetic Pulse).** This can be caused by lightning activity, and pulses can be induced into electrical wiring and aerials. The allowable surge is 100 volts for 50ms and 70 V for 100ms.

g. **Spikes.** Turn-on spikes result from the initial charging of input filters on power supplies. Turn-off spikes arise when reactive loads are switched, and the magnetic fields collapse on inductive loads, such as transformers, relay or contactor coils, solenoid coils, pump motors etc. Spikes can be as much as 500V peak-to-peak. MOV suppressors are often put across the coils.

h. **AC Transients.** Surges and transients on an AC shore power system can be carried through chargers to the boat DC system. Many chargers do not have any power filtering circuits.

21.2 Suppression Methods. There are a number of methods that can be used to reduce or hopefully eliminate interference. The use of shielded cables along with proper grounding is important but the use of proper equipment enclosures is also critical, as this minimizes electromagnetic radiation. A filter or capacitor is installed close to the "noisy" equipment, and effectively short circuits noise in the protected frequency range. Filters may take a number of forms:

a. **Filters.** The filter consists of either a capacitor, or a combination of capacitor and inductor connected across the power supply lines. STO-P use filters with a very low ground impedance, typically lower than 20 milliohms at 1 kHz, which cleans our ripple. An option is to supply sensitive equipment through a Navpac from NewMar. This is a supply-conditioning module that filters our spikes and noise, regulates supply voltage, and has an internal power pack to ensure supply continuity. The StartGuard from NewMar also protects against the surges that occur when the voltage drops when starting engines. This device is connected in parallel with the equipment, and sense circuit is connected with the starter switch or solenoid. The internal battery supplies the load when starting and recharges when in standby mode. The units are rated at 20 amps.

b. **Suppressors.** Suppression modules from Charles Marine use MOV technology and are available in AC and DC types. Many alternators do not have these fitted, so install them. Normally you will have noticed radio noise or interference on electronics equipment. A 1.0-microfarad is a starting point, but even experimentation with a couple of automotive types is simple and inexpensive.

c. **Ferrite Chokes.** Ferrites chokes are sleeves or rings that are placed over cables. They allow differential mode signals through but block common mode currents by interrupting RF ground loops and prevent RF from coupling into the cables. They are ideal for eliminating problems in e-mail connections to notebook, HF modem and SSB connections and are recommended by Sail-Mail in their installations. They can also be used on any cables such as autopilot cables or others exposed to interference. The Fair-rite chokes are available in the US from Newark Electronics (800-639 2759) and Amidon Electronics (714-850-4660). It is important when clipping on ferrites that no air gaps are left between the ferrite halves. Co-axial ferrite line isolators such as the T4 are available from The Radio Works (www.radioworks.com) and are used on the coaxial cable and placed on near the tuner unit. These block the stray RF ground path from the coaxial shield and transceiver grounds.

d. **Power System Stabilization.** In cases of high voltage induction, it is necessary to clamp voltages to a safe level, typically around 40 volts. One of the major causes of lightning strike damage is the failure of equipment power supplies to cope with high voltage transients. The most common method of achieving this is the connection of a Metal Oxide Varistor (MOV) across the power supply. As the voltage rises the resistance alters to shunt the excess voltage. A second method is the use of an avalanche diode across the supply. MOVs are designed for AC systems, and DC surges tend to have longer time durations. Also MOVs can be blown without warning.

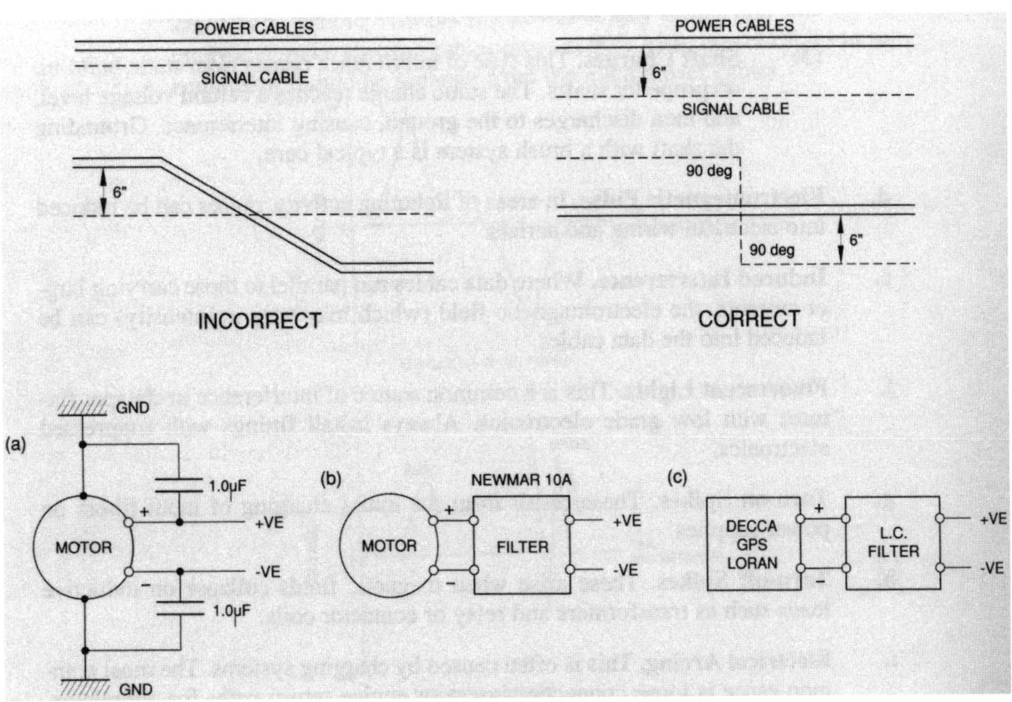

Figure 21-2 Noise Filtering and Stabilization

21.3 Screening. Screening masks sensitive equipment from radiated interference. Common sources include radio equipment, and high current carrying cables. The equipment or cables are covered and grounded by metal covers or screens (commonly called the Faraday cage). This may be a simple aluminum cover grounded to the RF ground point.

a. **Equipment Covering.** A useful product is the Sonarshield conductive plastic sheet. Simply cover the Loran, GPS, Radar or radio casing. (Southwall Technologies, 1029 Corporation Way, Palo Alto, CA 94303.) The total metal Faraday Shield approach is rarely required.

b. **Cable Covering.** Noisy power cables can be wrapped in noise tape, which is a flexible copper foil with an adhesive backing such as that from NewMar. I have used this product and method.

c. **Cable Shields.** Shields are designed to protect against interference from unknown or unspecified sources. The effectiveness of shields is measured in terms of transfer impedance. This is a measure of effectiveness in capturing the interference field and preventing it from reaching the conductor pairs inside. Data cables have shields made from a foil/polymer laminate tape or layers of brading. These also may have a drain wire installed to enable termination of the screen. Most manufacturers will also specify the termination of shields. Never ground at both ends, always ground one end only, typically the equipment end. In many cases shields are not connected at all, so check and connect them.

d. **Grounding.** The ground must be clean, which means that it should have a ground potential between equipment no greater than 1 volt peak-to-peak. A ground is capable of also conducting transients and emissions so it must be sound. Another grounding source on boats is the grounding of static causing equipment such as shafts and engine blocks. As discussed in alternators the negative connections to the engine block are a common source of problems. Ensure that the starter motor negative is attached close to the starter itself. Add an additional negative to the alternator. In many cases, interference is caused by arcing and sparking within the engine, as it is effectively part of the negative return conductor. Modifying the negative system eliminates this problem. Ensure that all ground connections are clean and tight.

e. **Cancellation.** The wires to a piece of equipment can be twisted together. This effectively causes cancellation as the electrical fields are reversed.

21.4 Noise Troubleshooting. Tracing the sources of noise is a matter of logic and systematic switching off of equipment to find the source. In some cases it may consist of two or more sources causing a cumulative effect. Some noise will be simply intermittent, such as static discharges, which may be synchronized with hot dry wind conditions, or lightning pulses, which may not even be visible locally. A cheap battery powered AM radio is a good tool for tracking down radiated sources on board, with static being easily picked up. Passing it close to equipment is the method used. Some noise is simply related to time of day. Interference from ionosphere factors on radios is well described in Chapter 19. This may affect GPS, HF, and satellite communications all simultaneously, giving the appearance of some greater problem.

21.5 Cable Planning. Cable planning is a major cause of problems. If you have problems cable routing will have to be assessed and possibly require re-routing of the sensitive cables. This factor is addressed in the wiring sections, and if run properly many problems can be reduced or eliminated.

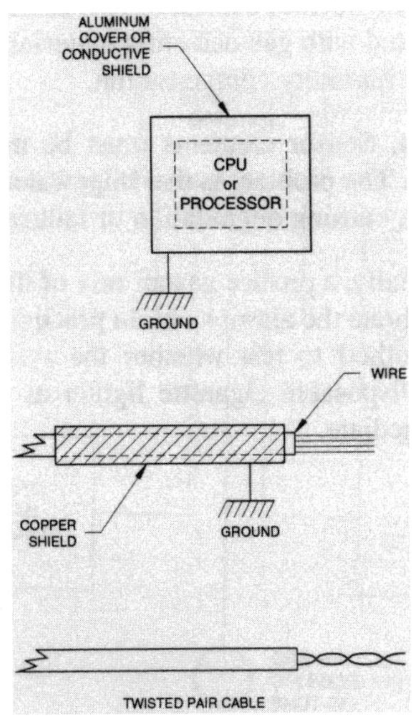

Figure 21-3 Grounding and Screening

Safety Systems

22.0 Gas Detectors. Propane gas is potentially lethal on a boat. If leaking gas accumulates in the bilges, once ignited it takes only a small amount of gas to destroy a vessel. If gas is installed, a quality gas detector is essential.

a. **Theory.** All gases have a lower explosion limit (LEL). As long as the gas/air ratio remains within this range, no explosion can occur. Once this level is exceeded, a significant explosion risk exists. A detector must indicate the presence of gas concentrations before the limit is exceeded, typically 50% of LEL. Better units have a sensitivity of 25% LEL.

b. **Detector Types.** Two types of gas detectors are in use in detection systems.

(1) The main commercial sensor is the catalytic type. On offshore installations, we recalibrated these units weekly to ensure precise operation.

(2) The most common type of sensor on small vessels is the semi-conductor type, which consists of a sintered tin oxide element. When gas is detected, the resistance alters and activates the alarm circuit. It takes several days of operation before the sensor stabilizes and final calibration can be made. Detectors may be subject to temperature drift in the sensing circuit. Good gas detectors incorporate a temperature sensor to correct this and ensure accuracy.

(3) Other detection devices use what is called the pellister principle. These devices consist of two heated platinum wire elements. One is coated with gas-detecting material, the other is used for temperature and humidity compensation.

c. **Installation.** Sensor elements must be mounted in areas where gas may accumulate. The problem is that bilgewater or moist salt air can contaminate the element, causing degradation or failure.

d. **Testing.** Ideally, a precise gas/air mix of the appropriate LEL ratio would be used to calibrate the alarm level. In practice, however, this is never done. The simplest method to test whether the system functions is by activating a butane or disposable cigarette lighter at the sensor. Activation should be almost immediate.

e. **Alarm Outputs.** All detectors should have a gas bottle solenoid interlock that closes when gas is detected. This function should be fail-safe in operation. An external alarm or exhaust fan can also be connected to the detector.

f. **Troubleshooting.** Note the following important factors:

(1) **Alarms.** If an alarm goes off, assume it is real. If the alarm proves to be false, you can normally readjust the alarm threshold. Do so only enough to compensate for the sensor drift causing the nuisance activation.

(2) **Sensor Element.** The principal cause of problems is a degraded sensor element. Carry a spare sensor for replacement. If after replacing the sensor the alarm still causes problems, have the electronic unit tested.

22.1 Fire Detection Systems. Smaller yachts should invest in self-contained units that have an integral battery. Larger vessels may have sophisticated addressable, multi-zone fire systems installed. The control unit processes sensor information and allows the setting of alarm thresholds, and time delays that activate alarms. The sensors or transducers are the detectors and manual call points that are processed. The various smoke types are different with respect to the smoke particle sizes. Hot fires tend to have very small and almost invisible particles; low temperature smoldering fires will have larger visible particles. Ion chamber detectors react quickly to small particles, but are slower on larger particles, and the reverse is true for photoelectric detectors.

a. **Optical Smoke Sensors.** These detectors are ideal for low levels of smoke. The Raleigh forward scatter principle uses the scattering properties of light from smoke particulates when they enter a light beam. The light sources use a narrow band gallium arsenide (GaAs) emitter and a silicon photodiode photo-detector, with a lens installed in front of each. They are aligned so that the optical axes of each will cross in the center of the sampled volume. Baffles are installed within the narrow light beam, so no light reaches the detector. When smoke enters the chamber, some light will get scattered and reach the photo-detector, and the quantity of light at the detector is proportional to the smoke density. This is processed within an amplifier and a 0 - 20mA analog signal is output to the control unit. Test response time is 6 to 22 seconds.

b. **Ion Chamber Detectors.** These operate by the air within a chamber being ionised by a very small radioactive source of Americium 241. This allows a small current to flow between the source and a cover, which has a fixed voltage between them. The collector is a perforated electrode, and has a nominal clean air potential relative to the outer electrode. When combustion particulates enter the chamber, the collector potential increases, and the level of charge can then indicate smoke density. The units operate best with invisible smoke materials released by fast burning fires. Test response time is 6 to 12 seconds.

c. **Heat Sensors.** There are two types: the first activates when a set temperature is reached; the second activates based on the rate of temperature rise above a threshold level. Many units combine both functions. The heat sensor uses a bridge consisting of two matched thermistors, which are arranged to respond on absolute temperature and rate of temperature change, and are fed to a differential amplifier. The thermistors are negative temperature coefficient types, one is exposed to air and the other is within the detector casing. The bridge voltage will track constant temperatures, when the temperature changes rapidly, the sense thermistor will be unable to follow, and generates an analog output.

d. **Carbon Monoxide Sensors.** These operate on the principle of oxidation of carbon monoxide gas to carbon dioxide. This conversion process takes place within a catalytic sensing cell. The process requires an exchange of electrons and the flow of electrons generates a small current within the cell. These are suited to slow burning fires. The 0–20mA output is normal at 7.5mA. When the carbon monoxide level increases a proportional output is also generated, and the alarm is activated before reaching the limit of 50ppm.

22.2 Security Systems. Trying to keep villains off your boat is always a major undertaking. You can never keep out a determined thief, but my approach has always been to make the exercise as difficult as possible. The big trend is now to wireless security systems, which means no wiring and even less trouble causing connections. A variety of detectors can be coupled with control units and alarms. Some detectors combine IR and microwave detectors in one unit.

a. **Ultrasonic Sensors.** These sensors are unsuited to vessel installation. They are easily set off by spurious signals and have a relatively high power consumption.

b. **Microwave Sensors.** These are often combined with PIR and use short K-band to reduce false alarm rates.

c. **Infrared Motion Detectors (IR).** These dual sensor units direct a pattern of infrared beams over a set area. When a heat source crosses a beam, the alarm is activated. Contrary to the theory that cats and other animals set them off, they can be calibrated to only react to human-sized heat sources. One unit, properly located, can cover a typical saloon, but the installation site must be carefully selected so that it is not easily visible. They are now available wireless.

d. **Magnetic Switches.** The most power-efficient security systems use magnetic switches on hatches and other access points. These are connected directly to the control unit. This system detects the thief before he enters the boat; so the alarm catches him on deck or in view. It is also fail-safe, so that the alarm still activates if a sensor cable is cut. These are now available wireless.

e. **Pressure Pads.** Pressure activated pads can be installed under carpets and mats.

f. **Deck Sensors.** These are installed above and below decks. They are effectively strain gauges that detect distortion when a person's weight is on the cockpit deck.

g. **Photoelectric Beams.** These miniature beams activate when a beam is broken.

22.3 Security Alarm Indication Systems. Once an intruder is detected, an alarm has to be activated to indicate his presence. The following alarm systems are recommended:

a. **Strobe Light.** A high-intensity xenon strobe light mounted on the stern arch or mast is the most common indication method. Many install a blue light, but you simply cannot see it easily. That is why police vehicles worldwide now use a red/blue light combination. I always fit an orange xenon strobe light, which is far more visible, but cannot not be relied upon on its own.

b. **Audible Alarm.** Install the highest output, two-tone siren you can find. Put one outside and one below. A high output unit wailing in a cabin is very painful and will cut short any intruder's stay. A number of audible alarms may also panic or disorientate a thief.

c. **Interlocking Systems.** Connecting various systems to the alarm is another popular method.

(1) Spreader and foredeck spotlights, as well as any spotlight on sternpost and arches can be interlocked to come on with alarm activation.

(2) One yachtsman I know was robbed so many times he attached a high-voltage, electric fence energizer to the pulpit of his fiberglass boat to give thieves a "rude shock." Such a system, however, may get you in legal trouble if it causes injury or death.

d. **Time Delays.** Entry and exit delays give you time to leave after you activate the alarm, or to disable the alarm when you return. I prefer to fit a remote isolator in a sail locker and have minimal delay. Generally, laws restrict alarm operation to 10 minutes. After that, the alarms must cease. Really ambitious thieves will set off the alarm and come back when the silence returns, so make sure that yours resets automatically.

e. **Back-to-Base Alarms.** This alarm method transmits a radio signal to a 24-hour monitoring station that can take corrective action. These systems can monitor all vessel alarms, including bilge levels, smoke and fire, gas, as well as security.

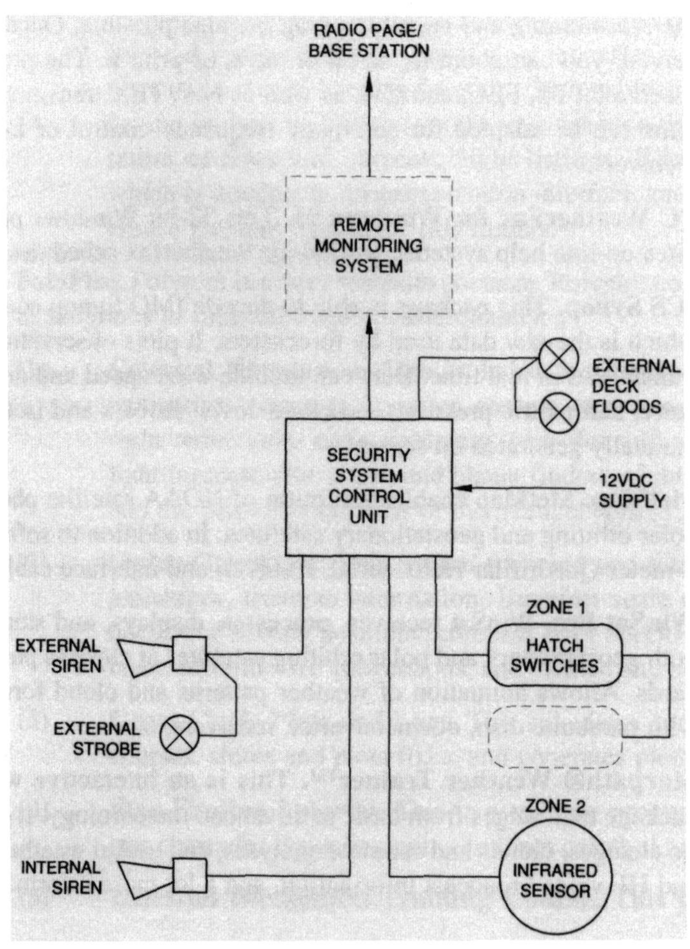

Figure 22-1 Security Systems

Computers

23.0 Notebook Computer Systems. The notebook computer is now a powerful mobile office work tool to incorporate major yacht management systems. There are several companies that specialize in mobile computing as well as a full range of options and marine software. I would recommend that you contact them and discuss the options. Mobile Computing Magazine (www.mobilecomputing.com) is a good start, also Home Office Computing Magazine (www.smalloffice.com) or (www.marinecomputing.com). The criteria in selecting a suitable computer are the same as for any other environment and makers such as Toshiba, Sony, IBM and Compaq all have acceptable models. Machines get smaller, lighter and slimmer and this is due to minimization of components, such as the CD-RW/DVD drive, or smaller battery or deletion of the disc drive. Also the display sizes shrink while the resolution is improved. For boat applications the fully integrated machine is probably the best way to go. There are specially prepared computers available for marine use. I have successfully carried notebooks around, on and off planes, helicopters and done most of my professional work on oil rigs and ships. I wrote this and other books aboard my own boat using a Sony Vaio. In the end if you treat them roughly they will fail; if you soak them they will fail. It's all a question of budget. If you are doing serious boating a rugged version may be ideal. Always choose wireless capability so you can utilize wireless hotspots and Internet access when you travel around.

a. **RAM and Processor.** Each notebook manufacturer has different processors, which affect the price. Whether it is an Intel Centrino or Pentium IV, Celeron, or AMD they are designed along with associated circuitry to maximize power efficiency. When selecting RAM (Random Access Memory), go for as much as you can afford. The more you have the faster applications will operate in particular some powerful graphics packages and chart plotting software.

b. **Hard Drive and Data Storage.** I would always opt for the largest hard drive possible within your budget, as so many software packages soon absorb capacity. The best thing going is a USB plug-in memory stick and I travel with a 1 Gb unit. You may choose to backup using a CD or get a plug-in external hard drive unit.

c. **Communications Ports and Modems.** My machine has 2 and some have 3 USB ports. Also you may need a serial port for some plug in applications to SSB units or inputs of navigation data.

d. **CD-RW/DVD Drive.** The majority of notebooks have integral drives. The drive speed is important and the higher the better if you play games or watch movies.

e. **Displays.** Desktop PCs used cathode ray tube screens, but now it's all flatscreens and the notebook uses a color LCD. The two main LCD display types available are the TFT Active Matrix (Thin Film Transistor) and the DSTN Passive Matrix (Dual Scan Twisted Neumatic). There are several variations on these types called HCAD, HPA and CB-DSTN. For performance the TFT has a sharper and brighter image with better color. If you view graphics, video and DVD it will refresh faster, and it allows viewing at wide angles. The downsides are that TFT consumes more power than DSTN screens, and DSTN units are cheaper. For chart plotting applications TFT is the better choice. Screen technology and operating software have improved graphics and video quality of LCD screens dramatically and on DVD applications I find them more than adequate.

f. **Power Supplies and Battery Life.** This is an important selection point for any considering boat applications. The two principal battery types are the LiON (Lithium-Ion) or the NiMH (Nickel Metal Hydride). The LiON batteries weigh less and hold charge for a longer period, with less memory effect than the NiMH, but they are more expensive. Carrying a spare battery is a good option, and also a battery conditioner will restore batteries suffering from memory effects. Cycling the computer down to full discharge and recharge is also a good practice. I have the Targus DC travel and car charge kit; it has a variety of other plugs including ones for aircraft. I usually plug into a cigarette lighter outlet on board and like the cellular phone my battery appears to recharge more fully when plugged into my boat DC supply.

g. **Keyboard and Mouse.** The smaller keyboard will never be as good as a full size PC but you do get used to them. While the keys have a different feel, which varies between makes you soon adapt. Key locations vary which is also about learning and adaptation. When using the pressure sensitive touch pad, it takes some practice to be efficient and move the cursor around the screen so it comes down to personal preferences. I always plug in an external optical mouse and full sized keyboard at the nav station.

23.1 Computer System Options. The following show some options.

a. **Printers.** The usual printer will be an inkjet. If there is the room and you operate a "boat office," a small laser is probably better. There are small and compact printers, and it will depend on the quantity and quality of printing planned.

b. **Scanners.** The scanner may suit those who file everything they get, and is a realistic option for quite a lot of paper based information storage.

23.2 Software Options. The following is a general guide, choose your favorites.

a. **Standard Software Packages.** MS Office is a good simple choice and includes Word and Excel. Many illustrations were done using Visio. Other packages to use are Adobe Acrobat and I use AutoCad LT. Of course I also have a CD ROM interactive sailing package, and run a few other demo packages from various marine electronics equipment manufacturers.

b. **Electronic Charting.** Refer to chapter on electronic charting.

c. **Entertainment.** In addition to games packages, the computer can be used with the CD player for music, and the DVD for movies. It is also possible to watch television on the laptop using the TV-to-Go package, which is a PCMCIA Card to plug in, with integral antenna or connect to an external one. It requires a Zoom Video (ZV) PC Card port.

d. **Weather Information.** There are many weather software options, and many are listed in the weatherfax chapter.

e. **Digital Photographs.** The digital camera (I use a Sony CyberShot model) and software allows easy image storage and e-mail of pictures to friends and relations.

f. **Tidal Predictions.** There are many packages available, and these include Tide Wizard, for UK, Europe or worldwide. It allows tidal predictions using the simplified harmonic method. Claimed accuracy is 0.1 meter. Results are tabulated or graphical. Another package is Tides and Currents, which uses the full harmonic method and data from the UK Hydrographic Office. A North American Region version is available.

g. **Astro and Celestial Navigation.** I have used non-windows DOS versions for years. The windows packages are really easy to use, and the Win Astro package typifies them and is a sight reduction program. It has a perpetual almanac for the sun, moon, planet and stars. The DOS packages I have used are AstroNav and Polystar.

h. **Boat Management.** There are log packages that import data via a NMEA interface from GPS, Log and Depth etc and automatically maintain a vessel log.

i. **E-mail.** There are a variety of methods, as discussed in the e-mail chapter, in particular for HF radio.

j. **Internet.** Wireless is making things less difficult when traveling, and changes are occurring rapidly and options for satellite access are improving all the time.

23.3 Computer Maintenance. The following basics will improve reliability.

a. **Disc Drive.** All hard disc drives are sealed and are maintenance free. Ensure that dust is vacuumed out regularly. Floppy disc (if you have one) head cleaning kits are also available.

b. **Printer.** Most use a small inkjet printer, and a few basic maintenance tasks are required to maintain optimum performance. Ensure you carry spare ink cartridges: Clean out the interior using a soft brush. Outside casings should be wiped using a damp cloth. Do not use solvents or abrasive cleaners. Never apply lubricants or electrical cleaners into mechanisms.

c. **Keyboard.** Brush out dust and particles regularly, and wipe the board with a slightly damp cloth. Keep wet fingers, coffee mugs, etc., well away. Small purpose-made, battery-powered vacuum cleaners are available to extract dust and particles.

d. **Screen.** Wipe using lint free cloth. Dampen with water, no excess though. Wipe off marks gently. You can get cleaners made for this purpose.

e. **Battery Packs.** All systems have battery packs. The major problem is that battery packs are not properly cycled and some can develop what is called memory effect. Discharge them properly on a regular basis and then recharge completely. Charging is also an issue, some opt for the use of a small inverter, and I use a Targus DC car charge kit that works very well.

23.4 Computer Troubleshooting. There are a few basic checks to make.

a. **Power Supply.** The majority of problems on computers on boats can be attributed to poor quality power supplies, both AC and DC carrying damaging voltage transients. Most externally powered laptops have a 15-volt maximum. If using an alternator fast charge device, disconnect the computer from the vessel DC system during charging.

b. **Printers.** The most common problems are as follows:

(1) **Paper Feed Jams.** In most cases it is caused by moisture being absorbed into the paper. Use only as much paper as required in sheet feeders, keep the rest packaged and dry.

(2) **Printer Communication Error Messages.** Use self-test functions first. Check that cables are properly inserted into sockets. Check correct default printer is selected.

Entertainment Systems

24.0 Music Systems. Without music, is your boat really ready for sea? On a new yacht, there is a definite psychological lift when the stereo goes on for the first time. When selecting or installing a music system, there are a number of important factors to consider. Stereo units specifically designed for vehicles are obviously the choice for smaller boats, while larger boats will put on quality shore type systems. Like radios, choose units that are designed for RVs and more rugged shore duty. A good option if you like your music is the use of a 10 CD cartridge pack from Sony, or Pioneer which give a lot of music without reloading, and of course there is the MP3.

a. **Power Output.** Power output is rated in watts, and is either specified in watts per channel (RMS), or total power output (PMPO). Watts per channel is the power through each speaker, total power output is combined power. There is no need for high rated units with 60 watts per channel simply because the area involved is relatively small, and the ear cannot distinguish between a 30-watt and a 60-watt system. Quality, not volume, is what counts.

b. **MP3 Players.** This is a technology just made for sailing boats. Very small, powerful, and with a large music storage capacity. These days an iPod has speaker attachments to give great versatility.

c. **Speakers.** A speaker produces sound when the cone moves back and forth, compressing air in waves that produce the sound. Best sound is produced along a line projecting outwards from cone center. Sound also radiates at around 45° of the main axis in a cone configuration. If not listening in this zone, sound reflects and distorts off other surfaces. The average speaker requires air space of at least one cubic foot. When the cone moves forward a vacuum is created, and as it moves back the air is compressed. Small cavities behind speakers will reduce the cone movement, and you will get a lower bass response, as low frequencies require more air movement. I installed Bose Environmental speakers that cope with damp air and still sound great. Also be aware that magnetics on the speaker cones can affect navigation instruments so install them well away from compasses and GPS. It is very important to install the best quality speaker cable you can. Long runs are normal in boats, and quality suffers accordingly. For speakers mounted in the cockpit or on the stern arch, use tinned cables.

d. **Graphic Equalizers.** The equalizer makes the difference between good music and great music. Its function is to divide the music into different frequency ranges, which you can adjust to suit your own tastes. A boat's shape and materials do not facilitate ideal acoustic reproduction; the equalizer overcomes this to a considerable extent.

e. **CD Players and Cassette Players.** The CD and cassette player is rapidly being consigned to oblivion by the MP3 player. If you want reasonable quality, buy from manufacturers such as Pioneer, Alpine, or similar. I always recommend a unit designed for RV or four-wheel drive vehicles. These units can stand severe vibration and are moisture resistant, making them ideal for yachts. Beware of the term "marine stereo." Many are relatively cheap compared to a car stereo system, and their advertised superiority in the marine environment is questionable.

24.1 Video and DVD Players. The video cassette has been replaced by the DVD as the main audio visual source on board. It is not recommended to attempt to play discs not rated for the player, as player damage is possible. Discs for other regions will be indicated on displays and not play or be ejected. Common faults are damaged or scratched discs that cause skipping, failure to play or disc rejection. Vibration can also cause skipping and location should be checked. Failure to play can also be caused by moisture on the laser pickup, which can be rectified by leaving the player on to warm up for an hour. Picture problems are usually caused by incorrect settings, poor cable connections, or the disc has incompatible copy prevention guards. Sometimes power off, removal of the plug for 30 seconds and restarting will solve abnormal operation problems, which are often caused by static, lightning, electrical noise or simple improper operation. Remember that any switch off causes a loss in settings. Current demands are relatively small, and are typically around 1 amp. It is a good idea to place a cover or bag over unused players with a bag of silica gel and perhaps a corrosion inhibitor, as players are not marine grade.

24.2 Satellite Television. There are some very sophisticated TV systems available including Applied Satellite Technology (AST), KVH TracVision and Raymarine 45 STV. Systems operate on the C-Band (3.7-4.2 GHz) and the Ku-Band (10.7-12.95 GHz), and some are dual frequency. The units use a control unit, and auto-tracking gyro-stabilized antenna dishes housed in a radome. Radomes have sizes in the range 0.6–1.5m. As satellite tracking has to be fast and accurate to remain locked on, units also have pitch and roll sensors, with 3-axis servos and track satellites while compensating for vessel pitch, roll and yaw. These sensors use rate gyro sensors and inclinometers. Companies are now investing heavily in research and development and AST has made significant advances in parabolic reflector efficiency, reduced weights and more precise tracking. They have a 4-axis design compensating for elevation, pitch and roll, azimuth and active skew to maximize signal quality. Some units also have integral GPS receivers that supplement the NMEA gyro data. At power-up typical satellite acquisition times are around 5 minutes. There are several search modes including auto, manual, search and scan modes. The servo is normally a high torque brushless motor. Installation principles are virtually the same as for Inmarsat antennas. As satellites transmit to limited areas reception is dependent on being within the footprint. Footprints are generally limited to coastal and landmass areas, and proper selection of the correct tracking frequency is necessary. Power consumption is typically 3–5 amps. There are over 300 TV channels available in Europe and many more in the US and elsewhere, CD quality music and high speed Internet downlinks are available. On many barges in Europe you will see small domestic satellite dishes. They are set up at the required azimuth and elevation each night. Some people I know have taken the subscription at home and transferred the demodulator and dish to the boat.

Some channels and elevations include Telecom 2B and Telekom 2B (5ºW); Telecom 2A (8ºW), Astra 1A– 1D (19.2ºE); Eutelsat II F1/Hotbird (13ºE); Eutelsat II F3/Hotbird (16ºE); Eutelsat II F2/Hotbird (10ºE). The dishes are small and easy to mount, and on inland waterways there are no motion problems to overcome.

24.3 Terrestrial Television. There are a range of TVs designed for 12- and 24-volt operation, and for long distance passage makers multi-system units with NSTC and PAL reception are available. Television aerials and their performance on vessels is a controversial subject along with often over-optimistic performance claims by some manufacturers They are as well relatively expensive. Comparable performance with home aerials should not be expected, and attempting to get a reasonable picture while under way is generally out of the question, and the off-watch should stick to DVDs. The principal problem is getting a good picture at anchorages, without the continual ghosting that occurs in varying degrees of severity as you swing around. The ghosting problem largely depends on the path of the transmitted signal, and the frequency characteristics of the transmission.

a. **Signal Distortion.** Transmission signals are essentially straight line and do not bend significantly when meeting obstructions. The effect is one of creating shadows and areas of low signal behind the obstruction. Reflection of the signal also alters the direction of propagation causing signals to arrive at the aerial from a direction other than the straight-line path from the transmitter. The receiver ultimately receives two signals that arrive at times different to each other. The effect is the reception of a distorted signal pattern. The distortion of signal occurs from a number of sources that include hills, other boats, rigging and reflection off the water surface.

b. **Signal Polarization.** Signal transmissions are generally horizontally polarized. When signal is reflected, the polarization is altered, causing distortion.

c. **Transmission Systems.** The variety of TV transmission systems in use is confusing. The main systems are M/NTSC (US, Canada, Guam, most of South America); B.G/PAL (Australia, Netherlands, Spain, Portugal, Italy, Canary Is.); I/PAL (UK, South Africa); L/SECAM (France).

24.4 TV Aerials. There various aerial types in use are as follows:

a. **Directional Aerials.** The aerials can be aligned with the transmitted signal. Aerials are of the domestic type, and may be of use if you live on board and rarely venture out from the marina, but on an anchorage are useless, as you must continually adjust the aerial.

b. **Omni-directional Aerials.** These aerials are able to receive transmission signals consistently and are not affected by the vessel swinging at anchor or mooring. This type of aerial does not discriminate between directly transmitted or reflected signals. The most common type of aerial is the ring or loop, which is hoisted when required. These aerials do have a problem with reception of reflected signal from any masts and rigging, and perform poorly in marinas.

c. **Active Aerials.** These units are typically a fiberglass or plastic dome, with an integral omni-directional loop aerial element inside. The signal is amplified to compensate for the smaller aerial and performance depends on a good gain value within the amplifier unit. These aerials are also designed for the reception of UHF signals as well as AM/FM radio transmissions.

d. **Installation Factors.** The following factors should be considered when mounting and installing aerials:

1. **Aerial Height.** Install the aerial as high as practicable.
2. **Aerial Cables.** Cables should always be low loss coaxial (RG59), which is normally 75-ohm impedance.

24.5 Satellite Radios. This great entertainment system from Sirius and XM uses digital satellite radio broadcast system, with 50 channels of music, along with 50 news, sports and entertainment channels. Contact them at www.siriusradio.com. The footprint is essentially the continental US and seawards around 200 nm.

Troubleshooting, Testing and Servicing

25.0 Troubleshooting. Troubleshooting has a definite philosophy that should be understood and followed if it is to be effective. Troubleshooting is a logical process of evaluating a system, and how it operates. It involves collecting evidence, such as burn marks or heat, unusual sounds, acrid smells, temperature variations etc. This can be supported by the correct use of instruments and analyzing the data displayed on them. This forms the basis for testing theories and assumptions, so that the precise fault can be identified and subsequently rectified. The following factors must be considered in any troubleshooting exercise.

a. **Systems Knowledge.** Understand the basic operations of the equipment. It is common to find that faults are in fact only improperly operated equipment. If there is a basic understanding of the system, it is considerably easier to break down the system into functional blocks, which makes the process much easier. A circuit diagram will show all of the components in a system.

b. **Systems Configuration.** Understand where all the system components are installed, where connections and cables are and where supply voltages originate.

c. **Systems Operation Parameters.** Understand what is normal during operation, and what are the parameters or operating range of the system. All too often expectations are very different from reality.

d. **Test Equipment.** Understand how to use a basic multimeter. Be able to make the simple tests of voltage and continuity of conductors.

25.1 Troubleshooting Procedure. The following approach should be used in troubleshooting systems.

a. **System Inputs.** Check that the system has the correct power input. Don't assume anything. For example there may be a voltage input but it may be too low. Check it with a multimeter.

b. **System Outputs.** Does the system have an output? Is the required voltage or signal being put out? If there is input and no output, you have already isolated the main area of the problem.

c. **Fault Isolation.** In any troubleshooting exercise split the system in two. This method is ideal when troubleshooting lighting circuits. It instantly isolates the problem into a specific and smaller area.

d. **Fault Complexity.** Most problems usually turn out to be rather simple. Start with the basics, and don't try to apply complex theoretical ideas you do not fully understand, as the result is a lot of wasted time, and embarrassment. Stand back and think first.

e. **Failure Causes.** When a fault has been isolated and repaired, ascertain why the failure has occurred, if possible.

25.2 How To Use a Multimeter. The majority of tests can be carried out using a multimeter. A multimeter as the name suggests is able to perform a range of electrical measurements. There are two types of multimeters, analog and digital. An analog meter has a needle to show the readings. The digital meter (DMM) displays the test values numerically on a display. Manual ranging meters require selection of measurement ranges, and auto-ranging automatically types select the best measurement range.

a. **Voltage Tests (AC & DC).** The volt is the unit of electrical pressure, and is the force required to cause a current to flow against a resistance. The basic equation is E = I x R. It is the most useful of all measurements, either to detect that it is present or to precisely measure voltage levels. I perform 95% of all my troubleshooting on complex oil rigs and commercial vessels with this function alone. The voltmeter is connected across the supply or equipment, which is negative probe to negative and positive-to-positive to measure the voltage potential between the two. Reversal of probes will simply show a negative reading. If the DMM is not auto-ranging, set the scale to the one that exceeds the expected or operating voltage of the circuit under test. To analyze results:

   (1) If the voltage is missing this indicates that the circuit supply is switched off, or the circuit is possibly broken, such as a connection or a wire (positive or negative), or a faulty switch or circuit breaker.

   (2) If the voltage is low, this indicates that the supply voltage to circuit from the battery is low, or that additional resistance is in the circuit, such as faulty connection.

b. **Continuity Tests.** The continuity test requires the use of the ohms setting. It is simply to test whether a circuit is closed or open. Many multimeters also incorporate a beeper to indicate a closed condition. Power must be switched off before testing. Set the scale to one of the meg ohm ranges. Touch the probes together to verify operation, and then place the probes on each wire of the circuit under test. What you are looking for is a simple over-range reading if the circuit is open, and low or no resistance if it is closed.

c. **Resistance Tests.** Resistance is resistance to the flow of electrons, and fundamental laws were formulated by George Ohm and called Ohms Law. The ratio of a voltage through a conductor to a current flowing in it is constant, and is equal to the resistance of the conductor. The basic equation is R = E/I. If DMM is not auto-ranging, set the range switch to the circuit under test, typically the 20-ohm range is used. Turn off circuit power, and discharge any capacitors. When testing, do not touch probes with fingers as this may alter readings. Prior to testing, touch the probes together to see that the meter reads zero.

d. **Current Tests.** Current is the rate of electron flow in a conductor in amperes, and the basic equation is I = E/R.

(1) **Direct Current (DC).** This is the movement of electrons through a conductor in a single direction only. The ammeter function of a multimeter is rarely used or required although some use it for measuring leakage currents. The switchboard ammeter normally can be used for all measurements. The ammeter is always connected in series with a circuit, as it is a measurement of current passing through the cable. The circuit should be switched off before inserting the ammeter in circuit. Most DMM have maximum DC measurement ratings of 10 amps only, and it is a little used function.

(2) **Alternating Current (AC).** This is the movement of electrons through a conductor in one direction, followed by a reversed movement in the other. At the start the voltage is zero, then at a quarter of the cycle it reaches maximum, at half way through the cycle it is zero, again at 3/4 through the cycle it again reaches maximum, then at completion of the cycle it attains zero again. The cycle is also referred to as a sinusoidal waveform. In normal practice all references to voltage, current and power values are what is termed Root Mean Squared (RMS) values. This is the maximum value multiplied by a constant, which in AC systems is 0.707. The frequency of AC is measured in cycles per second, and the measurement unit is the Hertz (Hz). In most systems this is nominally 50Hz or 60Hz, although aircraft operate at 400Hz. AC current measurements are made using a clamp-on current meter.

e. **Power.** The watt is the unit of energy or power. The basic equation is $P = E \times I$. In AC systems power factor is included so that $P = E \times I$ Cos. Power can be tested using a wattmeter, as it is in domestic situations using watt hour meters. Power meters also may be installed on generator panels and display in Kw. In most circumstances it is simply calculated using voltage and current readings.

f. **Capacitors.** When an insulator or electrolyte separates 2 metal plates, a potential difference exists. Excess electrons on the negative plate exert attraction on positive plate when the potential difference is removed, this reverses charge and then discharges. Unit of capacity is the farad (F) and microfarad (μF).

g. **Diode.** A diode is a semiconductor one-way valve, consisting of an anode and cathode. It allows electrons to flow one way only and has a high resistance the other way. To test a diode, where a diode test position is not included on the multimeter set, place a probe on each side of the diode. It will read low resistance in the forward direction and high resistance in the reverse or blocking direction. This is why they are often called blocking diodes. Diodes are used to form full wave bridge rectifiers that convert AC to DC. To test, each diode must be checked as a separate diode, in the forward and reverse direction.

25.3 Meter Maintenance. Look after your meter. Do not drop the meter or get it wet. There are a few basics that ensure reliability and safety:

a. **Probes.** Ensure that probes are in good condition. On many probes the tips sometimes rotate out, and a probe may come out and short across the terminals under test. Another problem is the solder connections of test leads break away due to twisting and movement.

b. **Cables.** Cables should be kept clean and insulation undamaged. Cables can age and crack. If a cable is damaged replace the cable. Do not attempt to test higher voltages, in particular AC voltages, if the cables are damaged. People have received severe shocks or been killed due to faulty leads.

c. **Batteries.** Replace the internal battery every 12 months, or at least carry a spare. Many meters will have a low battery warning function.

25.4 Before Calling Service Technicians. I remember an episode when a boat arrived in from a Pacific cruise and the skipper told me that his radar had been out for some months, and he could only get the display partially working but no picture. To really make his day I went to the stern mounted scanner, and simply switched on a local power switch. Imagine his reaction, he simply had forgotten to check it. Grown men do cry. Consider the following points before calling for serivce:

a. Did I operate the equipment properly? Read the manual again and go back to basics. Only when you are sure that you have operated the equipment properly and it doesn't work should you call the service technician.

b. Are all the plugs in and power on? It is amazing how many people forget to plug in an aerial or to put the power on. If the power is on at the breaker and not on at the equipment, double-check that the circuit connection on the back of the switchboard is not disconnected. Check that the equipment fuse has not ruptured. Check the obvious.

c. Ask yourself what you were doing immediately prior to the fault. A great number of faults occur immediately after working on often unrelated systems. The inadvertent disturbance of connections can and does occur regularly so go and check.

d. Write down clearly the fault and the situation when the unit failed. If a profile can be built up it may point to some other factors. Not only will it assist the service person but also it may assist you in solving the problem and avoiding an expensive call.

e. Keep a good technical file on board. If possible obtain copies of all the technical manuals, as service people cannot carry, or get every different manual. It will save you money if you give him the information, as time will be saved.

f. Clean up the area to be worked on. It is quite unfair to expect service people to work on filthy engines, and dirty bilges. If you want or don't mind grime tracked through the boat, then ignore this advice, and if the fellow is good he simply may decline to come back again.

g. Have a good tool kit ready. It is impossible to carry on to every boat a complete tool set. Any assistance like this is greatly appreciated. Make sure your flashlights work, and empty or clear any locker through which equipment is accessed.

25.5 Spare Parts and Tools. To maintain a reasonable level of self-sufficiency, the following tools and equipment should be carried on board every vessel. The list should be used as your itemized checklist.

Table 25-1 Tools and Spares List

Recommended Tools	Consumables & Spares
Electrical pliers	Self bonding tape (2 rolls)
Long nose pliers	Insulation tape (2 rolls)
Side cutters	Denso tape (2 rolls)
Cable crimpers (Ratchet type)	Heatshrink tubing
Set electrical screwdrivers	Spiral wrapping
Set Phillips head screwdrivers	Nylon cable ties, black (3 sizes)
Soldering iron (gas)	50m 2.5mm twin tinned cable
Soldering iron (12 volt)	100m 2.5mm single tinned cable
Roll solder	Circuit breaker (15A)
Adjustable wrench	Lamp-bicolor (2)
File, small half round	Lamp-stern/masthead (2)
File, small round	Lamp-anchor (2)
Socket wrench set	Switchboard indicator lamps (2)
Bearing puller set	Butt crimp connectors
Wire brush	Alternator regulator
Junior hacksaw	Alternator
Battery powered drill	Alternator warning light
Ring spanner set (offset)	Start relay (if fitted)
Set Allen keys	Alternator fuses (if fitted)
Digital multimeter	Fuses for electronics equipment
Meter battery	Anchor windlass fuses (2)
Set jumper wires & clips (Tandy)	Brushes-windlass (if fitted)
	Brushes-starter motor (if fitted)
	Brushes-refrigerator motor (if fitted)
	Brushes-alternator
	Coaxial connectors
	Battery terminals (2)
	Bearings for AC motors
	Engine fan belts (2 each)
	Electrical cleaner, CRC, (2 cans)
	Water dispersant, WD40 (2 cans)
	Silicon grease (1 tube)
	Silicon compound (1 tube)
	Petroleum jelly (Vaseline) (1 can)
	Anti-seize lubricant (Copper Slip)
	Distilled water (1 gal)
	Boeshield B-19 (3 cans)

Table 25-2 Basic Three Language Electrical Glossary

English	French	Spanish
Audible alarm	Avertisseur sonore	Bocina electronica
Alternator	Alternateur	Alternador
Alternator rating	Puissance de l'alternateur	Potencia del alternador
Alarm panel	Tableau des alarmes	Tarjeta instrumentos
Battery	Batterie	Bateria (Accumulador)
Bolt	Boulon	Perno, tornillo
Circuit breaker	Coupe-circuit or interrupteur	Fusible
Connection	Cablage	Connexion
Circuit diagram	Schema de cablage électrique	Eschema de connexiones
Current (electrical)	Courant	Corriente
Drive belt	Courroie de transmission	Correa de ventilador
Disconnect	Déconnecter, isoler	Desconectar
Electrician	Electricien	Electricista
Element	Elément	Resistencia
Fault	Faute	Defecto
Ignition switch	Contact (moteur)	Llave de contacto
Insulation	Isolement	Aislamiento
Current level	Intensité (amps)	Intensidad
Fuse	Fusible	Fusible
Ground (earth)	Mettre à la masse	Conectar con masa
Lights	Feux	Luz
Light bulb	Ampoule électrique	Bombilla, foco
Lightning	Eclair	Relampago, rayo
Navigation lights	Feux de position	Luz de navigación
Overheat	Surchauffe	Recalentarse
Oil Pressure sensor	Sonde de pression d'huile	Sensore pressione aceite
Pressure gauge	Manomètre d'huile	Monometero de aceite
Preheating glowplugs	Bougies de prechauffage	Bujia de precalentamiento
Relay	Relais	Rele
Recharge	Recharge de batterie	Recargar
Short Circuit	Court-circuit	Cortocircuito
Starter Motor	Démarreur	Motor de arranque
Sensor	Capteur/sonde	Sensor
Switch	Bouton poussoir/interrupteur	Pulsador
Tachometer	Compte-tours	Tacometro
Temperature sensor	Sonde de température	Sensore temperatura
Transmitter	Emetteur	Transmisor
Voltmeter	Voltmètre	Voltimetro
Voltage	Tension de système	Tensión del systema
Voltage drop	Chute de tension	Voltaje, bajar
Water pump	Pompe à eau	Bomba
Wire	Cable ou fil (électrique)	Alambre
Engine not starting	Le moteur ne démarre pas	El motor no arranca

SERVICE DIRECTORY

26.0 Installation and Service List. Following are lists of marine electricians, marine electronics technicians and companies, and other qualified electrical experts who come highly recommended. Marine electricians have a merchant marine or naval background, and will probably have many years of sea service behind them. They have an understanding of the environmental factors affecting marine electrical installations and are qualified to work on both AC and DC systems, as well as on many electronics. Beware of automotive electricians claiming to be marine electricians—they are not. Most good ones doing marine work do not hide that fact. There are some very good automotive electrical tradesmen doing marine work. Many are included in the lists. Go on recommendations, if at all possible. (Ask them if they own a boat!) Beware also of the domestic electrician who makes similar claims. Again, there are a few good tradesmen around who have an industrial background and can do a good job. If you are getting AC work done, ask to see a license or some qualification. Get references or check their backgrounds if at all possible. It's your life in the balance.

Table 26-1 Service Directory—United States/Canada

Port	Person/Company	Contact Numbers
Oakland, CA	Collins Marine Corp	(415) 957 1300
San Francisco, CA	Cal Marine Electronics	(415) 391 7550
Marina Del Rey, CA	Baytronics South	(213) 822 8200
Newport Beach, CA	Alcom Marine Electronics	(714) 673 1727
San Diego, CA	Power & Wind Marine Electrical	(619) 226 8600
Santa Barbara, CA	Ocean Aire Electronics	(805) 962 9385
Fort Lauderdale, FL	Avalon Marine Electronics	(305) 527 4047
Miami, FL	Electro Marine	(305) 856 1924
Houston, TX	Able Communication	(713) 485 8800
Portland, ME	Ross Marine Electronics	(207) 272 7737
Stamford, CT	Maritech Communications Corp	(203) 323 2900
Chesapeake, VA	Seaport Electronics	(804) 543 5600
Annapolis, MD	Coast Navigation	(301) 268 3120
Portsmouth, NH	Cay Electronics	(401) 683 3520
Seattle, WA	Seamar Electronics	(206)622 6130
Tacoma, WA	J & G Marine Supply	(206) 572 4217
Hawaii, HI (Honolulu)	Navtech	(808) 834 7672
Hawaii, HI (Kailua-Kona)	West Hawaii Electronics	(808) 329 1252
Vancouver, BC	Maritime Service	(604) 294 4444
Victoria, BC	Victoria Marine Electronics	(604) 383 9731
Halifax, NS	Gabriel Aero Marine	(902) 634 4004
Quebec, QUE	GAD Electronics	(418) 986 3677

Table 26-2 Service Directory—UK/European Channel Coast

Electrical		
Plymouth (England)	Ocean Marine Services	(01752) 500121
Plymouth (England)	Western Marine Power	(01752) 225679
Southampton (England)	Marinapower Electrical	(023) 8033 2123
West Sussex (England)	Keith Howlett	(01798) 813 831
Winchester (England)	Power & Air Systems	(01962) 841828
Brighton (England)	K McCallum	(01243) 775 606
Medway (England)	David Holden	(01795) 580 930
Jersey (Channel Islands)	Jersey Marine Electrics	(01534) 21603
Calais (France)	Bernus Shipyard	(21) 34 3040
Gothenburg (Sweden)	Wahlborgs Marina	(42) 29 3245
Stockholm (Sweden)	Linds Boatyard	(715) 624 211
Haringvliet (Holland)	Stellendam Marina	1879 2600
Ostend (Belgium)	Noordzee Jachtwerf	(59) 78100
Electronics		
Jersey (Channel Islands)	Jersey Marine Electronics	(01534) 21603
Guernsey (Channel Islands)	Radio & Electronic Services	(01481) 728837
Plymouth (England)	Tolley Marine	(01752) 222530
Hamble (England)	Hudson Marine Electronics	(023) 8045 5129
Southhampton (England)	Regis Electronics	(01983) 293 996
Lymington (England)	Regis Electronics	(01590) 679 251
Brighton (England)	DMS Seatronics	(01273) 605 166
Falmouth (England)	Western Electronics	(01326) 73438
Poole (England)	Fleet Marine	(01202) 6326 66
Ipswich (England)	R & I Marine Electronics	(01473) 659737
Boulogne (France)	Ocel	(0121) 317592
Honfleur (France)	La Barriere	(31) 890517
Dunkerque (France)	Marine Diffusion	(28) 591 819
St Valery-sur-Somme	Lattitude 50	(22) 26 82 06
Le Havre (France)	Electronique Equip.	(35) 546070
Carentan (France)	Gam Marine	(33) 711 702
St Vaast La Hougue	Marelec	(33) 546 382
Granville (France)	Nautilec	(33) 500 496
Cherbourg (France)	Ergelin	(33) 532 026
La Rochelle (France)	Pochon	(45) 413 053
Port Camargue (France)	Y.E.S.	(66) 530 238
Nieuwpoort (Belgium)	Sea Trade & Service	(058) 237230
Sneek (Holland)	Jachtwerf Rimare	(05150) 12396
Hamburg/Wedel (Germany)	Yachtelektrik Wedel	(04103) 87273
Cork (Ireland)	Rider Services	0002 841176

Table 26-3 Service Directory—Mediterranean

Port	Person/Company	Contact Numbers
Valetta (Malta)	S & D Yachting	54 Gzira Rd, Gzira
Bodrum (Turkey)	Motif Yachting	(6141)2309 VHF 71
Marmaris (Turkey)	ATC Yacht Service	(612) 13835
Kusadasi (Turkey)	Dragon Yachting	(636)12257 VHF71
Fethiye (Turkey)	Alesta Yachting	(615)11861 VHF 16
Venice (Italy)	Cantiere Zennaro	71 7438
Larnaca (Cyprus)	IMF Marine Elect.	(04) 655 377
Larnaca (Cyprus)	KJ Electronics	(41) 636 360
Rhodes (Greece)	Y Paleologos	(0241) 25460
Syros (Greece)	E Bogiatzopoulos	(0281) 22254
Corfu (Greece)	P Mavronas	(0661) 26247
Piraeus (Greece)	General Electronic Repairs	461 4246
Mallorca (Spain)	Euro Marine Services	(71) 676141
Torrevieja (Spain)	Torrevieja Int'l Marina	571 3650
Benalmadena (Spain)	M Blenkinsopp	(52) 560906
Gibraltar	H Shephard & Co.	75148
Monaco	Electronic Services	(93) 26 76 67
Antibes (France)	GMT Maritime (electronics)	(93) 34 23 87
	Georges Electricite (electrical)	(93) 33 73 72
Nice	Electronique Marine	(93) 56 58 73
St Laurent du Var	International Marine Tecnic	(93) 07 74 55
Cannes La Bocca	Radio-Ocean	(93) 47 72 15
Port de Ste Maxime	Express Electronique	(94) 96 53 48
	Paulo Services (electrical)	(94) 96 67 15
Port Grimaud	Electronic Services	(94) 56 44 60
Les Marines de Cogolin	Electronique Marine	(94) 56 05 69
Mandelieu	Elec Marine Napouloise	(93) 49 06 02
	Mandelieu Electricite Marine	(93) 49 57 07
Menton	Mar-Elec	(93) 41 62 62
Beaulieu	CRM	(93) 01 09 09
Cap Ferrat	Electro Mechanique	(93) 01 67 57
Hyeres	Monmartre Boulanger(electronics)	(94) 38 88 84
	SIARI (electrical)	(94) 38 73 71
Toulon	Pro Electronique	(94) 03 00 50
Marseille	Marseille Marine (electronics)	(91) 91 31 42
	Electric Auto Yachting (electrical)	(91) 73 30 14

Table 26-4 Service Directory—Caribbean

Port	Person/Company	Contact Numbers
Electrical Services		
Castries (St Lucia)	Rodney Bay Marina	(809) 452 9922
	Columban Ellis	(809) 452 7749
St Maarten (Neth Antilles)	Bobbys Marina	Call on VHF 16
St Georges (Grenada)	Moorings Marina (Gill Findlay)	4402119 Ch66/71
Jolly Harbor (Antigua)	Electro Tek	(809) 462 7690
Electronics Services		
St Maarten (NL Ant.)	Radio Holland	(599) 525414
English Harbour (Antigua)	Signal Locker	(809) 463 1528
Falmouth Harbour (Antigua)	Cay Electronics	(809) 460 1040
Hamilton (Bermuda)	Marionics Caribbean	(809) 460 1780
San Juan (Puerto Rico)	Electronic Communications	(809) 295 2446
St. Thomas (U.S.V.I.)	Marine Comms	(809) 295 0558
Tortola (B.V.I.)	Master Marine Electronics	(809) 788 6888
Castries (St Lucia)	Caribbean Radio & Telephone	(809) 724 2035
	Geary Electronics	(809) 776 1444
St Barth (French WI)	Cay Electronics	(809) 494 2400
Guadeloupe	Cay Electronics	(809) 452 9922
Martinique	Morne Doudon	(809) 452 2652
Bridgetown (Barbados)	GME Int. (Port de Gustavia)	(590) 27 89 64
Point Cumana (Trinidad)	Marina	(590) 908919
La Guaira (Venezuela)	Samafon	(596) 660564
	Williams Electrical	(809) 425 2000
	Goodwood Marina	(809) 632 4612
Refrigeration		
Castries (St Lucia)	Rich Electronics (SEA)Caraballeda YC	(31)941 789
	Mars Refrigeration Services	(809) 452 2994

Table 26-5 Service Directory—Australia

Port	Person/Company	Contact Numbers
Electrical Services		
Sydney	Malbar Marine Electrics	(02) 476 4306
Pittwater	Barrenjoey Marine Electrics	(02) 997 6822
NSW South Coast	Ken King (Seaboard Electrics)	(044) 465 012
Lake Macquarie	Hunter Marine Electrics	(049) 532 353
Southport	Southport Industrial & Marine	321 167
Runaway Bay	Runaway Bay Marine Electrics	572 188
Mooloolaba	Lawries Marina	(071) 441 122
Townsville	Manlin Electrical & Marine	(077) 796 231
Cairns	All Marine Electrics	517 219
Hobart	M & K Madden	295 195
Melbourne	Goaty's Marine Electrical	(018) 327 403
Darwin	Percy Mitchell Electrical	(89)814 288
Adelaide	John Yandell	(08)47 5660
Fremantle	Cully's Electrical Service	430 5181
Electronics Services		
Sydney (NSW)	Peter Morath (Radar Specialist)	(02) 883959
	Ted McNally	(02) 522 8235
	Olympic Instruments (VDO)	(02) 449 9888
Port Macquarie (NSW)	Peter Long (Computel)	(018) 653 128
Whitsundays (QLD)	Phillip Pleydell	(079) 467 813
Runaway Bay (QLD)	Micro Logic	(075) 37 1455
Mooloolaba (QLD)	Mooloolaba Radio	(074) 44 4707
Bundaberg (QLD)	Rampant Marine Elec.	(071) 534 994
Gladstone (QLD)	Rigneys Electronics	(079) 727 839
Cairns (QLD)	Pickers Marine	(070) 511 944
Townsville (QLD)	Breakwater Chandlery	(077) 713 063
Melbourne (VIC)	John Powell	(018) 591 780
Port Adelaide (SA)	International Comms.	(08) 473 688
Fremantle (WA)	Maritime Elect. Services	(09) 335 2716
Darwin (NT)	NavCom	(089) 811 311
Refrigeration		
Sydney	Dave Bruce Moorebank Marine	(02) 602 9571

Table 26-6 Service Directory—Pacific

Port	Person/Company	Contact Numbers
Electrical Services		
Noumea (New Caledonia)	Gerard Destaillets	(687) 25 43 00
Suva (Fiji)	Yacht Help (VHF 71)	(679) 311982
Port Moresby (PNG)	Marine & Industrial Elec. Eng.	(675) 25200
Lae (PNG)	Lae Battery Services	(675) 421125
Auckland (NZ)	MJJ Electrics Afloat	(09)473 871
Whangarei (NZ)	Orams Marine	(89)489 567
Opua (NZ)	Elliots Boat Yard	4027705
Honiara (Solomons)	Pacific Electrics	22454
Apia (Western Samoa)	Samoa Marine	(685) 22721
Nuku'alofa (Tonga)	Flemming Electric	(676) 21095
Nuku Hiva (Marquesas)	Alain Bigot	920 334
Kowloon (Hong Kong)	Islander Yacht Basin	3719 1336
Electronics Services		
Fukuoka-Shi (Japan)	Nakamura Sengu	(092) 531 4995
Majuro (Marshall Islands)	Mariscom	(692) 9 3271
Noumea (New Caledonia)	Marine Corail	(687) 275 848
Guam (Marianas)	Pacific Isle Communications	(671) 649 9797
Nuku'alofa (Tonga)	Tait Electronics	Tungi Arcade
Nuku Hiva (Marquesas)	Alain Barbe	920 086
Bay of Islands (NZ)	Rust Electronics	(09) 403 7247
Whangarei (New Zealand)	Ray Roberts Marine	(09) 438 3296
Auckland (New Zealand)	Seaquip Marine	(09) 424 1260
Bay of Plenty (NZ)	Bay Marine Electronics	(07) 577 0250
Papeete (Tahiti)	Marine Corail	(689) 428 222
Vina del Mar (Chile)	Nauticos Mauricia Opazo	(32) 66 31 50
Santiago (Chile)	Parker y Cia	(41) 740 730
Callao (Perù)	Marco Peruana	(14) 659 497
Panama	Marco Panama	(27) 3533

Table 26-7 Service Directory—Atlantic & Indian Oceans

Port	Person/Company	Contact Numbers
Las Palmas (Canary Islands)	Internatica Gran Canaria	(28) 246 590
Las Palmas (Canary Islands)	Servicios Electronicos	(28) 243 935
Tenerife (Canary Islands)	Heinemann Hennanos	(22) 680 859
Funchal (Madeira)	Maria-Faria	(91) 368 858
Horta (Azores)	JBN Electronica	(96) 23781
Ponta Delgada (Azores)	Mid Atlantic Yachts Serv.	(92) 31616
Montevideo (Uruguay)	Electromaritima Uruguaya	(02) 203 857
Buenos Aires (Argentina)	Sistemas Electronicos	(01) 343 0069
Galle (Sri Lanka)	Windsor Yacht Services	(09) 22927
Mombasa (Kenya)	Comarco Communications	(11) 318 778
Antanarive (Madagascar)	Landis Madagaskar	(02) 25151 55
Cape Town (South Africa)	Sea Gear	(21) 448 3777
Cape Town (South Africa)	Wilbur Ellis Co.	(21) 448 4517
Durban (South Africa)	Durban Yacht Services.	(31) 0 1953/4

SOURCES AND LITERATURE

American Boat and Yacht Council (ABYC) Recommendations

Battery Service Manual, Battery Council International, 1982

Donat, Hans. *Engine Monitoring on Motor Boats.* VDO Marine, 1985

International Regulations for Preventing Collisions at Sea, 1972

Lloyd's Rules for Yachts and Small Craft

Warren, Nigel. *Metal Corrosion in Boats,* 3E. Sheridan House, 2006

ACKNOWLEDGMENTS

Many thanks to Grahame MacCleod for the Visio illustrations, Paul Checkley for the Autocad illustrations, as well as the following friends and companies for their assistance, and to acknowledge the use of various drawings, service manuals and other reference material and information on related websites of many other companies.

Quinn's - Port Adelaide; Olympic Instruments (VDO Agent); Dr. Steve Bell; Kenneth Parker; The Cruising Association UK; Index Marine; ABYC; USCG; MSA UK; AMSA; Simrad; Raymarine; Ingram Corporation; Bosch; Derek Barnard, Penta Coomstat; Telstra Maritime; IPS Radio & Space Service; Collins Marine; Datamarine International; BP Solar; Solarex; Whale Pumps; SeaFresh Watermakers; Guest Corp; Hubble; Trevor Scarratt (Adverc BM); ICS Electronics; M.G. Duff & Co; Marlec Engineering; Lewmar; Ampair; Lestek; Balmar; Motorola; Ample Power Company; Marinco; Glacier Bay; New-Mar; Northern Lights; DIY-Boat Magazine; Cruising World; Practical Boat Owner; Yachting Monthly; Cruising Helmsman Magazine; Ocean Navigator Magazine; PassageMaker Magazine.

INDEX